BOYCE, JEFFERSON C
DIGITAL LOGIC: OPERATI
       000376903

621.381   B78

DIGITAL LOGIC
Operation and Analysis

DIGITAL LOGIC

Operation and Analysis

Second Edition

JEFFERSON C. BOYCE
Allan Hancock College

PRENTICE-HALL, INC. Englewood Cliffs, New Jersey

Library of Congress Cataloging in Publication Data
BOYCE, JEFFERSON C.
    Digital logic.

    Rev. ed. of: Digital logic and switching circuits.
1975
      Bibliography: p. 485
      Includes index.
      1. Digital electronics. 2. Logic circuits.
3. Switching circuits.   I. Title.
TK7868.D5B69 1982   621.3819′582   81-15805
ISBN 0-13-214619-3   AACR2

Editorial production/interior design: Margaret McAbee

Manufacturing buyer: Gordon Osbourne

© 1982   by Prentice-Hall, Inc., Englewood Cliffs, N.J.   07632

All rights reserved. No part of this book
may be reproduced in any form or
by any means without permission in writing
from the publisher.

Printed in the United States of America

10   9   8   7   6   5   4   3   2   1

Prentice-Hall International, Inc.,   *London*
Prentice-Hall of Australia Pty. Limited,   *Sydney*
Prentice-Hall of Canada, Ltd.,   *Toronto*
Prentice-Hall of India Private Limited,   *New Delhi*
Prentice-Hall of Japan, Inc.,   *Tokyo*
Prentice-Hall of Southeast Asia Pte. Ltd.,   *Singapore*
Whitehall Books Limited,   *Wellington, New Zealand*

To
Jay, Dan, Margi, and Kathi (who gave up her room);
they made this effort necessary:
but especially to Betty,
whose love and understanding
made it possible.

And now to Emma,
Matthew, and Zachary;
they made *this* effort
necessary!

Contents

Chapter 1   **An Introduction to Digital Logic**   1
- 1-1   An Historical Background   1
- 1-2   Analog (Linear) vs Digital Techniques   3
- 1-3   Analog (Linear) Circuit Analysis   7
- 1-4   Digital Circuit Analysis   11
- 1-5   Logic Analysis Techniques   15

Chapter 2   **Numbers and Counting**   19
- 2-1   The Decimal Number System   19
- 2-2   The Binary Number System   22
- 2-3   Number System Conversions   24
- 2-4   The Counting Process   29
- 2-5   The Flip-Flop (FF)   33
- 2-6   A Binary Counter   38

Chapter 3   **Basic Combinational Logic**   45
- 3-1   Fundamentals of Combinational Logic   45
- 3-2   The OR Function in Digital Logic   57
- 3-3   The AND Function in Digital Logic   60
- 3-4   The NOT Function in Digital Logic   62
- 3-5   Simple Combined Functions   65
- 3-6   Derived Functions   73
- 3-7   A Simple Application of Basic Combinational Logic Functions   79
- 3-8   Summary   84

Chapter 4   **Advanced Combinational Logic**   86
- 4-1   Basic Laws of Boolean Algebra   87
- 4-2   Common Identities of Boolean Algebra   108
- 4-3   Summary of Algebraic to Logic Conversion   118
- 4-4   A Combinational Logic Circuit (Binary Half-Adder)   119

Chapter 5   Algebraic Simplification Methods   124

- 5-1 What is Logic Simplification? 124
- 5-2 Algebraic Simplification Procedures 126
- 5-3 1-Level Logic Simplification 128
- 5-4 2-Level Logic Simplification 130
- 5-5 3-Level Logic Simplification 136
- 5-6 4-Level Logic Simplification 137
- 5-7 Algebraically Simplifying the Binary Adder 141

Chapter 6   Graphic Simplification Methods   150

- 6-1 Types of Graphs 151
- 6-2 Preparation for the Use of Maps 156
- 6-3 2-Variable Map Simplification 161
- 6-4 3-Variable Maps 164
- 6-5 4-Variable Maps 171
- 6-6 Simplifying the Binary Adder Graphically 173
- 6-7 Comparison of Simplification Methods 177
- 6-8 Summary 178

Chapter 7   Analysis and Troubleshooting of Combinational Logic   181

- 7-1 A Proposed Analysis Procedure 181
- 7-2 A Typical Analysis 183
- 7-3 Troubleshooting Combinational Logic 195
- 7-4 A Typical Troubleshooting Problem 219

Chapter 8   Sequential Logic   227

- 8-1 An Example of Sequential Logic Use 227
- 8-2 The General Sequential Element 228
- 8-3 The R-S Flip-Flop 238
- 8-4 The Clocked R-S Flip-Flop 241
- 8-5 The Master-Slave Principle 243
- 8-6 The T Flip-Flop 244
- 8-7 The D Flip-Flop 247
- 8-8 The J-K Flip-Flop 248
- 8-9 Sequential Logic Analysis 250
- 8-10 Summary 256

Chapter 9   Troubleshooting Combinational/ Sequential Logic   255

- 9-1 A Proposed Procedure 255
- 9-2 A Typical Mixed Function 256
- 9-3 Troubleshooting Mixed Functions 267
- 9-4 A Typical Troubleshooting Problem 274
- 9-5 Summary 277

Contents   ix

Chapter 10   **Counters and Their Applications**   **279**
     10-1   The Asynchronous Binary Counter   279
     10-2   The Synchronous Binary Counter   287
     10-3   Special Purpose Counters   293
     10-4   Counter Circuit Analysis   299
     10-5   Counter Applications   307

Chapter 11   **Registers and Their Applications**   **312**
     11-1   The Storage Register   313
     11-2   The Shift Register   320
     11-3   Shift Counters   329
     11-4   Application and Analysis   332
     11-5   The Future   339

Chapter 12   **Miscellaneous Logic Functions**   **342**
     12-1   The Exclusive-OR Operation   342
     12-2   Arithmetic/Logic Circuits   349
     12-3   Multiplexers   353
     12-4   Memory Elements   358

Chapter 13   **Binary Codes**   **368**
     13-1   Binary Related Radix Codes   368
     13-2   Binary Coded Decimal (BCD) Codes   372
     13-3   Alphameric Codes   379
     13-4   Summary   386

Chapter 14   **Code Converters and Display Devices**   **388**
     14-1   Code Converters   388
     14-2   Display Devices   405
     14-3   Summary   418

Chapter 15   **Microprocessors**   **421**
     15-1   Introduction   421
     15-2   Microprocessor Concepts   423
     15-3   The Microprocessor as Hardware   428
     15-4   The Microprocessor as Software   436
     15-5   Summary   447

Chapter 16   **Digital Applications**   **449**
     16-1   Telemetry   449
     16-2   Data Communication   454
     16-3   Digital Test Equipment   457
     16-4   Microwave Oven Control   460
     16-5   Summary   466

Appendix A   Postulates, Theorems, and Laws of Boolean Algebra   **467**

Appendix B	MIL-STD-806B Extracts	469
Appendix C	Equivalent Logic Symbols	474
Appendix D	Logic Families	478
Appendix E	Number Tables	483
Bibliography		485
Index		487

Preface

Much has happened since the first edition of this book was prepared. No longer is the digital age in its infancy. It is an adult, strongly asserting itself and affecting society. Engineers and technicians no longer have to learn from "trial-and-error" or from highly complex digital computer texts. A plethora of books ranging from over-simplifications to over-complexities now exist. This book bridges the gap by exploring the *general* field of digital techniques. It is a unique blend of the fundamentals of digital logic and modern implementations. It prepares the reader to recognize, understand, analyze, and troubleshoot digital logic equipment *without* having to relate to a more complex type of digital device such as a computer. However, the digital influence has resulted in much simplification in the computer field. Therefore, material has been added to acquaint the reader with microprocessors and their applications.

The title of this book, DIGITAL LOGIC: OPERATION AND ANALYSIS, appropriately describes its contents. As will be observed as this book is used, the terms *operation* and *analysis* are closely allied. Following development of the basic concepts of digital logic in the first few chapters, it becomes apparent that separation of the two terms is almost impossible. No interconnection of the hardware elements of digital logic can be made that does not require *analysis* to determine the *operation* of the resulting circuit. *Analysis in this book is used to describe application of the relatively few rules of digital logic to circuits to determine the operation of those circuits.* The circuits may be as simple as two or three interconnected logic elements, or as complex as a digital computer. The same techniques apply.

As a result of reading and applying the principles and exercises provided in this book, a reader should be able to follow the logic diagram data flow from input to output and be able to use simple Boolean Algebra, truth tables, logic maps, and waveforms to analyze and understand the operation of digital logic equipment. Since this approach to logic tends to be hardware-oriented rather than theoretical, minimal electronic and mathematical background is required. The mathematical and electronic principles needed are developed in the book.

Chapters 1 and 2 are introductory and discuss the basic concepts of digital logic, numbers, and counting. The fundamentals of combinational logic (gates) are presented in Chapters 3 and 4; Chapters 5 and 6 discuss algebraic and graphical simplification of logic expressions. The reader is invited to use as little or as much of Chapters 4, 5, and 6 as needed to gain the required knowledge and skills. Combinational logic is summarized in Chapter 7, where procedures for analysis of this type of logic circuit are developed by introducing the use of troubleshooting flow charts.

The knowledge gained during this preliminary study of combinational logic is applied in Chapter 8 where sequential (time-dependent) logic is investigated. Chapter 9 integrates combinational and sequential logic by developing the analysis and troubleshooting procedures (again using flow charts) applicable to circuits that contain both types of logic.

If any division of this book is to take place, it should be after Chapter 9. All previous chapters are somewhat theoretical in nature, although presented from a practical viewpoint. The remaining chapters discuss specific types of logic circuits. For example, Chapter 10 is devoted entirely to counters. Implementation of the counting function is discussed from both the individual logic element packaging method to the integrated circuit (IC) packages currently in use.

Registers are discussed in the same manner in Chapter 11, while miscellaneous logic functions, such as adders, comparators, multiplexers, and the like are covered in Chapter 12. Chapters 13 and 14 are somewhat related, since they present the number codes used in digital equipment, the converter devices used with such codes, and the devices used to display information.

Chapter 15 introduces the concepts of microprocessors and considers both the hardware and the software aspects. An overview of the complete digital field is presented in Chapter 16 by discussing a number of common applications at the block diagram level.

It is impossible to list all of the individuals whose influence and suggestions have resulted in the organization of, and the material in, this book. Edward Francis, David Boch's and Margaret McAbee of

Prentice-Hall and Professor Kosow of Staten Island Community College deserve recognition for their untiring efforts during the evolution of this book. My colleagues in The Boeing Company and at Allan Hancock College, whose encouragement and advice has been invaluable, must receive much of the credit for this effort. The many students who suffered through early versions of portions of this book also should not be overlooked. Readers have contributed many suggestions that are incorporated in the second edition and, most of all, the patience of an understanding family, all of whom gave up so much so that this book could be developed, must be acknowledged.

JEFFERSON BOYCE

DIGITAL LOGIC
Operation and Analysis

1

AN INTRODUCTION TO DIGITAL LOGIC

Revolution is the mainstay of mankind's survival—not violent revolution, but a steadily progressive change in our way of life. We shall devote this chapter to a discussion of such basic changes as the transportation revolution, the revolution in energy production, and, most importantly, the revolution in machines designed to aid our intellectual pursuits.

1-1 AN HISTORICAL BACKGROUND

The Transportation Revolution

We have progressed rapidly in this century. In the early 1900s, one could travel cross-country only by rail at the convenience of the railroad. Today the same trip can be routinely made by auto, at the individual's convenience—a revolutionary accomplishment!

Still more fantastic has been the transportation revolution brought about by the airplane. The first powered flight by the Wright brothers in 1902 led to the beginning of the first passenger service in 1919 between London and Paris. Today hundreds of·millions of passengers and a large volume of freight traffic are carried yearly.

Even "the sky is the limit" no longer. We now travel outside the earth's atmosphere. It is almost commonplace for astronauts (or cosmonauts) to orbit the earth. A few individuals have even travelled to the moon (¼ million miles), landed, explored its surface, and returned safely with samples of another planetary body. For many of us, perhaps, travel within our planetary system may become a reality in this century. Can anyone deny the "transportation revolution?"

The Energy Revolution

Similar progress has been achieved in the application of energy. The 1900 census showed that the largest percentage of power for industrial use was furnished by steam engines. Today electrical power has almost completely taken over.

Considerable ingenuity has gone into meeting our needs for electrical power. As energy requirements grew, coal and oil-fired generating plants have been supplemented by hydroelectric systems. Nuclear power plants have begun to fill the ever-expanding need for electrical energy as fossil fuel shortages have appeared. Solar energy supplies electrical power in remote areas, and further supplements already-existing systems. The energy revolution is just beginning.

Intelligent Machines

Last—but not least—in our revolutionary progress is the growing partnership between man and machine. The eighteenth century Industrial Revolution may soon take second billing to an "intellectual revolution." The concept of applying "intelligent machines" to many fields of endeavor has existed for centuries; the evolution of electronic techniques has made this a reality.

One of the earliest intelligent machines built was the *abacus*. This digital machine appears in history prior to the birth of Christ. The example shown in Fig. 1-1 is a modified form that represents the alterations made by the Japanese and the Chinese civilizations. A skilled abacus operator can successfully compete with mechanical desktop calculators, but lags behind electronic calculators.

Little progress was made in the development of machine aids until the appearance of Blaise Pascal's desk calculator in 1642. Pascal's device, using simple gears, could add and subtract. Due to a lack of mechanical precision, other mathematicians could not improve Pascal's machine beyond achieving multiplication.

The next major milestone associated with the development of digital machines occurred in the early 1800s. Charles Babbage envisioned a mechanical device that incorporated many principles of the modern digital computer. His "Difference Engine" was developed to calculate and print mathematical tables. Again, imperfect materials, a shortage of precision tools, withdrawal of government support (after 1843), and a lack of understanding among his associates resulted in the abandonment of his project after several incomplete models had been constructed.

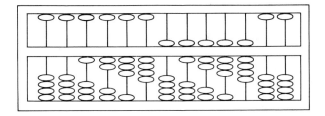

Figure 1-1 Japanese abacus, or Soroban

As scientists developed techniques that applied electrical and electronic principles to record storing, account handling, and bookkeeping, new and more versatile machines appeared. Again, the impetus of wartime activities (World War II) spurred development; in 1940 a computer using relays was devised, and pulse techniques developed for radar were incorporated into applied mathematics. Thus automation requirements, arising from production needs, resulted in machines that could perform routine tasks without human intervention. Soon these production machines were capable of making decisions concerning quality, quantity, and so forth.

The general development of electronic computers in World War II has been followed by an almost endless activity in the design and application of these "intelligent machines." Textbooks, papers, and magazines abound on both the technical and non-technical aspects of the "care and feeding" of computers. *The computer merely uses techniques that have existed for years.*

In this text our primary concern is to explore these basic techniques and their diverse applications; however, we shall pay particular attention to the study of that broad field of digital techniques which can be applied to data communication and to data processing. Since the term *digital techniques* may be unfamiliar, we shall begin by comparing them with common analog techniques.

1-2 ANALOG (LINEAR) VS. DIGITAL TECHNIQUES

Analog (Measuring) Devices

An analog (linear) *device* is basically a measuring instrument. The measured property can take on an infinite number of possible values. For example, consider temperature. The liquid thermometer (an analog device) measures and displays an approximate value. With sufficient precision, a thermometer reading of +70°F. might actually appear as +69.92987°F. There are an infinite number of possible temperatures between +69°F. and +70°F. The only limitation on resolution is the accuracy of the measuring device.

With a liquid thermometer, however, the *actual* temperature is not really being measured. Instead, the height of an enclosed column of liquid (mercury or alcohol) is made *analogous* to the temperature. The thermometer is a *transducer* that converts temperature into the expansion of a liquid column. Analog devices can also transduce physical phenomena into electrically measurable quantities using magnitudes or phases of voltages and currents, electrical resistances, shaft rotation, or the like. In such cases, the primary factor in the analog device* is the *magnitude* of the electrical output, called the electrical *signal*. That is, how large or how small is the signal?

Digital (Counting) Devices

Whereas an analog signal is usually a *smooth, continuously varying signal,* a digital signal has a *discrete, discontinuous character*. A major criterion for any digital device is the presence (or absence) of a signal—not its actual value. With digital conversion we want to know *if* something has changed—not how much it has changed. Numerical values are represented directly as digits (not as continuously varying quantities). *The digital device counts, while the analog device measures.*

In a digital device, the presence, absence, and repetition rate of the signals are the important factors—not the signal amplitude. A digital system transmits only discrete (step) changes and is discontinuous. (An analog system transmits physical quantity in a smoothly and *continuously* varying manner.)

An ordinary, hand-operated tally counter is a simple example of a digital device (Fig. 1-2). The counting lever must be pushed all the way down to register a count. There is no middle position of the lever that can register a half count. *It either counts one digit, or it does not count.*

Today all digital systems are based on whether something is actually there, whether it is true or false, whether there is a hole in a punched card, and so forth. Even mammoth, multimillion dollar computers are based on this simple yes/no decision. Complexity results when the equipment is instructed to make such decisions millions of times a second.

Digital Applications

Digital techniques are persistently invading fields that formerly used analog techniques. For example, remote measuring equipment, which once depended on analog methods, is now being replaced by

---

*Electronic components used with analog devices or transducers are mentioned in this text only where outputs from such devices are to be converted to digital operation.

Analog (Linear) vs. Digital Techniques 5

Figure 1-2 Tally counter (courtesy of Veeder-Root, Hartford, Conn.)

digital equipment. Analog measurements are converted at their source to digital signals, then transmitted, processed, stored, and acted upon at some remote point prior to readout and action (Fig. 1-3).

Telephone conversations are being broken up by digital equipment, combined into a composite signal, transmitted, and decoded at some distant terminal. Some approaches to telephone transmission actually convert conversations to digital values instead of just "breaking-up" information (Fig. 1-4).

Inaccuracies in reading conventional moving-pointer measuring instruments are being overcome by applying digital techniques. Digital voltmeters are common today, and so are the number of digital readout instruments available to the technical and scientific disciplines. Figure 1-5 shows a typical digital multimeter used for general testing and measurement.

Many functions formerly performed by analog devices in missile and aircraft guidance and control now use digital techniques (Fig. 1-6).

Figure 1-3 Data Acquisition Subsystem
Courtesy of Pheonix Data Inc.

Figure 1-4 Digital telephone system (courtesy of Lynch Communications Systems, San Francisco, CA.)

Greater accuracy results, and tremendous reduction in weight and volume is effected.

Digital techniques are also making inroads in medical electronics, the control of industrial processes and railroad trains, material handling, the monitoring and operating of ocean-going ships, the supervision of industrial plant functions, the centralized control of electrical power transmission and distribution, automobile servicing, machine control (Fig. 1-7), supermarkets (Fig. 1-8), the stock exchange (Fig. 1-9), and

Figure 1-5 Digital Multimeter (courtesy RACAL-DANA)

Analog (Linear) Circuit Analysis 7

Figure 1-6 F-111 Navigation display unit (left) and computer control (right) (courtesy Singer-General Precision, Inc., Kearfott Division)

other activities. Of course, many of these functions require the use of a digital computer, but again, the digital computer is just another device that performs tasks with digital assistance.

1-3 ANALOG (LINEAR) CIRCUIT ANALYSIS

The electronic circuits used in digital data handling and processing systems are subject to many of the same analysis techniques as those used for conventional analog (linear) circuits. To establish the portions

Figure 1-7 Dynapath System 4 Numerical Controller (courtesy Bendix Industrial Controls Division)

Figure 1-8 Digital Supermarket Control (courtesy Digitronics Corp.)

of the techniques that are similar, therefore, we shall use the analysis of a simple radio receiver as an example. *This is only an example.* If you are not familiar with radio receivers, don't worry. You won't have to analyze them in your digital work.

Block Diagram Analysis

A common way to begin the analysis of an electronic device is to develop an overall understanding of the complete device. This is usually accomplished by using a *block diagram.* Such a diagram indicates what is functionally necessary to meet the specifications for a particular piece of equipment. For example, assume that this device has the following specifications.

FM Tuning Range: 152 to 174 mHz
Bandwidth: 100 kHz at 6 dB down
Sensitivity: 6 μV at 5 kHz deviation
Squelch Sensitivity: Opens for 5 μV signal
Audio Output: 1 watt to 8 ohm speaker

Analog (Linear) Circuit Analysis

Antenna Impedance: 50 to 75 ohms
Power Requirements: 110 to 130 Volts, 60 Hz

The block diagram shown in Fig. 1-10* shows the structure of these specified functions.

Figure 1-9 Stock Exchange securities terminal (courtesy of SYCOR, Inc., Ann Arbor, Michigan).

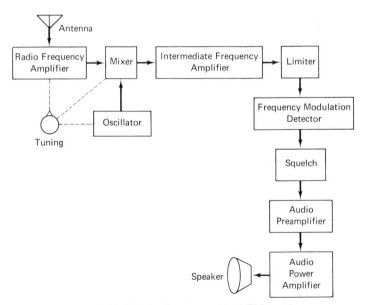

Figure 1-10 Block diagram, typical FM receiver

*Power connections are not shown, but are assumed to be present and properly connected.

Once the functions of individual blocks are determined and all of the functions are integrated, the operation of the radio receiver can be explained. Although many readers may be familiar with the operation of this device, such familiarity is not a prerequisite. The block diagram level of analysis purposely avoids using electronic circuit details. Only the *functions* of the blocks and the *flow of data* from block to block are important.

Here the antenna receives an incoming signal and passes it to the radio frequency (RF) amplifier which is tuned so that only the desired signal is made clearer and louder. In the mixer, the amplified signal is combined with a local oscillator signal to form the sum and difference frequencies. The RF amplifier, mixer, and oscillator are tuned simultaneously, thus maintaining constant sum and difference frequencies at the output of the mixer. The intermediate frequency (IF) amplifiers are tuned to select only the sum or difference frequency, which is amplified in these stages. Since this device is a frequency modulation (FM) receiver, limiter stages are incorporated to remove all amplitude modulation. The FM detector develops an audio output that is a close replica of the original modulation applied to the transmitter. To prevent constant background noise (always present without an incoming signal) from reaching the speaker, a squelch circuit is included. This circuit effectively blocks input to the audio preamplifier until a signal of a predetermined and preset level is reached. The squelch circuit then opens and allows the desired audio signal to be applied to the preamplifier for amplification. Sufficient power to operate the speaker is developed by the audio power amplifier stage. Finally, the speaker converts the electrical variations from this power amplifier into audible sounds.

Schematic Diagram Analysis

If further detail is required, the next level of explanation is usually accompanied by a *schematic diagram.* The schematic diagram (Fig. 1-11) shows such individual discrete components as transistors, capacitors, and resistors—all interconnected to perform the functions of individual blocks in the block diagram.* Although complete schematic diagrams without any correlation to block diagrams are commonly provided, schematic diagrams of individual blocks (circuit design permitting) should be furnished. The schematic diagram in Fig. 1-11 shows a typical mixer stage.

---

*Discrete component circuit analysis is beyond the scope of this text, but readers with sufficient electronic background should experience no difficulty with this circuit.

Digital Circuit Analysis

Figure 1-11 Schematic diagram, mixer stage

1-4 DIGITAL CIRCUIT ANALYSIS

A block diagram may also be used with digital equipment to functionally describe the *system operation*. Due to the complexity of most digital systems, however, it is impractical to provide a complete circuit schematic diagram. Besides, the use of only a few basic circuits is well suited to an intermediate level of analysis. The complexity of digital systems is due not to "exotic" circuits but to the multiple use of a few simple circuits. As you read this section you will see some examples of the techniques used in digital circuit analysis.

Block Diagram Analysis

Figure 1-12 shows a simple digital measuring system in block diagram form. Although greatly simplified, this digital measuring/recording system is a typical application of present digital techniques to industrial processes. Transducers, described in Sec. 1-2, provide inputs to this digital system. The electrical outputs of transducers vary considerably, depending on the conversion method used. Therefore a "linear" amplifier, called a *signal conditioner*, is commonly included in series with the transducer output either to amplify or limit to a predetermined range the input to the digital measuring/recording system.

The system shown in Fig. 1-12 has enough transducers and signal conditioners to measure all of the parameters required in the monitored process. Upon command from the *control unit*, the *multiplexer* selects one of the many input signals and feeds it to the *analog/digital converter*. Here the analog signal is converted by one of several methods to a pure

Figure 1-12 Digital measuring/recording system

digital signal and then routed to either the *data display* or the *data recorder*, or both, depending on the requirement. As soon as the conversion of one channel is complete, the multiplexer is stepped to the next channel, and the conversion process is repeated. An internal timing device, known as a *clock,* establishes the basic multiplexing cycle for the complete system. Many variations of sampling times, display times, and recording media exist, and provision is usually made for simple adjustment and selection of both input and internal parameters.

Schematic Diagram Analysis

Like the radio receiver, the functions of individual blocks are determined and then these block functions are integrated before a full explanation of the operation of the digital system is achieved. In digital systems, circuit schematics result in endless repetition of the same circuits, however. For example, Fig. 1-13, a partial schematic diagram, represents only a small portion of the clock in Fig. 1-12. For simplicity, only those discrete components that enter directly into the explanation of the operation are defined.

Figure 1-13 shows a circuit employed in digital clocks that divides by 6. For each 6 "input signals" received, 1 "output signal" is generated. A complete electronic circuit would obviously be quite lengthy. However investigation of Fig. 1-13 reveals some interesting aspects. The Q_1-Q_2 circuitry is almost identical to that of Q_3-Q_4 and Q_5-Q_6. The Q_1-Q_2, Q_3-Q_4, and Q_5-Q_6 circuits perform the same functions.

Logic Diagram Analysis

If the operation of these circuits can be described by sets of rules and represented by special symbols, then the complete circuit or system can be described by its own set of symbols. This new level of diagram is called a *logic diagram,* and its symbols are called *logic symbols*.

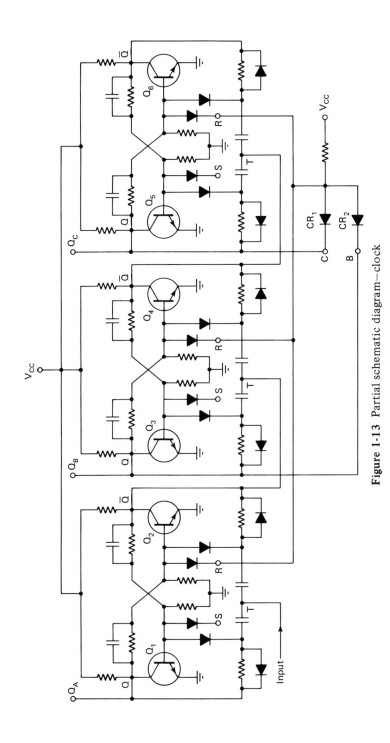

Figure 1-13 Partial schematic diagram—clock

Figure 1-14 Logic diagram, divide-by-six circuit

The logic diagram for the circuit diagram in Fig. 1-13 is shown in Fig. 1-14. Note that Fig. 1-14 shows a considerable decrease in complexity compared with Fig. 1-13. Only two symbols appear in the logic diagram. Both the AND GATE symbol (Fig. 1-15a) and the FLIP-FLOP symbol (Fig. 1-15b) represent *functions**—not the specific arrangement of discrete components shown in Fig. 1-13.

Referring to Figs. 1-13 and 1-14, the AND GATE corresponds to CR_1-CR_2 and associated components. The Q_1-Q_2, Q_3-Q_4, and Q_5-Q_6 circuitry correspond to FF_A, FF_B, and FF_C, respectively. But since we need to understand only the functions in logic diagrams, it is sufficient to realize that the AND GATE *combines signals* according to predetermined rules and that the FLIP-FLOP *stores information.* Actually, the "divide-by-6" device *counts* the number of times the input signal changes and also provides a change in output signal for each sixth input-signal change. Counting is accomplished through the binary number system (see Ch. 2); the condition of FLIP-FLOPS A, B, and C at any

(a) AND GATE

(b) FLIP-FLOP

Figure 1-15 Logic symbols (a) AND gate, (b) Flip-flop

*Numerous examples of electronic implementation of logic functions appear in Appendix D.

specific time is determined by how many input-signal changes have taken place. Thus the FLIP-FLOPS store numbers (at least temporarily) in the binary number system. The AND GATE determine what numbers are stored and route the control signals so that the device recycles back to the 0 count after detection of the sixth input change and generation of the output signal.

The digital circuit analyses performed throughout this text use the MIL-STD-806B logic symbols,* a portion of which appears in Appendix B. As these symbols appear in later chapters, the complete rules associated with their operation are developed. IEEE 91-1973 Standard also shows commonly used symbols. (See Appendix C.)

1-5 LOGIC ANALYSIS TECHNIQUES

HIGH and LOW

In addition to the new symbology and a new level of diagram, we must develop a new approach to signal analysis. As we mentioned earlier, a radio receiver functions by selecting incoming signals that have a particular frequency and amplitude, but a digital measuring device is activated by the *presence* (or *absence*) *of signals* and the *rate* at which they arrive.

Most digital devices operate on the principle of *bistable circuits.* This means that the devices are stable when they are in *either* of two states: *one state is represented by the presence of a signal, and the other state is represented by the absence of a signal.* In logic diagram analysis, *the relative level of the signal is the criterion.* A signal may be either high or low. The actual value of the voltage or current that will represent *HIGH* is determined by the device designers. Likewise, the actual value of voltage or current that represents *LOW* does not really affect the analysis of a logic diagram. *The important consideration is that* HIGH *represents one of the stable states and* LOW *represents the other.*

Binary Number System

As we mentioned briefly, counting is not performed by the familiar *decimal system.* Ten separate stable states would be needed to represent the decimal digits 0 through 9, inclusively; electronically, this becomes an almost insurmountable task. A number system using only 2

---

*MIL-STD-806B (Graphic Symbols for Logic Diagrams) is the established standard of notation. However, minor changes may occur in the future, and the latest revision may therefore be obtained from the Superintendent of Documents, U.S. Government Printing Office, Washington, D.C. 20402.

stable states does exist. This system is called the *binary number system* and has only *two values*, 0 and 1. The binary number system is vitally important to any logic circuit analysis, so it is discussed in detail in later chapters.

Boolean Algebra

Mathematics is a useful tool in digital circuit analysis. Algebra, trigonometry, and calculus are commonly used to analyze *analog circuits*. Because all logic operations in *digital circuits* depend on either the existence or absence of a signal, a variable may have only *one* of the two possible values. Reduced to a two-valued system, both mathematical operations and the analysis of digital circuits are greatly simplified. *The mathematics of two-valued logic is called Boolean algebra.* The basic postulates and theorems of Boolean algebra appear in Appendix A, but their applications are explained in detail as we encounter them throughout the text.

Maps and Truth Tables

The use of graphic methods for linear and nonlinear systems analysis is widespread. The characteristic curves of active devices such as semiconductors and vacuum tubes are applied to the design of electronic circuit operations. Experimental data are derived, tables manufactured, and graphs of responses of both active and passive devices developed. In the digital field, graphic techniques* are not as complicated, but they are still quite useful. Such techniques include truth tables, Venn diagrams, Veitch diagrams, Karnaugh maps, and so forth.

At this point, it might be useful to examine the application of a truth table to the analysis of the AND GATE. *A truth table is a tabular listing of all possible combinations of input conditions to a logic device, and a resultant output for each case.* As such, a truth table is a useful method of condensing all information concerning the functional characteristics of a logic device. Figure 1-16 is a simplified truth table for any 2-input AND GATE. Logic devices and combinations thereof may be analyzed by truth tables or by other techniques mentioned earlier.

Waveforms

Many of the most common measuring instruments used in analog and linear applications are not useful in digital system analysis, since the information most needed relates to the absence or presence of a signal

---

*The most common graphic techniques are examined in subsequent chapters.

Logic Analysis Techniques   **17**

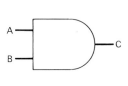

Input		Output
A	B	C
Low	Low	Low
Low	High	Low
High	Low	Low
High	High	High

Figure 1-16 Truth table, 2-input AND gate

and the rate at which it occurs. The *cathode-ray oscilloscope* (CRO) is, however, a useful measurement device. Within all digital devices, signals are constantly changing from HIGH to LOW, and in general, digital devices are *time-sequenced*. By using the CRO the time relationships between internal and external signals may be obtained. The commonly encountered signal changes (waveforms) of linear and analog circuits seldom appear in digital equipment. Instead, constantly changing HIGH and LOW signals are routed throughout the circuitry and may be observed on the CRO or other special logic-oriented test devices. These signal changes are unique to each type of circuit being investigated. Figure 1-17 shows the waveforms that might be encountered in the divide-by-6 circuit.

Waveforms are used in conjunction with logic diagrams, Boolean algebra, and graphic methods to integrate the analysis of digital equipment. All of these methods of analysis are discussed in detail later.

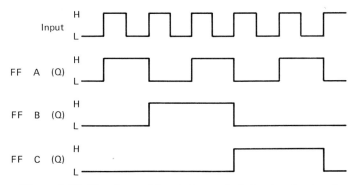

Figure 1-17 Waveforms (signal changes), divide-by-six circuit

PROBLEM APPLICATIONS

1-1. Trace the evolution of digital computers from ancient times to the present day.
1-2. Define the term *analog*.
1-3. Define the term *digital*.
1-4. List as many applications as possible where digital equipment is replacing or supplementing analog equipment.
1-5. Compare analog and digital signals and their characteristics.
1-6. What is the purpose of a block diagram?
1-7. What is the purpose of a schematic diagram?
1-8. What is the purpose of a logic diagram?
1-9. Why are logic diagrams used in conjunction with digital equipment?
1-10. Why is the binary number system used in digital equipment?
1-11. What is a truth table?
1-12. Why are waveforms used in troubleshooting digital circuits?
1-13. What is the most practical digital-equipment troubleshooting instrument?

2

NUMBERS AND COUNTING

2-1 THE DECIMAL NUMBER SYSTEM

Our constant, almost casual use of the decimal number system may prevent us from utilizing some of its more important features. Therefore review of certain properties and characteristics of the decimal system would be helpful.

The *radix* or *base* of a number system is the total number of *unique* (different) *symbols* available in that system. The largest-valued symbol always has a magnitude of *one less than the radix*. For example, in the decimal number system there are *ten* different symbols: 0, 1, 2, 3, 4, 5, 6, 7, 8, and 9, where the largest, 9, is one less than 10 (the base).

Counting

A table of progression of decimal number system symbols is shown in Fig. 2-1. Note especially the treatment of numbers greater in value than the basic symbols of the number system. The 0 symbol follows the highest-valued symbol each time all of the allowable symbols have been used in the counting sequence. When the 0 symbol appears, the results of accumulating the counts from 1 through 9 must be "shifted" into the column (position) immediately to the left of the 0 symbol. The basic counting sequence is then resumed in the present column. The 0 symbol is called a *shift marker*. This action can be seen at numbers 10, 20, and 100 in Fig. 2-1.

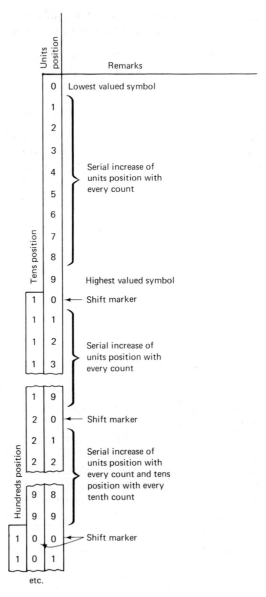

Figure 2-1 Decimal counting concepts

The number 006 (commonly written as 6) represents 6 units, while the number 060 (commonly written as 60) represents the quantity sixty. This method of using Arabic numeral symbols to indicate a magnitude or quantity is actually a type of shorthand. Each position in a decimal number has a value that is ten times the value of the next posi-

The Decimal Number System

tion to the right. *In other words, every positional value is a multiple of ten and can be expressed as ten raised to some exponent.*

Example 2–1. Express 123 in exponential form.
Solution: $123 = 1 \times 100 + 2 \times 10 + 3 \times 1$
$= 1 \times 10^2 + 2 \times 10^1 + 3 \times 10^0$

Note: Any number raised to the exponent 0 equals 1.

This progression of increasing exponents can be continued as far to the left of the decimal point as desired. The same progression can also be extended to the right of the decimal point, but then the exponents are *negative.* For example, the first position to the right of the decimal point is the "tenth's" position; it has a value of 10^{-1} or $1/10^1$.

Any number in the decimal number system can be represented like those in Fig. 2-2. The symbol placed in any position indicates how

Decimal Number	Thousands	Hundreds	Tens	Units	Decimal Point	Tenths	Hundredths	Thousandths		
	10^n	10^3	10^2	10^1	10^0	.	10^{-1}	10^{-2}	10^{-3}	10^{-m}	Power
	n	3	2	1	0		−1	−2	−3	−m	Position
		1000	100	10	1	.	0.1	0.01	0.001		Base 10 Equiv. value
1					1	.					
10				1	0	.					
123			1	2	3	.					
1234.56		1	2	3	4	.	5	6			
0.789						.	7	8	9		
10.10				1	0	.	1	0			
53				5	3	.					
45.312				4	5	.	3	1	2		
146.64			1	4	6	.	6	4			
2927.975		2	9	2	7	.	9	7	5		

Figure 2-2 Representation of decimal numbers

many *multiples* of that power of 10 are in the total quantity represented by the number.

2-2 THE BINARY NUMBER SYSTEM

Radix and Symbols

Now that some of the concepts and characteristics of the decimal number system have been established, we can ask the next logical question. Is it possible to have a number system with a radix less than 10? Let's look at the requirements for *any* number system. To begin: a *position marker* is needed to indicate positional shift; hence the zero is required. In addition to zero, at least one other symbol with a different value is required, thus establishing the need for the one.

Now comes the next question. Are two symbols sufficient to completely define a number system? Perhaps this question can best be answered by attempting to count in the decimal-system manner.

Counting

With a radix of 2 (just two symbols), the maximum number that can be shown with a single digit is the number 1 (one). If we want to show the decimal number two, the 1 must be moved one place to the left and the 0 must be used to indicate that a shift has taken place. So the decimal number two (in binary) is 10 (read as "one-zero" not "ten"). See Fig. 2-3 for the development of the *binary* (base 2) *counting table*.

Decimal	*Binary*	*Comments*
0	0000	Identical
1	0001	Identical
2	0010	Shift to the left required
3	0011	A 2 and a 1 makes 3
4	0100	Another shift to left required
5	0101	A 4 and a 1 makes 5
6	0110	A 4 and a 2 makes 6
7	0111	A 4 and a 2 and a 1 makes 7
8	1000	Another shift to left required
9	1001	An 8 and a 1 makes 9
10	1010	An 8 and a 2 makes 10
11	1011	An 8 and a 2 and a 1 makes 11
12	1100	An 8 and a 4 makes 12
13	1101	An 8 and a 4 and a 1 makes 13
14	1110	An 8 and a 4 and a 2 makes 14
15	1111	An 8 and a 4 and a 2 and a 1 makes 15

Figure 2-3 Binary counting table

The Binary Number System

Obviously, the ability to count exists in number systems other than base 10. Figure 2-3 has not been extended beyond count 15, but, for all practical purposes, it is entirely possible to count as *high* in a binary (base 2) number system as in a decimal (base 10) number system.

Figure 2-4 shows how any number in the binary system may be represented. Note that this figure appears similar to Fig. 2-2, which represented the decimal system. The only change is that *multiples of powers of* 2* (instead of powers of 10) are now being used. This is only reasonable, since the base of the binary system is 2!

Example 2-2. Express decimal number 1234.56 in the form of Eq. (2-1).
Solution: $n = 3$; $m = 2$

$N = 1 \times 10^3 + 2 \times 10^2 + 3 \times 10^1 + 4 \times 10^0 + 5 \times 10^{-1} + 6 \times 10^{-2}$
$N = 1000 + 200 + 30 + 4 + 0.5 + 0.06$
$N = 1234.56$

Decimal Number	2^n	Eights 2^3	Fours 2^2	Twos 2^1	Ones 2^0	Binary Point	Halves 2^{-1}	Fourths 2^{-2}	Eighths 2^{-3}	2^{-m}	
	n	3	2	1	0	.	-1	-2	-3	$-m$	Position
		8	4	2	1	.	$\frac{1}{2}$	$\frac{1}{4}$	$\frac{1}{8}$		Base 10 equiv. value
1					1	.					
2				1	0	.					
5			1	0	1	.					
15		1	1	1	1	.					
10.5		1	0	1	0	.	1	0			
0.875					0	.	1	1	1		
7.125		0	1	1	1	.	0	0	1		
8.625		1	0	0	0	.	1	0	1		
1.25		0	0	0	1	.	0	1	0		
5.375		0	1	0	1	.	0	1	1		

Figure 2-4 Representation of binary numbers

*A table of powers of 2 is found in Appendix E.

Any Number System

Actually, *any number* may be used as the base for a number system. Figure 2-5 shows the skeleton of a general number system with a radix r. Note that the same format is used as had been used for the base 2 and base 10 systems.

Need for the Binary Number System

The binary system eliminates the extreme difficulty of *accurately* representing decimal digits in electronic equipment. Ten discrete levels are required to allow use of all the symbols in the decimal system. These ten discrete levels must be sufficiently distinguishable from each other to assure that no mistakes in symbol recognition occur. Such criteria are overly restrictive in most electronic data-handling devices.

In the binary system, however, only *two discrete levels* are needed. Fortunately, decimal values can be coded into equivalent binary values so that the binary system may be used. Two arbitrary levels of voltage or current can be established, thus leaving no doubt about which binary value is represented. The value 0 is assigned to one of the arbitrary levels and the value 1 to the other level.

Nearly all of the data handling and processing equipment discussed in this text uses the binary system. Nevertheless one of the basic requirements in working with and understanding digital equipment is the ability to convert from decimal to binary and vice versa.

2-3 NUMBER SYSTEM CONVERSIONS

It is very easy to confuse a number in the binary system with a number in the decimal system. For example, binary 1010 is equivalent to decimal 10. To minimize confusion, therefore, we have devised a *system of subscripts*. Thus the *base* of the number system appears as a subscript— binary 1010 becomes 1010_2 (base 2) and decimal 10 becomes 10_{10} (base 10). In most cases, however, the number system being used is obvious, so it is unnecessary to employ subscripts.

Decimal to Binary (D/B) Conversion

Decimal-to-binary (D/B) *integer conversion* is initially approached on a "reasoning" basis. After establishing that such conversions are easily performed, a general solution can then be presented. For example, consider D/B conversion of 14_{10} (decimal 14). The only "tools" needed are a table of powers of 2 (Appendix E) and a knowledge of the principles of positional notation described in Eqs. (2-1), (2-2), and (2-3). Just as 14_{10} is really $1 \times 10^1 + 4 \times 10^0$, the equivalent binary numbers

Number System Conversions

Figure 2-5 General number system

may be made up of sums of products of admissible binary symbols and powers of 2. Examination of the table of powers of 2 shows that 14_{10} is less than 1×2^4 (16_{10}), so 1×2^4 cannot be used. However 1×2^3 (8_{10}) is less than 14_{10}, so 14_{10} must include 1×2^3. The remainder of 6_{10} must be examined in the same manner as was 14_{10}. 1×2^3 has already been used, so 1×2^2 (4_{10}) must be the next value. Note that 6_{10} includes 1×2^2 with a remainder of 2_{10}. Continuing the process, we discover that 2_{10} includes 1×2^1 with a remainder of 0. Therefore

$$14_{10} = 1 \times 2^3 + 1 \times 2^2 + 1 \times 2^1 + 0 \times 2^0 \text{ or } \mathbf{1110_2}$$

This operation is called the *direct method of D/B conversion*. Its formal, step-by-step procedure is listed below.

1. Successively subtract the largest possible power of 2 from the decimal number.
2. Continue performing step 1 until a zero remainder occurs. (Negative remainders are not allowed.)
3. Write a 1 for each power of 2 subtracted.
4. Write a 0 for each power of 2 not used.
5. Write the 1s and 0s in the order in which the subtractions are performed. All powers of 2 (from the largest through and including 2^0) must be accounted for by using either 1 or 0.

Sample conversions are shown in Examples 2-3 and 2-4.

Example 2-3. Convert decimal number 14 to a binary number using the direct method.

Solution:
```
      14
    -  8  (2³)    1   Most Significant Digit (MSD)
       6
    -  4  (2²)    1
       2
    -  2  (2¹)    1
       0
    -  0  (2⁰)    0   Least Significant Digit (LSD)
       0
```

$$1 \times 2^3 + 1 \times 2^2 + 1 \times 2^1 + 0 \times 2^0 = 14_{10}$$
$$\mathbf{1110_2 = 14_{10}}$$

Example 2-4. Convert 25_{10} to a binary number using the direct method.

Solution:
$$\begin{array}{r}25\\-16\ (2^4)\quad 1\ \text{(MSD)}\\\hline 9\\-8\ (2^3)\quad 1\\\hline 1\\-1\ (2^0)\quad 1\ \text{(LSD)}\\\hline 0\end{array}$$

$\left\{\begin{array}{l}\text{Note that 4 } (2^2)\text{ and}\\ 2\ (2^1)\text{ cannot be subtracted from 1;}\\ \text{they must be included (Rule 5).}\end{array}\right.$

$$1 \times 2^4 + 1 \times 2^3 + 0 \times 2^2 + 0 \times 2^1 + 1 \times 2^0 = 25_{10}$$
$$11001_2 = 25_{10}$$

It is possible (and sometimes quite practical) to perform decimal binary conversion by the *successive division method*. A table of powers of 2 is not required—all we have to know is how to divide by 2! Here is the detailed step-by-step procedure and some typical examples.

1. Successively divide the decimal number by 2.
2. Place the quotients directly beneath the dividend.
3. Place the remainders opposite the quotients.
4. The equivalent binary number becomes the remainders—the *final* remainder being the *most significant digit* (MSD) and the *first* remainder being the *least significant digit* (LSD).

Example 2-5. Convert 14_{10} to a binary number using the successive division method.

Solution:
$$\begin{array}{rl}2\,\lfloor\underline{14}&\\ 2\,\lfloor\underline{\ 7}&0\ \text{(LSD)}\\ 2\,\lfloor\underline{\ 3}&1\\ 2\,\lfloor\underline{\ 1}&1\\ 0&1\ \text{(MSD)}\end{array}$$

$1110 = 14_{10}$

Example 2-6. Convert decimal number 25 to a binary number using the successive division method.

Solution:
$$\begin{array}{rl}2\,\lfloor\underline{25}&\\ 2\,\lfloor\underline{12}&1\ \text{(LSD)}\\ 2\,\lfloor\underline{\ 6}&0\\ 2\,\lfloor\underline{\ 3}&0\\ 2\,\lfloor\underline{\ 1}&1\\ 0&1\ \text{(MSD)}\end{array}$$

$11001_2 = 25_{10}$

Number System Conversions

Decimal-to-binary conversion of fractional numbers can be performed by applying the following steps.

1. Successively multiply the decimal fraction by 2.
2. Ignore any integer results when performing the multiplication.
3. Place the products directly below the multiplicand.
4. Work toward zero or until the number of digits required for degree of precision is reached.
5. The first integer part obtained is the MSD.

Some typical examples of this method of conversion are shown below.

Example 2-7. Convert 0.8125_{10} to a binary number.

Solution: $\boxed{0}$. (8125) x 2
(MSD) $\boxed{1}$. (6250) x 2
         $\boxed{1}$. (2500) x 2 $0.8125_{10} = 0.1101_2$
         $\boxed{0}$. (5000) x 2
(LSD) $\boxed{1}$. (0000)

Example 2-8. Convert 0.7893 to a binary number.

Solution: $\boxed{0}$. (7893) x 2
(MSD) $\boxed{1}$. (5786) x 2
         $\boxed{1}$. (1572) x 2
         $\boxed{0}$. (3144) x 2
         $\boxed{0}$. (6288) x 2 $0.7893_{10} = 0.1100101_2$
         $\boxed{1}$. (2576) x 2
         $\boxed{0}$. (5152) x 2
(LSD) $\boxed{1}$. (0304)

Binary to Decimal (B/D) Conversion

Conversion from decimal to binary provides the most efficient method of conditioning inputs to digital data processing and computing equipment. Once the required operations are performed, however, a problem arises. Since the binary system is not employed by everyone, the results of the performed operations may not be easily readable. Therefore it may be desirable to reconvert binary numbers into decimal numbers. Fortunately, decimal-binary (D/B) conversion processes have binary-decimal (B/D) inverses.

The *direct method of* B/D *conversion* is accomplished by using either a B/D conversion table (Appendix E) or a powers of 2 table (Appendix E). Two basic rules apply.

1. A 1 in any digit position indicates that the corresponding power of 2 is to be used.
2. A 0 indicates that the power of 2 is not required.

Some examples are presented below.

Example 2-9. Convert 11011_2 to a decimal number using the direct method.

Solution: $11011_2 = 1 \times 2^4 + 1 \times 2^3 + 0 \times 2^2 + 1 \times 2^1 + 1 \times 2^0$
$= 1 \times 16 + 1 \times 8 + 0 \times 4 + 1 \times 2 + 1 \times 1$
$= 16 + 8 + 2 + 1$
$11011_2 = 27_{10}$

Example 1-10. Convert 0.1101_2 to a decimal number using the direct method.

Solution: $0.1101_2 = 1 \times 2^{-1} + 1 \times 2^{-2} + 0 \times 2^{-3} + 1 \times 2^{-4}$
$= \frac{1}{2} + \frac{1}{4} + 0 + \frac{1}{16}$
$= 0.5 + 0.25 + 0.0625$
$0.1101_2 = \mathbf{0.8125_{10}}$

Although it is not absolutely necessary, B/D conversion of mixed numbers is usually performed separately for the integer and for the fractional portions. The individual parts of the number are then combined, as Example 2-11 illustrates.

Example 2-11. Convert 101101.0101_2 to a decimal number using the direct method.

Solution: $101101 = 1 \times 2^5 + 0 \times 2^4 + 1 \times 2^3 + 1 \times 2^2 + 0 \times 2^1 + 1 \times 2^0$
$= 1 \times 32 + 0 \times 16 + 1 \times 8 + 1 \times 4 + 0 \times 2 + 1 \times 1$
$= 32 + 0 + 8 + 4 + 0 + 1$
$= 45_{10}$
$0.0101 = 0 \times 2^{-1} + 1 \times 2^{-2} + 0 \times 2^{-3} + 1 \times 2^{-4}$
$= 0 + \frac{1}{4} + 0 + \frac{1}{16}$
$= 0.3125$

Therefore $101101.0101_2 = \mathbf{45.3125_{10}}$

B/D conversion of integers may be approached by the *successive multiplication method,* sometimes referred to as the *double-dabble method,* which is the inverse of the successive division method used in D/B conversion. This method of B/D conversion can be performed by applying the following steps.

Number System Conversions

1. Double the highest order binary digit.
2. Add this doubled value to the next lower order binary digit and record the sum.
3. Double the sum obtained.
4. Add this doubled value to the next lower order binary digit and record the new sum.
5. Continue Steps 3 and 4 until the last (lowest-order) binary digit has been added to the previously doubled sum and a final sum has been obtained. The final sum is the decimal equivalent of the binary number.

Examples 2-12 and 2-13 demonstrate the double-dabble method.

Example 2-12. Convert 10100_2 to a decimal number using the double-dabble method.

Solution:
Step 1.	$1 \times 2 = 2$
Step 2.	$0 + 2 = 2$
Step 3.	$2 \times 2 = 4$
Step 4.	$1 + 4 = 5$
Step 3.	$5 \times 2 = 10$
Step 4.	$0 + 10 = 10$
Step 3.	$10 \times 2 = 20$
Step 4.	$0 + 20 = 20$
Step 5.	Final Sum = 20

Example 2-13. Convert 11100_2 to a decimal number using the double-dabble method.

Solution:
Step 1.	$1 \times 2 = 2$
Step 2.	$1 + 2 = 3$
Step 3.	$3 \times 2 = 6$
Step 4.	$1 + 6 = 7$
Step 3.	$7 \times 2 = 14$
Step 4.	$0 + 14 = 14$
Step 3.	$14 \times 2 = 28$
Step 4.	$0 + 28 = 28$
Step 5.	Final Sum = 28

2-4 THE COUNTING PROCESS

Both the ability to convert from one number system to another and the ability to count are useless without a means to mechanize the counting process. Paper and pencil sufficed for centuries, but now high-speed counting requirements demand new methods. Once again, since the decimal number system is relatively well understood, the mechanics of decimal counting have been used to develop a method of mechanization.

Figure 2-6 Odomoter mechanism (courtesy Veeder-Root, Hartford, Conn.)

Decimal Counting

The concepts of decimal counting were explained in Section 2-1. Now consider, as an example, the odometer on an automobile (Fig. 2-6). As the miles go by, the units dial increases. Then, as the units dial rolls past 9 to 0, the tens dial increases by 1, as shown in the decimal counting table in Fig. 2-1. The process will continue, following the procedure described in Sec. 2-1.

Another example is the simple adding machine. When adding 5 to 6, the sum produced exceeds the highest-valued symbol available in the number system. As in the odometer example, passing from 9 to 0 on the units dial causes the tens dial to increase by 1. Thus increasing the units dial on the adding machine one more time results in a sum of 11. The drawing in Fig. 2-7 shows how the "carry" is accomplished.

Binary Counting

Binary counting is easier to mechanize than decimal counting, but examples of binary counters are not common. The device* shown in Fig. 2-8, however, does verify the Binary Counting Table in Fig. 2-3. Continuous reference should be made to Fig. 2-8 while reviewing the binary counting operation.

All "rockers" are initially positioned so that the 1s are covered and the 0s are uncovered. This presets the counter to 0000. The first ball dropping through the chute is deflected to the *left* by the Least Significant-Digit (LSD) rocker separator. The weight of the ball causes the rocker to tip to the 1 position, and the ball is then routed to the storage area via the return chute. One ball has been used (Fig. 2-9a), so the binary counter now displays 0001 (1_{10}).

The second ball through the chute strikes the *right* side of the LSD rocker, causing it to tip to the right which covers the 1 and uncovers

---

*All major parts of the mechanical binary counter are labeled to conform with nomenclature used in our explanation.

Figure 2-7 "Carry" mechanism

the 0. The ball rolls from the right side of the LSD rocker to the left side of the next rocker. Ball-weight causes this rocker to tip to the left, thus covering the 0 and uncovering the 1. The second ball is now routed to the storage area via the return chute. Two balls have been used (Fig. 2-9b), so the binary counter now displays 0010 (2_{10}).

The third ball through the chute strikes the left side of the LSD rocker, tipping it to the 1 position. The ball is routed to the storage area via the return chute. The binary counter display (Fig. 2-9c) is now 0011 (3_{10}).

The fourth ball through the chute strikes the right side of the LSD rocker, tipping it to the 0 position. The ball rolls from the LSD rocker to the right side of the next rocker, tipping it to the 0 position. It then rolls from this rocker to the left side of the next rocker. This tips to the 1 position, and the fourth ball returns to the storage area. The binary counter (Fig. 2-9d) now displays 0100 (4_{10}).

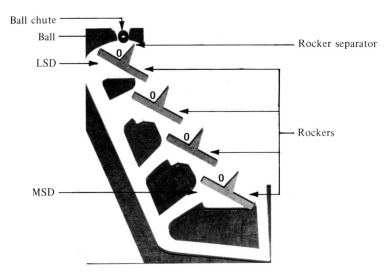

Figure 2-8 Mechanical binary counter

Figure 2-9 Mechanical binary counter, counts 1, 2, 3, and 4

As balls continue to drop through the chute, the counter is actually reproducing the Binary Counter Table in Fig. 2-3. We recommend that the reader continue this exercise until decimal count 15 is reached —then proceed to one additional count (16_{10}) to gain a comprehensive understanding of the mechanics of binary counting.

2-5 THE FLIP-FLOP (FF)

The mechanical binary counter performs its required functions adequately but slowly. If operational speed is needed, a *bistable electronic circuit,* called a *flip-flop* (FF), should be used. This bistable FF duplicates the action of each of the bistable rockers on the mechanical counter. The following discussion shows how *electronic counting* is accomplished. More detail appears in chapters 8 and 10.

Physical Form

Flip-flops are available in many different physical forms. Early digital devices used electromechanical relay, vacuum tube, and transistor FFs. In recent years *integrated circuit* FFs have almost entirely replaced the other types. Semiconductor technology has developed subminiature assemblies that contain all the functions of a FF on a single chip of silicon less than 0.05 inch square and 0.006 inch thick. These assemblies, called *integrated circuits* (ICs), package complex logic functions into a space no larger than a single transistor (about 0.1 inch square). Modern technology builds complete digital systems containing tens of thousands of gates and FFs on a single IC. Figure 2-10 shows various digital logic packages.

Symbology

Because we are not concerned with the individual electronic parts and their interconnections, a single symbol is sufficient to represent the FF. Although many different symbols are used, this text will adhere to the United States military standard containing graphic symbols for use with logic diagrams (MIL-STD-806B). As each symbol* is introduced, its use is explained.

The FF symbol that we have selected was first shown in Fig. 1-15b. Its shape** (and that of most logic symbols) denotes the logical operation performed by the device it represents. The standard FF has *three inputs,* labeled S, C, and T. The S is the SET *input*; the C is the CLEAR or RESET *input*; and the T is the TRIGGER or TOGGLE *input*. In many cases, only the inputs required to perform the FFs operation in the logic circuit are shown. Two outputs are provided, named the Q *output* and the $\overline{Q}$ *output*. ($\overline{Q}$ is spoken and read as "NOT Q.")

---

*Appendices B and C contain all symbols used in this text and their relationship to other commonly used symbols.
**It is important in drawing symbols to use the proper size and shape. Templates conforming to MIL-STD-806B are readily available and should be used when drawing logic diagrams.

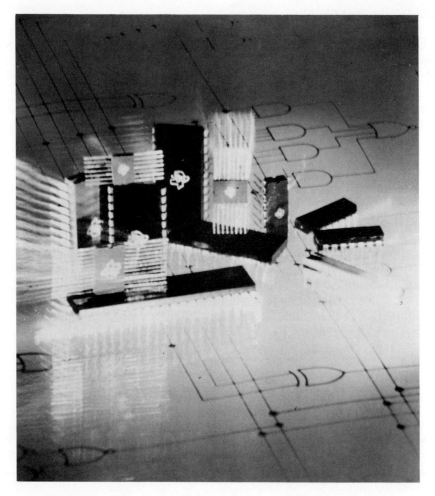

Figure 2-10 Digital logic packaging (courtesy Texas Instruments)

Rules of Operation

The reaction of a FF to the original application of power is both unimportant and indeterminate. Most analog devices assume a specific input/output (I/O) relationship when power is initially applied. But since the FF is bistable, it may take on *either* of its two states upon application of power. Many different terms are used to describe these states: ON and OFF, SET and RESET, SET and CLEAR, ONE and ZERO, 1 and 0, and Q and $\bar{Q}$. The basic rules of FF operation found within the framework of MIL-STD-806B provide a means of determining the state of a FF.

The Flip-Flop (FF)

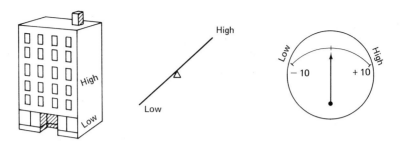

Figure 2-11 HIGH and LOW analogies

If the flip-flop is in the set state, the Q output is HIGH and the $\bar{Q}$ output is LOW.

Now it becomes necessary to define what we mean by HIGH and LOW. Since there are many ways of implementing the FF, different combinations of circuit voltages or currents can be used to indicate HIGH and LOW. So, we must again paraphrase MIL-STD-806B.

A HIGH output is relatively high in respect to the LOW output.

Imagine a physical analogy: if you are on the "up" side of a teeter-totter, you are *relatively high* with respect to the "down" side; or if you are on the third floor of a building, you are relatively high compared with the ground floor. For the electronically oriented reader, 0 volts might be LOW, and +10 volts might be HIGH; while –10 volts might be LOW and 0 volts might be HIGH. It is the *separation between two discrete values of some parameter* that determines HIGH and LOW. In any case, HIGH is *always* higher than LOW! Figure 2-11 presents a visual representation of this concept.

Flip-Flop Operation

A common method of illustrating FF operation is to draw diagrams representing HIGHs and LOWs. (*H* normally denotes HIGH, and *L* denotes LOW.) These diagrams are called *waveforms* and are nothing more than the "pictures" that would appear on a cathode-ray oscilloscope (CRO) when the FF is operating. All waveforms are labeled to firmly define which level is HIGH and which is LOW.

As we stated earlier, if the Q output is HIGH, the FF is in the SET or the 1 state. If, however, the $\bar{Q}$ output is HIGH, the FF is in the CLEAR, RESET, or 0 state. A very interesting aspect of the FF is that the Q output and the $\bar{Q}$ output are always *opposite* to each other. When the Q output is HIGH, the $\bar{Q}$ output is LOW and vice-versa. This is

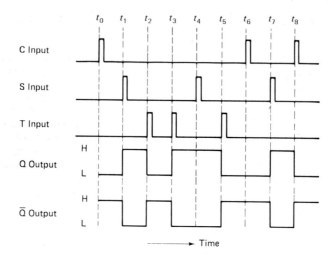

Figure 2-12 Flip-flop operation

equivalent to having available *both* a binary number and its ones complement. That availability is quite advantageous in performing arithmetic operations in the binary number system.

No matter what state the FF is in, it can be put into the SET state by applying a HIGH-going signal to the SET input; similarly a HIGH-going transition on the CLEAR input places the FF into the CLEAR state. A HIGH-going transition on the TRIGGER input also causes the FF to change state; if it is in the CLEAR state it will go SET, and if it is in the SET state it will go CLEAR. All of the operations of the FF may be seen in Fig. 2-12, which is a waveform diagram of FF states based on specific input signals.

This diagram represents the conventional presentations encountered in instruction handbooks for digital data processing equipment. Such diagrams are *time-oriented,* where the time starts on the *left* side of the page and progresses to the right. Time intervals are represented by subscripted letters (t_0, t_1, etc.) with t_0 usually representing the start of an operation, as it does in Fig. 2-12. The FFs response to various inputs (keyed to time intervals) is listed in Table 2-1.

One specific combination of input signals in Fig. 2-12 is conspicuous by its absence. When actuating signals are applied to 2 or more inputs *simultaneously,* the results are *indeterminate*—the FF *may* or *may not* change state! Such a situation is unacceptable, and precautions must be taken to prevent such *multiple-input* signals. Methods of preventing indeterminate operation will be discussed in later chapters.

Often it is inconvenient to use either CROs or other test instruments capable of displaying waveforms. An alternate method of deter-

The Flip-Flop (FF)

TABLE 2-1 FF response to various inputs at selected time intervals

Time	Input Terminal	Response
t_0	CLEAR	FF goes to CLEAR state
t_1	SET	FF goes to SET state
t_2	TRIGGER	FF changes state (CLEAR)
t_3	TRIGGER	FF changes state (SET)
t_4	SET	FF stays in SET state
t_5	TRIGGER	FF changes state (CLEAR)
t_6	CLEAR	FF stays in CLEAR state
t_7	SET	FF goes to SET state
t_8	CLEAR	FF goes to CLEAR state

mining the state of a FF employs *indicator lamps*. For example, by using electronic circuitry called *lamp drivers* in conjunction with indicator lamps, it is possible to cause the lamp to be turned on fully when the input is HIGH and turned completely off when the input is LOW. Connection of the lamp driver to the Q output of the FF enables visual indication of the state of the FF. Thus when the FF is in the SET state and the Q output is HIGH, the indicator lamp is ON; when the FF is in the CLEAR state and the Q output is LOW, the indicator lamp is OFF.

The waveforms in Fig. 2-12 can be modified, as in Fig. 2-13, to show visual indications correlated with waveforms and FF states. The

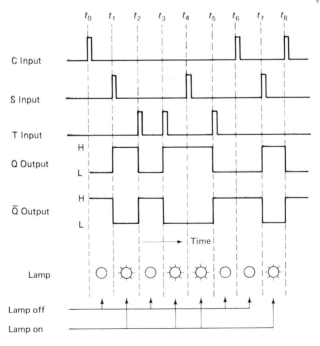

Figure 2-13 Flip-flop operation with visual indication

Figure 2-14 Digital techniques training device
Courtesy Heath Co.

use of visual indicators in place of waveform display is commonly encountered not only in such digital devices as computers but also in training devices (Fig. 2-14).

2-6 A BINARY COUNTER

Logic Diagram

An examination of our explanations of FF operation and binary counting should clearly show that it is possible to count only 0 and 1 with a single FF. To count to numbers greater than 1 in the binary system, we must devise a way to count the second, third, and fourth orders (and so on). If properly connected, *one* FF may be used to represent *each* binary order. A logic diagram to accomplish this counting function is shown in Fig. 2-15.

Note that the logic diagram and waveforms in Fig. 2-15 do not agree with conventional drawing practices. However, to simplify the explanation of circuit operation, a certain amount of "artistic license" was exercised. Each FF was given a name that corresponds to the binary

A Binary Counter

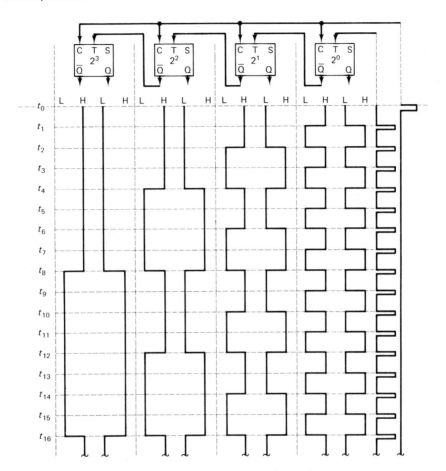

Figure 2-15 Binary counter and waveforms

order that it represents. The right-most FF is labeled 2^0; the FF immediately to the left is labeled 2^1. Note the correspondence between FF positions, names, and order when compared with conventional presentation of binary numbers, i.e., LSD on the right and MSD on the left.

Also note that the accompanying waveforms are "sideways," occurring from top to bottom instead of from left to right. Required power to each FF is assumed to be applied, and it is further assumed that FF states prior to t_0 either are unimportant or are to be disregarded.

To assure that counting begins with 0 at t_0, an actuating signal is applied simultaneously to the C (CLEAR) input of all FFs. Note that all C inputs are connected together in Fig. 2-15 and that the CLEAR pulse occurs on the time scale at t_0. Thus all FFs start out in the 0 state, and the binary number represented is 0000_2. The T (TRIGGER) input* of the LSD (2^0) FF is connected directly to the source of actuating signals to be counted. The $\bar{Q}$ output of the LSD FF connects directly to the T input of the next-most-significant digit FF (2^1). In a similar manner, $\bar{Q}$ outputs of the other FFs connect to the T inputs of next-most-significant digits, until the MSD FF is reached.

Analysis

At t_1 the first actuating signal to be counted is applied to the T input of the 2^0 FF, causing it to change state. The Q output (which was LOW) goes HIGH, and $\bar{Q}$ output (which was HIGH) goes LOW. Since the $\bar{Q}$ output of the 2^0 FF is directly connected to the T input of the 2^1 FF, a changing signal is coupled to the 2^1 FF. However, it is a LOW-going signal and causes no change. If the 2^1 FF doesn't change state, no change is coupled to the T input of the 2^2 FF and it will not change state. (Similarly, no change occurs in the 2^3 FF.) With the 2^3, 2^2, and 2^1 FFs in the RESET state and the 2^0 FF in the SET state, the binary number represented is 0001.

No change occurs in any of the FFs until the arrival of the actuating signal at t_2. The leading edge (HIGH-going portion) of the pulse causes the 2^0 FF to change from the SET to the RESET state, resulting in a HIGH-going $\bar{Q}$ output. This HIGH-going transition, which is connected to the T input of the 2^1 FF, causes the 2^1 FF to go from the RESET to the SET state.

The $\bar{Q}$ output of the 2^1 FF, which is connected to the T input of the 2^2 FF, has no effect on the 2^2 FF state, since the change is LOW-going—not HIGH-going. Since no change occurs in the 2^2 FF, no change occurs in the 2^3 FF and the binary number 0010 is established. Note that this corresponds to decimal number 2, which represents the number of input-actuating signals applied to the counter.

The arrival of the third actuating trigger pulse, t_3, will once again cause the 2^0 FF to change state. In going from RESET to SET state, the $\bar{Q}$ output goes from HIGH to LOW. Again, according to the rules for operation of the FF, a LOW-going signal transition will not actuate

---

*Methods of generating trigger inputs are discussed later. At this point, it is assumed that such methods do exist and that the trigger inputs are arriving at specific times, as pointed out in the diagram.

A Binary Counter

the next FF; therefore the 2^1 FF will not change state. Consequently, both the 2^2 and the 2^3 FFs remain unchanged. The binary number now stored is 0011_2, which corresponds to 3_{10}.

Application of the t_4 trigger pulse causes the 2^0 FF to RESET, and the resulting HIGH-going transition of the $\bar{Q}$ output triggers the 2^1 FF from SET to RESET state. Since the $\bar{Q}$ output of the 2^1 FF then changes from LOW to HIGH, it also activates the T input of the 2^2 FF, causing it to be SET. An evaluation of the FF states now shows the following results.

2^0 RESET, storing a binary 0
2^1 RESET, storing a binary 0
2^2 SET, storing a binary 1
2^3 RESET, storing a binary 0

The counter stores the number 4_{10}, represented as 0100_2.

The reader should continue logic analysis of the counter in order to gain valuable experience in this method of analysis. In particular, the reaction of the counter at t_{16} should be critically analyzed—a number of interesting conclusions may be reached. For example, in this particular logic circuit, a higher order FF can change state only when the next-lowest-order FF changes from SET to RESET. Examination of the waveforms for the 2^0 FF shows that it changes state *every time* an input trigger pulse is received. The 2^1 FF changes state for *every two* input triggers; the 2^2 FF changes state for every *four* input triggers; and the 2^3 FF changes state for every *eight* input triggers. Note the correspondence between FF names and the number of times they change state. Obviously, the name of the FF (which represents the positional order in the binary system) not only defines the decimal value of that position but also defines how many inputs may be counted prior to change of state. This implies a division capability, which will be discussed in detail later.

The State Table

A table of FF states (Fig. 2-16) correlated with input triggers may be derived using a complete logic analysis of the counter and the resulting waveforms. The table starts with all FFs in the RESET state (all 0s). Then, when input triggers are applied to the T input of the 2^0 FF, the individual FFs change state, as described earlier. The *state table* therefore describes counter operations in a manner as meaningful as the waveforms. The reader should carefully compare Fig. 2-16 with Fig. 2-15 to establish this correlation.

2^3	2^2	2^1	2^0
0	0	0	0
0	0	0	1
0	0	1	0
0	0	1	1
0	1	0	0
0	1	0	1
0	1	1	0
0	1	1	1
1	0	0	0
1	0	0	1
1	0	1	0
1	0	1	1
1	1	0	0
1	1	0	1
1	1	1	0
1	1	1	1

Figure 2-16 Binary counter state table

EXERCISES

Convert the following decimal numbers to exponential form.

2– 1.   7
2– 2.   206
2– 3.   5280
2– 4.   0.6836
2– 5.   914.254

Convert the following exponential numbers to decimal numbers.

2– 6. 5×10^0
2– 7. $1 \times 10^2 + 3 \times 10^1 + 7 \times 10^0$
2– 8. $3 \times 10^3 + 6 \times 10^2 + 4 \times 10^1 + 2 \times 10^0$
2– 9. $1 \times 10^{-1} + 2 \times 10^{-2} + 4 \times 10^{-3} + 3 \times 10^{-4}$
2–10. $4 \times 10^2 + 7 \times 10^1 + 6 \times 10^0 + 8 \times 10^{-1} + 3 \times 10^{-2} + 9 \times 10^{-3}$

Convert the following binary numbers to exponential form.

2–11.   1
2–12.   101
2–13.   11001
2–14.   0.1101
2–15. 1100111.1011

Convert the following exponential numbers to binary numbers.

2–16. $0 \times 2^1 + 1 \times 2^0$
2–17. $1 \times 2^2 + 1 \times 2^1 + 0 \times 2^0$
2–18. $1 \times 2^5 + 1 \times 2^4 + 0 \times 2^3 + 1 \times 2^2 + 0 \times 2^1 + 1 \times 2^0$
2–19. $0 \times 2^{-1} + 1 \times 2^{-2} + 1 \times 2^{-3} + 0 \times 2^{-4} + 1 \times 2^{-5}$
2–20. $1 \times 2^3 + 1 \times 2^2 + 0 \times 2^1 + 1 \times 2^0 + 0 \times 2^{-1} + 1 \times 2^{-2}$

A Binary Counter

Convert the following decimal numbers to their binary equivalents.

2–21.	3	2–31.	126
2–22.	7	2–32.	400
2–23.	8	2–33.	2048
2–24.	12	2–34.	3000
2–25.	13	2–35.	4000
2–26.	16	2–36.	0.5
2–27.	24	2–37.	0.125
2–28.	38	2–38.	0.03125
2–29.	79	2–39.	0.42
2–30.	84	2–40.	3.1416

Convert the following binary numbers to their decimal equivalents.

2–41.	101	2–51.	110101
2–42.	111	2–52.	0.0011
2–43.	1101	2–53.	0.0101
2–44.	01101	2–54.	1.01
2–45.	10111	2–55.	1.011101
2–46.	11010	2–56.	11.101
2–47.	0011101	2–57.	0101.1001
2–48.	100000	2–58.	1101.0001
2–49.	101101	2–59.	11011.1101
2–50.	110001	2–60.	10000100000.010001

PROBLEM APPLICATIONS

2-1. Define the term *radix*.
2-2. Why is the binary number system used in digital logic?
2-3. What is the decimal value of the largest number that can be represented by a 10-digit binary number?
2-4. How many binary digits would it take to represent the decimal number 1000?
2-5. Explain the general concept of decimal counting, i.e., the use of symbols and shift-markers.
2-6. Explain the general concept of binary counting, i.e., the use of symbols and shift-markers.
2-7. What is a *flip-flop* (FF)?
2-8. Draw the standard symbol for a FF and label all inputs and outputs.
2-9. Define HIGH and LOW as used in digital logic.
2-10. How can the state of a FF be determined?
2-11. Why are waveform drawings used to show FF operation?
2-12. Define the activating signal that causes FF transition.
2-13. What is the function of the CLEAR input on a FF?
2-14. What is the function of the SET input on a FF?
2-15. What is the function of the TRIGGER input on a FF?
2-16. Completely describe the rules of operation for a standard FF.

2-17. Draw a timing diagram showing the Q output terminal of a FF with 5 sequential pulses applied to the TRIGGER input. Assume that the FF was RESET at t_0.

2-18. Assume that a FF is in the RESET state. What state will the FF assume after a HIGH-to-LOW transition is applied to the
   (a) CLEAR input?
   (b) TRIGGER input?
   (c) SET input?

2-19. A FF used as part of a binary counter is labeled as "2^4." Explain what this name means and what can be determined about the FF from its label.

2-20. Draw a timing diagram showing the Q output of a 3-stage binary counter, assuming the following sequence.
   (a) The counter is initially cleared.
   (b) Seven pulses are applied to the T input of the first stage.
   (c) A LOW-to-HIGH transition is applied to the CLEAR input.

2-21. Draw a logic diagram of a 5-stage binary counter and explain its operation for each input pulse to be counted.

2-22. What is the maximum count attainable from a 5-stage binary counter?

2-23. How many FFs are required to store decimal number 10 in the binary scale?

2-24. Using your imagination, list at least five potential applications for a FF.

3

BASIC COMBINATIONAL LOGIC

3-1 THE FUNDAMENTALS OF COMBINATIONAL LOGIC

Most digital equipment consists of two general types of logic circuits: *combinational* and *sequential.* In a *combinational logic circuit,* the output depends on the *input combination* present when the output is examined. But a *sequential logic circuit* contains memory elements, so the output depends not only on the state of the input at any given instant but also on the *entire input history* of the circuit. A complete logic system can contain both types of circuits.

Combinational circuits can make *decisions* which range from the very simple to the extremely complex. This capability is one of the primary reasons for the application of digital techniques to many fields formerly dominated by analog techniques. Understanding how these decisions are made requires some knowledge of the mathematics of logic (Boolean algebra). It also requires the application of such "tools" as *truth tables, symbology,* and *waveforms.* This chapter will explore the fundamentals of combinational (*decision-making*) elements.

Basic Logic Operations

Logic operations are quite similar to algebraic operations. We will therefore begin by reviewing one of the simplest algebraic operations—*addition.* Practically everyone uses addition daily, but how often is the structure of this operation considered?

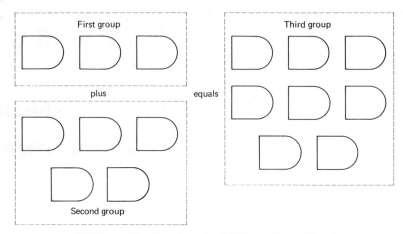

3 AND gates (First Group) <u>added to</u> 5 AND gates (Second Group) results in a new grouping (Third Group) of 8 AND gates.

Figure 3-1 Concept of addition

Addition is frequently considered to be a *primitive concept* because it is defined only by the properties assumed for it. The addition operation combines two or more numbers to obtain a new number, called the sum. Figure 3-1 illustrates this concept.

Algebraic operators are similar to grammatical verbs, since they indicate "doing something." Thus the arithmetic statement 3 + 5 = 8 may be translated to 5 *must be added* to 3 to obtain 8. Note the *action meaning* of the + symbol.

In algebra, letter symbols represent variables whose numerical values range from plus infinity to minus infinity and whose number form can be real, imaginary, complex, rational, irrational, integral, or fractional. The addition operation (+) in algebra can therefore be expressed in the following manner.

$a + b = c$

1	1	2
1	2	3
2	2	4
3	5	8
5	4	9

The other algebraic operations, subtraction, multiplication, division, exponentiation, and root extraction, may be approached in a similar manner—definitions* are developed from whatever is to be done

---

*Once the definition of subtraction is developed, the other four operations may be derived from the addition and subtraction definitions.

The Fundamentals of Combinational Logic 47

to a specified group of numbers. Hence the crucial facts for the reader to remember are that an algebraic operator indicates *what is to be done* and that its meaning is *strictly a matter of definition.*

This system of mathematics, using the six fundamental operations, can provide a solution for most of our everyday problems. For example, both the electronics technician and the engineer employ the system as a tool in understanding and designing complex equipment. In the specialized field of digital equipment, however, a simplified system of mathematics is applied. It is called *Boolean algebra* and utilizes the principles of *symbolic logic.*

The basic operations in symbolic logic are CONJUNCTION, DISJUNCTION, and NEGATION. Each of these operations, like those in algebra, can be described in terms of the effect it produces when it is used with specified variables and constants. In symbolic logic, letter symbols represent variables that are always *two-valued* (binary);* these may be expressed as one or zero, true or false, pulse or no pulse, plus voltage or minus voltage, and so forth.

The first logic operation, CONJUNCTION, can be defined as follows.

c is true only when a and b are simultaneously true.

The symbol for CONJUNCTION** is ($\cdot$), so the logic function just defined is written as $a \cdot b = c$ or $ab = c$. $ab = c$ is read as "a AND b is equal to c." Thus the logic operation AND has a mathematical name, CONJUNCTION, and a symbol ($\cdot$), just as the algebraic operation "times" uses the mathematical name multiplication and the symbol ($\cdot$). Table 3-1 shows all possible conjunctive combinations of the variables a and b having a resultant value of c.

Table 3–1 CONJUNCTION

$a \cdot b = c$

a	b	c
0(F)	0(F)	0(F)
0(F)	1(T)	0(F)
1(T)	0(F)	0(F)
1(T)	1(T)	1(T)

No other possibilities exist

The second logic operation, DISJUNCTION, can be defined in the following manner.

---

*For the purpose of our discussion, 1 and true are used synonymously, as are 0 and false; i.e., **true = 1** and **false = 0**.

**Due to its similarity to conventional algebraic multiplication, this operation is sometimes called *logical multiplication.*

> c is true if either a is true or b is true, or if both a and b are true at the same time.

The symbol for DISJUNCTION* is (+), so the logic function just defined is written as $a + b = c$. $a + b = c$ is read as "a OR b is equal to c." Thus the logic operation OR has a mathematical name, DISJUNCTION, and a symbol (+), just as the algebraic operation "plus" uses the mathematical name addition and the symbol (+). Table 3-2 shows all possible disjunctive combinations of the variables a and b having a resultant value of c.

Table 3–2 DISJUNCTION
$a + b = c$

a	b	c
0(F)	0(F)	0(F)
0(F)	1(T)	1(T)
1(T)	0(F)	1(T)
1(T)	1(T)	1(T)

No other possibilities exist

The last of the fundamental logic operations is NEGATION and can be defined as follows.

> If b is the negation of a, then a is true if b is false and a is false if b is true.

The symbol for NEGATION** is an overbar placed above a variable ($\bar{b}$), so the logic operation just defined is written as $a = \bar{b}$, $\bar{a} = b$. $a = \bar{b}$ is read as "a is equal to NOT b." Conversely, $\bar{a} = b$ is read as "NOT a is equal to b." Table 3-3 shows all combinations of variables a and b under the operation of NEGATION.

Table 3–3 NEGATION

a	b
0(F)	1(T)
1(T)	0(F)

Other types of logic operations exist. However, these can always be derived from the three fundamental operations that we have just discussed.

---

*Due to its similarity to conventional algebraic addition, this logic operation is sometimes called *logical addition*.

**No *direct* correlation to this operation exists in conventional algebra.

Logic Variables and Constants

In Chapter 1 Boolean algebra was defined as the mathematics of logic. Thus the basic logic operations of Boolean algebra link existing techniques and knowledge with the new field of digital logic. Just as in other branches of mathematics, certain postulates and theorems* are used to define the characteristics of Boolean algebra. We will employ only a few of these rules here, but as additional rules are required, they will be developed and explained.

Since the basic operations of digital logic (CONJUNCTION, DISJUNCTION, and NEGATION) have been defined, we can now turn to an examination of the elements to which they will be applied—Boolean *constants* and *variables*. In ordinary algebra all integers and fractions may be used as constants. Since a *constant* is something (a value, quantity, etc.) that has a *fixed meaning*, conventional algebra contains a large number of constants. Fortunately, Boolean algebra contains only *two* possible constants, 0 and 1.

A *variable*, on the other hand, is a quantity that can assume the value of *any constant* at *any time*. Since there are only two constants in Boolean algebra, Boolean variables may assume only one of the two values (i.e., 0 or 1). Variables are conventionally denoted by capital letters—A, B, Q, X, and so forth.

The NEGATION operation is also commonly called INVERSION. Table 3-3, which defined NEGATION, showed that when the variable a is 0, the NEGATION of a is 1; also when a is 1, the NEGATION of a is 0. Since variables may assume the values of constants, this NEGATION statement could also be written as

$$a = 0, \bar{a} = 1 \quad \text{and} \quad \bar{a} = 0, a = 1$$
therefore $\quad\quad\quad 0 = \bar{1} \quad \text{and} \quad \bar{0} = 1$

When a Boolean constant is *inverted*,** it is made equal to the other constant.

The definition of the AND operation was included in our discussion of the CONJUNCTION operation. So, from Table 3-1, it becomes apparent that if two or more constants are ANDed together, all of the constants must be 1 for the result to be 1. Since 1 has been previously equated with the "true" form of a variable and since variables can assume the values of constants, if two or more constants are ANDed together, all must be true for the result to be true. For example, if $A \cdot B = 1$, both A and B must be equal to 1. If either of the variables were 0, then a 0 would be ANDed with a 1, and the result would have to be 0.

---
*Appendix A is a composite listing of these postulates and theorems.

**In many texts, a variable without an overbar (e.g., A) is considered to be in its "true" form, while the "inverted" form of the variable has an overbar (e.g., $\bar{A}$).

Truth Tables

As indicated in Chapter 1, a *truth table* for a logic device is a tabular listing of all possible combinations of inputs and the resultant output for each combination. It represents a useful method of condensing all information concerning the functional characteristics of a logic device and provides a visualization that makes these characteristics more readily perceivable. The name "truth table" comes from formal *symbolic logic*,* where the truth or falsity of statements is investigated.

In symbolic logic, English-language phrases are combined into sentences, and the truth value of these sentences is then investigated. Of course, many rules govern the types and forms of phrases allowed. But one or two examples should be sufficient to help us bridge the gap between symbolic logic and the digital applications of logic.

Phrases used in symbolic logic are called propositions and are usually in the *declarative* form. Since these phrases are assertive in nature, they contain an inherent truth value; i.e., *they are either true or false*. A phrase such as "the FF is in the SET state" is a good example. If the FF is SET, the phrase is true; if it is not SET (CLEAR), the phrase is false. "The TRIGGER pulse is present" represents another phrase with inherent truth value. If the TRIGGER pulse is present, the phrase is true; if it is absent, the phrase is false.

Phrases may be combined by using *logical connectives* (previously called logical operators) to form more complex phrases. When two phrases are joined by the logical operator AND, however, new meanings may appear. Thus the preceding phrases become "the FF is in the SET state AND the TRIGGER pulse is present." The truth value of this new phrase now depends not only on the truth value of each of the individual phrases but also on the definition of the logical operator AND. An analysis, using English-language statements, might appear as follows.

> If the FF is not SET (CLEAR), the truth value of the first phrase is false. If the TRIGGER pulse is not present, the truth value of the second phrase is false. Therefore, if the FF is CLEAR and the TRIGGER pulse is absent, the truth value of the complex statement is false.

All such possible combinations in symbolic logic may be so examined—at the obvious expense of a great many words!

The application of a form (mathematical) of symbolic logic in which a special set of symbols replaces the language statements would

---
*Many texts have been written about formal symbolic logic, so the reader who desires to delve into its intricacies will have no difficulty in locating the appropriate material.

The Fundamentals of Combinational Logic

greatly simplify the presentation. Thus letting the variable A represent "the FF is in the SET state" and the variable B represent "the TRIGGER pulse is present," the complete statement can be written as "$A \cdot B$." All possible combinations of A and B can now be placed in matrix or table form, and the truth value of $A \cdot B$ examined quite easily. Table 3-4 defines the original logical statement.

Table 3-4 Truth table defining the logical statement $A \cdot B$

A	B	$A \cdot B$
F	F	F
F	T	F
T	F	F
T	T	T

This discussion may be carried one step further—to investigate the relationships of dependent and independent variables. A small addition to the previous statement will demonstrate these relationships.

"The FF is CLEARed if the FF is SET AND the TRIGGER pulse is present."

A has already been defined as "the FF is SET" and B as "the TRIGGER pulse is present." The additional phrase "the FF will be CLEARed" is *dependent* on the condition of A and B, just as it would be in conventional algebra. By letting X represent this dependent variable, the complete statement becomes $X = A \cdot B$. Modifying the truth table to reflect this new statement results in Table 3-5. Note that each row of the last column gives the truth value of the function that is determined by the truth values entered in the preceding columns of that row.

Table 3-5 Truth table defining the logical equation $X = A \cdot B$

A	B	X
F	F	F
F	T	F
T	F	F
T	T	T

By applying techniques borrowed from its "parent"—symbolic logic—digital logic becomes a more understandable and therefore more useful tool.

An Introduction to Boolean Algebra

Having discussed constants, variables, and basic operations, we shall now investigate some simple *theorems* (laws) of Boolean algebra. The tables used to define CONJUNCTION, DISJUNCTION, and NEGA-

TION provide some insight into a method of proof for certain theorems. Basic operation tables have been reproduced in Fig. 3-2, along with a reference designation for each row. The information contained in each of these rows constitutes the *basic postulates* of Boolean algebra. A list of the postulates (keyed to Fig. 3-2) follows:

①	$0 \cdot 0 = 0$	(P–1)
②	$0 \cdot 1 = 0$	(P–2)
③	$1 \cdot 0 = 0$	(P–3)
④	$1 \cdot 1 = 1$	(P–4)
⑤	$0 + 0 = 0$	(P–5)
⑥	$0 + 1 = 1$	(P–6)
⑦	$1 + 0 = 1$	(P–7)
⑧	$1 + 1 = 1$	(P–8)
⑨	$0 = \bar{1}$	(P–9)
⑩	$1 = \bar{0}$	(P–10)

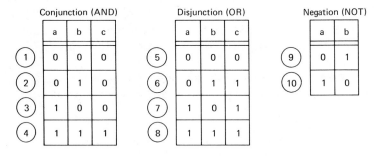

Figure 3-2 Basic logical operations

From these postulates, we have developed thirteen basic theorems of Boolean algebra.

Because we defined a variable as a quantity that can assume the value of any constant, the variables in our theorems are either 0 or 1 (they are the *only* constants available in the binary system).

The Fundamentals of Combinational Logic

THEOREM 1: $\boxed{A \cdot 0 = 0}$ (T–1)

Proof: If A is equal to 0, then Postulate 1 is proof.

$A = 0, \quad \therefore \ 0 \cdot 0 = 0$

If A is equal to 1, then Postulate 3 is proof.

$A = 1, \quad \therefore \ 1 \cdot 0 = 0$

THEOREM 2: $\boxed{0 \cdot A = 0}$ (T–2)

Proof: If A is equal to 0, then Postulate 1 is proof.

$A = 0, \quad \therefore \ 0 \cdot 0 = 0$

If A is equal to 1, then Postulate 2 is proof.

$A = 1, \quad \therefore \ 0 \cdot 1 = 0$

THEOREM 3: $\boxed{A \cdot 1 = A}$ (T–3)

Proof: If A is equal to 0, then Postulate 2 is proof.

$A = 0, \quad \therefore \ 0 \cdot 1 = 0$

If A is equal to 1, then Postulate 4 is proof.

$A = 1, \quad \therefore \ 1 \cdot 1 = 1$

THEOREM 4: $\boxed{1 \cdot A = A}$ (T–4)

Proof: If A is equal to 0, then Postulate 3 is proof.

$A = 0, \quad \therefore \ 1 \cdot 0 = 0$

If A is equal to 1, then Postulate 4 is proof.

$A = 1, \quad \therefore \ 1 \cdot 1 = 1$

THEOREM 5: $\boxed{A \cdot A = A}$ (T–5)

Proof: If A is equal to 0, then Postulate 1 is proof.

$A = 0, \quad \therefore \ 0 \cdot 0 = 0$

If A is equal to 1, then Postulate 4 is proof.

$A = 1, \quad \therefore \ 1 \cdot 1 = 1$

THEOREM 6: $$\boxed{A \cdot \overline{A} = 0}$$ (T–6)

Proof: If A is equal to 0, then $\overline{A}$ is equal to 1 by Postulate 9.

$$\left. \begin{array}{l} A = 0 \\ \overline{A} = 1 \end{array} \right\} \quad \therefore \quad 0 \cdot 1 = 0$$

If A is equal to 1, then $\overline{A}$ is equal to 0 by Postulate 10.

$$\left. \begin{array}{l} A = 1 \\ \overline{A} = 0 \end{array} \right\} \quad \therefore \quad 1 \cdot 0 = 0$$

THEOREM 7: $$\boxed{A + 0 = A}$$ (T–7)

Proof If A is equal to 0, then Postulate 5 is proof.

$$A = 0, \quad \therefore \quad 0 + 0 = 0$$

If A is equal to 1, then Postulate 7 is proof.

$$A = 1, \quad \therefore \quad 1 + 0 = 1$$

THEOREM 8: $$\boxed{0 + A = A}$$ (T–8)

Proof: If A is equal to 0, then Postulate 5 is proof.

$$A = 0, \quad \therefore \quad 0 + 0 = 0$$

If A is equal to 1, then Postulate 6 is proof.

$$A = 1, \quad \therefore \quad 0 + 1 = 1$$

THEOREM 9: $$\boxed{A + 1 = 1}$$ (T–9)

Proof: If A is equal to 0, then Postulate 6 is proof.

$$A = 0, \quad \therefore \quad 0 + 1 = 1$$

If A is equal to 1, then Postulate 8 is proof.

$$A = 1, \quad \therefore \quad 1 + 1 = 1$$

THEOREM 10: $$\boxed{1 + A = 1}$$ (T–10)

Proof: If A is equal to 0, then Postulate 7 is proof.

$$A = 0, \quad \therefore \quad 1 + 0 = 1$$

The Fundamentals of Combinational Logic

If A is equal to 1, then Postulate 8 is proof.

$$A = 1, \quad \therefore \ 1 + 1 = 1$$

THEOREM 11: $\boxed{A + A = A}$ (T-11)

Proof: If A is equal to 0, then Postulate 5 is proof.

$$A = 0, \quad \therefore \ 0 + 0 = 0$$

If A is equal to 1, then Postulate 8 is proof.

$$A = 1, \quad \therefore \ 1 + 1 = 1$$

THEOREM 12: $\boxed{A + \overline{A} = 1}$ (T-12)

Proof: If A is equal to 0, then $\overline{A}$ is equal to 1 by Postulate 9.

$$\left. \begin{array}{l} A = 0 \\ \overline{A} = 1 \end{array} \right\} \quad \therefore \ 0 + 1 = 1$$

If A is equal to 1, then A is equal to 0 by Postulate 10.

$$\left. \begin{array}{l} A = 1 \\ \overline{A} = 0 \end{array} \right\} \quad \therefore \ 1 + 0 = 1$$

THEOREM 13: $\boxed{\overline{\overline{A}} = A}$ (T-13)

Proof: If A is equal to 0, then $\overline{A}$ is equal to 1 by Postulate 9. If $\overline{A}$ is equal to 1, then another negation operation $(\overline{\overline{A}})$ makes $\overline{\overline{A}}$ equal to 0 by Postulate 10.

$$\left. \begin{array}{l} A = 0 \\ \overline{\overline{A}} = 0 \end{array} \right\} \quad \therefore \ 0 = 0, \ \text{proving} \ \overline{\overline{A}} = A.$$

Similarly, if A is equal to 1, then $\overline{A}$ is equal to 0 by Postulate 10. If $\overline{A}$ is equal to 0, then another negation operation $(\overline{\overline{A}})$ makes $\overline{\overline{A}}$ equal to 1 by Postulate 9.

$$\left. \begin{array}{l} A = 1 \\ \overline{\overline{A}} = 1 \end{array} \right\} \quad \therefore \ 1 = 1, \ \text{proving} \ \overline{\overline{A}} = A.$$

Note that in each of these proofs, all possible values were examined for each variable. The fact that only *two* values exist greatly simplifies the task of proving theorems in Boolean algebra. Since most readers have labored wearily in deriving proofs for conventional algebraic and geometric theorems, where an infinite number of values must be assumed, the digital system offers an obvious mathematical advantage.

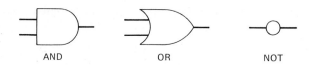

Figure 3-3 Combinational logic symbols

Symbology

Earlier, we devised a mathematical shorthand as a tool for digital-equipment analysis and troubleshooting. Now we will develop a *symbolic shorthand*. To begin, we must define the term *Boolean algebra expression*. Mathematically, an *expression* is a "string" of constants and/or variables connected by one or more arithmetic/algebraic operators. By revising this definition slightly, a Boolean algebra expression becomes *a string of constants and/or variables connected by one or more logical operators*.

Boolean algebra expressions can be represented graphically by symbols.* Such symbols can be directly related to the electrical/electronic gate and inverter circuits used in digital equipment. With such close correlation to the actual hardware, troubleshooting and analysis become much easier.

Figure 3-3 shows the symbols that define the AND, OR, and NOT operations. In the AND gate, signal flow is always from the flat to the round side—the *round* side is the *output*. Signal flow in the OR gate is always from the concave to the convex side—the *convex* side is the *output*. The symbol defining NEGATION (the NOT operation) is unique because *it never appears by itself*. It must be connected to some other symbol (e.g., an AND gate or an OR gate). Signal flow is conventionally from *left* to *right*. In various combinations, these three (AND, OR, NOT) symbols describe most of the combinational logic expressions commonly encountered.

Combinational logic elements are commonly called *gates*. In linear (analog) circuits, if a signal passes through or is blocked, depending on the characteristics of the controlling signal, one speaks of "linear gates," where output would be an approximate replica of at least one of the inputs. But for digital purposes, a *gate* is defined as a *circuit having two or more input terminals and one output terminal, where an output is present when and only when prescribed inputs are present*. The output of the digital (logic) gate may bear no resemblance to any of the inputs, except that the output occurs during an interval selected by the control voltage(s).

---

*Although numerous groups of symbols exist, MIL-STD-806B tends to be the standard governing a majority of the digital-equipment diagrams used by military and commercial organizations. See Appendices B and C.

The term *output* must be redefined for digital equipment. The output of an analog device might vary from minus infinity through zero to plus infinity. But, remember, digital equipment using the binary system can have only one of two outputs at any given time, and for all practical purposes, either of these two output values could be useful. Within this context, therefore, *a combinational logic device* (i.e., a *gate*), *produces an output whenever the gate performs the function denoted by the symbol.*

3-2 THE OR FUNCTION IN DIGITAL LOGIC

One aid to understanding a new concept (such as digital logic) is to construct a physical analogy that would be encountered in an everyday situation. Thus it is convenient—even necessary—for us to initially consider each logical operation as a physical entity. By using this approach, the reader should better comprehend the close correlation between the logic operations and the hardware that performs these operations.

A Physical Analogy

The conventional water faucet (Fig. 3-4) found in many homes provides an excellent analogy for a logic operation. Almost everyone has observed that if *both* the hot and cold water controls are turned

Hot water control	Cold water control	Faucet output
OFF	OFF	No water (OFF)
OFF	ON	Water (ON)
ON	OFF	Water (ON)
ON	ON	Water (ON)

Figure 3-4 A physical analogy of an OR gate

OFF, no water comes from the faucet. If, however, *either* the hot or the cold water control (or both) are turned ON, a flow of water results. A table of all possible control combinations and the resultant flow of water or lack thereof is also shown in Fig. 3-4. Comparison of this table with the table for the logic operation DISJUNCTION (OR)* shows a remarkable resemblance. Apparently, the faucet performs the logic operation of DISJUNCTION.

---

*The symbol for a device that performs the OR operation is shown in Fig. 3-3.

Terms Associated with the OR Operation

Since this is a non-electronic text and since so many possible methods of implementing logic functions exist, actual values of voltage and/or current are not used descriptively when outputs and inputs are discussed. Instead, we use the HIGH (H) and LOW (L) notation. The H and L approach not only greatly simplifies logic circuit analysis but also makes it unnecessary to use actual numerical values for inputs and outputs.

The Boolean algebra expression for OR and the associated truth table have already been described. In addition, waveforms are commonly employed in the analysis of OR operations.*

Applications of the OR Operation

The device that performs the OR operation is called an OR *gate*. Physically, it may assume many forms. Most modern OR gates are in integrated circuit (IC) form. The number of inputs is not restricted to two, as implied in our earlier discussion. Input limitations are a matter of electronic design, and OR gates are commonly packaged with two, three, four, or eight inputs (an attempt to standardize production and logic design). Some manufacturers provide special configurations of OR gates with greater input capability. Physical packaging is similar to the examples shown in Fig. 2-10.

Many early applications of digital logic were in the area of *switching circuits*, although relays were used in place of actual switches. A *relay* is an electrically operated device that mechanically switches electrical circuits. Its symbol appears in Fig. 3-5a, which is a schematic diagram showing a relay-implemented OR circuit. The truth table in Fig. 3-5b defines the four possible combinations of inputs that can exist with two variables. Note that this table is identical to the truth table for the OR operation.

Relay logic is still used in some industrial and communication control applications. Such circuitry is rapidly being replaced by solid-state, digital-logic modules, since modern technology requires extreme reliability and small physical size in its complex system functions.

Theorems Using the OR Operation

Our Boolean algebra theorems using the OR operation can now be applied to the actual hardware. Again, a "true" (1) is considered to be HIGH and a "false" (0) is LOW. Theorem 7, $A + 0 = A$, can be viewed

---

*To summarize the OR operation, the English-language statement, the Boolean algebra expression, the truth table, and typical waveforms are reproduced at the end of this chapter.

The OR Function in Digital Logic

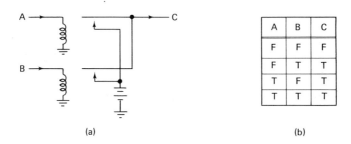

Figure 3-5 An OR circuit using relays

as two inputs to an OR gate. By definition, one input is a constant binary 0 (LOW). The other input, A, is a variable and can take on either binary 0 or binary 1 as a value. If A is a binary 0, it is at a LOW level; the OR truth table shows that when both inputs are LOW, the output is LOW. However if A is binary 1, then it is at a HIGH level; the OR truth table shows that these input conditions result in a HIGH output.

Actual demonstration of this theorem requires a *logic-signal source,* which has the capability of generating on command either a HIGH or LOW output. Such a device is relatively easy to implement, but for now, we will represent it by a simple rectangle. Control is provided for each signal source so that the output of the device will be either HIGH or LOW, as required. A logic diagram of the complete test set-up required to demonstrate Theorem 7 is shown in Fig. 3-6a with its associated waveforms and truth table.

One of the most valuable applications of the truth table is proving Boolean *identities*. If we can show that one column in a truth table (representing a variable) has a one-to-one correspondence with another column in the truth table (representing another variable or the results of a logical operation), then an "identity" exists between the variables or the results represented by each column. *One-to-one-correspondence* means that for each entry in a column in the truth table, an identical entry exists in the same row in another column. Thus in the short truth table used to define $A + 0 = A$, each entry under the A column has a corresponding entry in the OUT column, and identity is thereby established. That is, regardless of the logical value of A, when it is ORed with binary 0, the result will be the logic value of A.

Theorem 8, $0 + A = A$, may be easily demonstrated by interchanging the two inputs to the OR gate used in Theorem 7. Demonstration of Theorem 9, $A + 1 = 1$, is shown in Fig. 3-6b.

Theorem 10, $1 + A = 1$, may also be demonstrated by interchanging the two inputs to the OR gate used in Theorem 9. Theorem 11 is demonstrated in Fig. 3-6c.

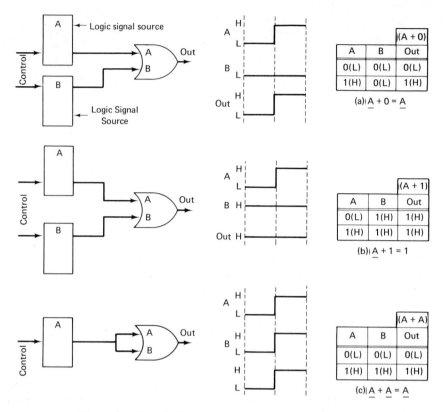

Figure 3-6 Proof of theorems using the OR operation

3-3 THE AND FUNCTION IN DIGITAL LOGIC

A Physical Analogy

Another logic function, the AND operation, is seen in the form of a physical analogy in practically all homes. To illuminate a light both the associated switch *and* the circuit breaker must be on. If either the switch or the circuit breaker is off, the light will not illuminate. A simple schematic diagram showing this *switch implementation* of the AND operation is found (with its associated table of combinations) in Fig. 3-7. Comparison of this table with Table 3-1 (for the logic operation CONJUNCTION) shows that the circuit breaker/wall switch combination performs the AND* operation.

---

*The symbol for the hardware that performs the AND operation is shown in Fig. 3-3.

The AND Function in Digital Logic 61

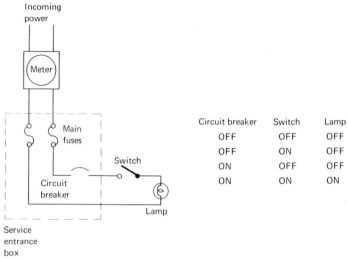

Figure 3-7 An analogy of an AND gate

Terms Associated with the AND Operation

The Boolean algebra expression and the related truth table for the AND operation have already been discussed. The third method of logic analysis, the waveform, is also adaptable to this operation. For easy reference, a synopsis of the important points relating to the AND operation is given in Sec. 3-8.

Applications of the AND Operation

The device that performs the AND operation is called an AND *gate*. Physically, it may assume many forms. Externally, the IC AND gate resembles the "packages" shown in Fig. 2-10. Like the OR gate, the number of inputs is restricted only by design requirements.

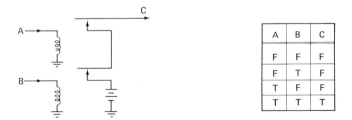

Figure 3-8 Relay AND circuit

Basic Combinational Logic

Switch implementation of the AND operation was shown in Fig. 3-7. Relay logic can also be employed in this operation. A typical relay operated AND circuit with its attendant truth table is shown in Fig. 3-8.

Theorems Using the AND Operation

Basic Boolean algebra theorems also provide a means of demonstrating the AND gate in operation. Theorem 1, $A \cdot 0 = 0$, is implemented with hardware in Fig. 3-9a. Theorem 2 may be demonstrated by merely interchanging the two inputs to the AND gate.

Theorem 3, $A \cdot 1 = A$, is shown in Fig. 3-9b. Theorem 4, $1 \cdot A = A$, is represented simply by interchanging the inputs in Theorem 3.

Theorem 5, $A \cdot A = A$, is shown in the logic diagram in Fig. 3-9c.

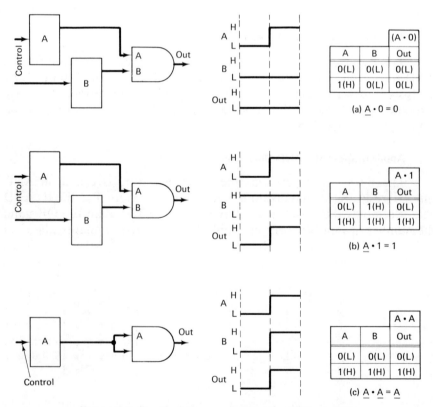

Figure 3-9 Proof of theorems using the AND operation

3-4 THE NOT FUNCTION IN DIGITAL LOGIC

A Physical Analogy

The last of our fundamental logic operations, NEGATION (the NOT function), is not as easily visualized with a physical analogy as were the OR and the AND functions. Perhaps the engine oil pressure warning indicator system on a modern automobile comes as close as possible to a physical analogy. If oil pressure is within limits, the warning light is off. If oil pressure falls below required values, the warning light is on. All possible combinations of this system are shown here.

Oil Pressure	Warning Light
NORMAL	OFF
LOW	ON

Comparison of this list with Table 3-3 (for the logic operation NEGATION) indicates that the oil-pressure warning system performs in a manner similar to the NOT* function.

Terms Associated with the NOT Operation

The Boolean Algebra expression and the related truth table for the NOT operation have been discussed previously. Waveform analysis is also adaptable to this operation. A summary of the important points relating to the NOT operation is given in Sec. 3-8.

The Inverter

Before continuing the discussion of the NOT operation, we should introduce another logic symbol. Within digital logic systems, the need often arises to increase the current or power capabilities of a logic signal. A device called an *amplifier* does this (its symbol is shown in Fig. 3-10a). The amplifier's output signal may have greater power capabilities, but no *logic operation is performed* on the signal applied to the amplifier.

Depending on design, an amplifier** may or may not "invert" the incoming signal. The NOT symbol is commonly attached to the amplifier symbol (Fig. 3-10b, c) to show the inversion process; when this combination of symbols is used, it represents a device that performs the

---

*The symbol for the NOT operation is shown in Fig. 3-3.
**Means are provided within MIL-STD-806B to denote the inverting characteristics of amplifiers used in logic systems.

(a) Amplifier   (b) Output LOW   (c) Input LOW

Figure 3-10 The NOT symbol redefined

NOT operation. Such a device is commonly called an *inverter*. Physically, the inverter resembles other logic devices, so identification of the inverter is usually made by part number from the manufacturers' specification sheets.

Since the NOT symbol *cannot* appear alone, it *must* be physically connected to the amplifier symbol—*where* it connects is extremely important. The NOT symbol is placed adjacent to other logic symbols to indicate that a LOW signal either activates the logic operation on input or is a desired result of the logic operation on output. If the NOT symbol is connected at the output side of the amplifier symbol (Fig. 3-10b), it means that when the logic operation (NEGATION) is performed, the *output* will be LOW. LOW has tentatively been correlated with "false" (0), so the input must be HIGH ("true" or 1). If the NOT symbol is attached to the input side of the amplifier (Fig. 3-10c), then the *input* has to be LOW to provide activation of the NEGATION function. The NOT symbol,* therefore, is a *clue*, indicating what can be expected of the symbol to which it is attached. A schematic diagram of the engine oil-pressure warning system (Fig. 3-11a) that we discussed earlier shows one application of the NOT operation. When there is no oil pressure the pressure switch is closed; a complete circuit exists and the warning light is ON. As oil pressure builds up during normal operation the switch opens; the warning light then goes OFF.

Oil pressure warning system

Figure 3-11 Implementation of the NOT function

---

*Placement of the NOT symbol on the amplifier symbol has little meaning in the construction of hardware, but as we have just demonstrated, it does mean a great deal in terms of logic.

Simple, Combined Functions

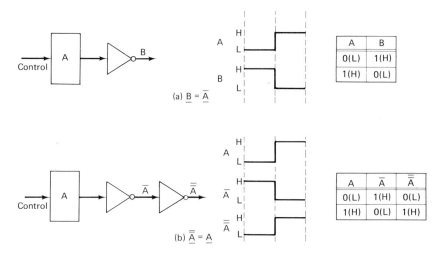

Figure 3-12 Proof of theorems using the NOT operation

Theorems Using the NOT Operation

By the digital-logic definition of the NOT function, the output of the inverter is the inverse or, in Boolean algebra, the "other" value, when compared with the input. For example (see Fig. 3-12a), if the input to an inverter is the variable A, the output is $\bar{A}$. Extension of this application results in the proof for Theorem 13, $\bar{\bar{A}} = A$ (Fig. 3-12b).

3-5 SIMPLE, COMBINED FUNCTIONS

Logic operations are not limited to a *single* function (e.g., OR by itself). Nor does a limit exist on the number of simple logic operations that can be combined to form complex logic functions. We shall examine various combinations of logic operations and their functions in this section.

Multiple OR Operations

OR operations may be combined on more than one level. For example, consider the Boolean expression $A + (B + C)$. The parentheses imply that A is to be ORed with the result of ORing B with C. Thus one OR gate is needed to disjunctively combine (OR) B with C, and another OR gate is needed to disjunctively combine (OR) A with the results. The logic diagram describing these operations is shown in Fig. 3-13a.

If we want to find the logic value for this expression, the equation $X = A + (B + C)$ can be written. A truth table can then be developed,

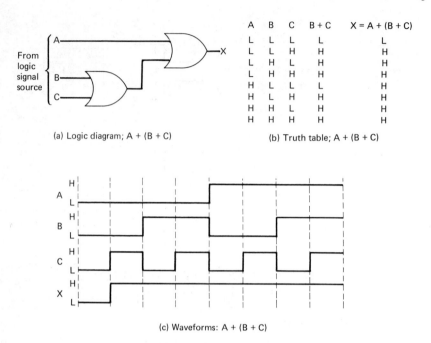

Figure 3-13 A + (B + C)

and the results of the total logic operation depicted in the X column. Note that in the truth table (Fig. 3-13b), eight combinations of the three variables exist. When evaluating logic functions, the total number of combinations of variables is determined by taking the base of the binary number system (2) and raising it to the power of the total number of variables being considered. Thus $2^3 = 8$ – the total number of combinations to be examined. This concept may be generalized by the following equation:

$$N = 2^M \qquad (3\text{-}1)$$

where N = the total number of combinations to be examined and M = the total number of variables being considered.

Waveforms (Fig. 3-13c) showing all possible combinations of the three variables may also be drawn. Like the truth table, all eight possible combinations must be examined.

Multiple AND Operations

AND operations may also be combined on more than one level. For example, consider the Boolean expression $A \cdot (B \cdot C)$. The parentheses imply that A is to be ANDed with the result of ANDing B

The NOT Function in Digital Logic

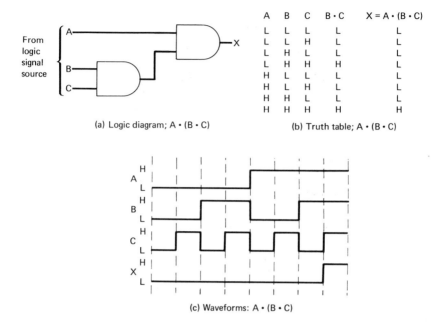

(a) Logic diagram; A · (B · C)

(b) Truth table; A · (B · C)

(c) Waveforms: A · (B · C)

Figure 3-14 A·(B·C)

with C. Thus an AND gate is needed to conjunctively combine (AND) B with C, and another AND gate is needed to AND A with the results. The logic diagram describing these operations is shown in Fig. 3-14a.

If we want to find the logic value for this expression, the equation $X = A \cdot (B \cdot C)$ can be written; a truth table can be developed; and the results of the total logical operation can be shown in the X column. Note that in the truth table (Fig. 3-14b), eight combinations of the three variables exist. Applying Eq. (3-1), $2^3 = 8$, which is the total number of combinations to be examined.

Waveforms showing all possible combinations of the three variables may also be drawn. Like the truth tables, all eight possible combinations must be examined (see Fig. 3-14c).

Until now, we have considered waveforms to be present for a given period of time. However one of the most common applications of multilevel gate logic is the *selection of given time periods* in which a relatively short "pulse" of information is allowed through. A good example would be a digital system that requires a "clock" which provides timing signals of a specific nature (usually short pulses at regular intervals). Some operations may not require pulses at the basic clock interval but only at specific times, depending on logic conditions elsewhere in the system. By ANDing these logic conditions with clock pulses, specific

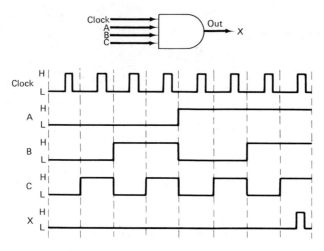

Figure 3-15 A "clocked" AND gate

pulses can be selected and routed to perform their functions. The logic diagram and representative waveforms shown in Fig. 3-15 represent such a circuit. Here, clock pulses are ANDed with *A, B,* and *C* to select only every eighth pulse.

AND/OR Operations

Both AND and OR operations may be combined. The Boolean expression $A \cdot (B + C)$ is an example; it implies that A is to be ANDed with the results of ORing B with C. Thus an OR gate is needed to OR B with C, and an AND gate is needed to AND the results with A. The logic diagram and truth table resulting from changing the expression to an equation $[X = A \cdot (B + C)]$ is shown in Fig. 3-16. Typical waveform analysis is also provided. Inputs are from a "logic-signal" source.

OR/AND Operations

Another simple example may be illustrated by the Boolean equation $X = A + (B \cdot C)$. A is to be ORed with the results of ANDing B with C. An AND gate is needed to AND B with C, and an OR gate to OR A with the results. The logic diagram, truth table, and waveforms are provided in Fig. 3-17. Again, inputs are from a "logic-signal" source.

Order of Operations and Grouping Signs

The *order* in which Boolean algebra operations are *performed* is also very important. It affects the *logic value of an expression*—just as the order of arithmetic operations in conventional arithmetic affects

Simple, Combined Functions

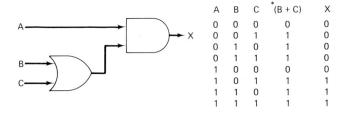

A	B	C	*(B + C)	X
0	0	0	0	0
0	0	1	1	0
0	1	0	1	0
0	1	1	1	0
1	0	0	0	0
1	0	1	1	1
1	1	0	1	1
1	1	1	1	1

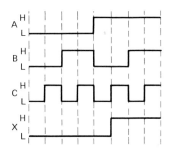

*Note the use of 0 as LOW and 1 as HIGH

Figure 3-16 $X = A \cdot (B+C)$

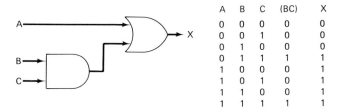

A	B	C	(BC)	X
0	0	0	0	0
0	0	1	0	0
0	1	0	0	0
0	1	1	1	1
1	0	0	0	1
1	0	1	0	1
1	1	0	0	1
1	1	1	1	1

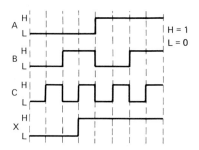

H = 1
L = 0

Figure 3-17 $X = A + (B \cdot C)$

the arithmetic value of an expression. If the total value of the expression 100 x 25 + 2000 is to be determined, the procedure involves first the multiplication of 100 x 25 and then the addition of 2000. The result is the answer 4500. A review of arithmetic rules discloses that, unless otherwise indicated, *multiplication and division of numbers within terms is performed before addition and subtraction.* However if the above expression is written as 100 x (25 + 2000), then 202,500 is the correct answer. Arithmetic rules state that *operations within parentheses are performed before operations outside of parentheses.*

Similar discrepancies arise in Boolean algebra. Consider the three variables A, B, and C. Combining them in the expression $A + B \cdot C$ can result in different logic values, depending on the order in which the operations are performed. Comparison of the effect of performing the OR operation prior to the AND operation $[(A + B) \cdot C]$ with performing the AND operation prior to the OR operation $[A + (B \cdot C)]$ is shown in Table 3-6.

Table 3–6 $(A + B) \cdot C \neq A + (B \cdot C)$

A	B	C	A + B	(A + B) · C	B · C	A + (B · C)
L	L	L	L	L	L	L
L	L	H	L	L	L	L
L	H	L	H	L	L	L
L	H	H	H	H	H	H
H	L	L	H	L	L	H
H	L	H	H	H	L	H
H	H	L	H	L	L	H
H	H	H	H	H	H	H

Not equal

The order of priority in Boolean algebra is NOT (*inversion*) first, AND second, and OR last, *unless otherwise indicated by grouping signs,* such as parentheses, brackets, braces, or the vinculum. Following these rules, the previous expression $A + B \cdot C$ can be evaluated only as $A + (B \cdot C)$ with the existence of the parentheses implicitly understood. In fact, knowing this, the expression can be further simplified to $A + BC$. (As a matter of convention, the AND operator is usually omitted in Boolean algebra.) BC is also considered to be grouped naturally because the letters are written together and are separated from A by the OR operator. Both of the conventions ($B \cdot C$ and BC) are used, but BC is preferred (unless clarity demands showing the logic operator).

When working with logic expressions, misuse of grouping signs is a common occurrence. Too many grouping signs do not usually generate problems, as long as they occur in pairs. But this is valid *only* when the

Table 3–7 $(A \cdot B) + C \neq A \cdot (B + C)$

A	B	C	$A \cdot B$	$(A \cdot B) + C$	$B + C$	$A \cdot (B + C)$
L	L	L	L	L	L	L
L	L	H	L	H	H	L
L	H	L	L	L	H	L
L	H	H	L	H	H	L
H	L	L	L	L	L	L
H	L	H	L	H	H	H
H	H	L	H	H	H	H
H	H	H	H	H	H	H

Not equal

grouping signs have been placed properly. Inserting a pair of parentheses to enclose $A + B$ in the previous expression obviously changes the complete meaning of the original expression. However enclosing $B \cdot C$ in parentheses has no effect and is, in fact, an unnecessary step—grouping signs are *not* necessary when ANDed variables or constants are combined with other variables or constants by the OR operator. An insufficient number of parentheses usually results in the wrong interpretation of logic expressions. Thus if we wish to AND $B + C$ with A, we must enclose $B + C$ in parentheses. Otherwise, the expression would be evaluated as $A \cdot B + C$, and the rules for order of priority of operations would AND B with A—prior to ORing with C. Table 3-7 clearly shows that $A \cdot B + C$ is not the same as $A(B + C)$.

If an expression already contains *parentheses* () and additional signs of grouping are required, first *brackets* [ ] and then *braces* { } are used. The *vinculum* (‾) is also a grouping sign—it groups the portion(s) of an expression to be inverted. Remember, letters and expressions are also considered to be *grouped* when they are written without a connecting operator. Here is a complex expression containing various examples of common grouping practices.

$$Y + W \cdot [W(X\overline{Z} + YZ)] + [WX\overline{YZ} + \overline{X + Y + WZ}]$$

AND/OR/NOT Operations

The NOT operation may be combined with the AND operation, the OR operation, and the combined AND/OR operations. Logic functions using combinations of OR, AND, and NOT range from the very simple to the very complex. Certain Boolean algebra theorems are well suited to a demonstration of such logic functions. Theorem 12, $A + \overline{A} = 1$, for example, requires the use of an OR gate and the NOT operation. When A is applied to the input of an inverter (NOT), the output is $\overline{A}$.

The same variable, A, may also be applied as an input to an OR gate. Thus both required inputs, A and $\bar{A}$, are available. The logic diagram in Fig. 3-18a illustrates one method of proving that $A + \bar{A} = 1$; the attendant truth table also provides proof of the theorem.

Theorem 6, $A \cdot \bar{A} = 0$, combines the NOT operation with the AND operation. Both A and $\bar{A}$ can be made available in the same manner as they were in Theorem 12. The logic diagram and accompanying truth table appear in Fig. 3-18b.

Figures 3-18a and b are typical examples of single-level, relatively simple logic circuits using AND/OR/NOT operations. Extending the inputs to two variables, A and B, does not especially complicate matters, as Fig. 3-18c shows.

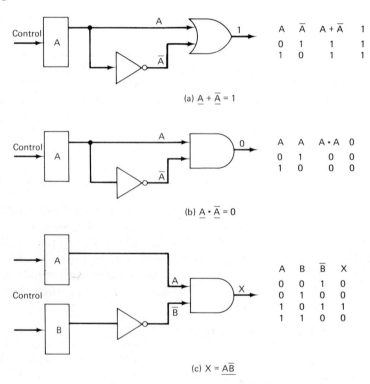

Figure 3-18 Simple AND/OR/NOT operations

With two variables, only four possible combinations exist, so no real problem should be encountered in analyzing the operation. Also, all four locations of input inverters should be examined, and the effect of each configuration on the truth table should be carefully noted by comparing the information in Fig. 3-19.

Derived Functions

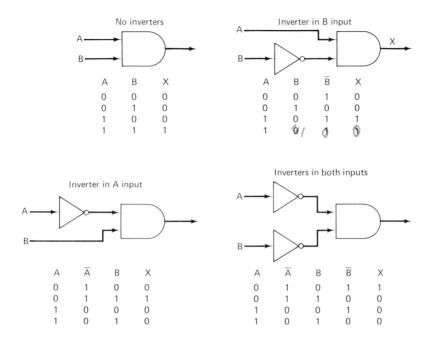

Figure 3-19 Variations of A and B combined conjunctively

3-6 DERIVED FUNCTIONS

In our preceding discussion, AND/OR operations were combined with the NOT operation. In each case, the inverter, which implemented the NOT operation, extended the capabilities of either the AND or the OR gate. Two special cases exist, however, that we will now introduce to illustrate the versatility of simple digital logic.

The NAND Function

The first case is one in which both inputs to an OR gate are fed through inverters. Under these circumstances, the OR gate functions in an *opposite* (complementary) manner. With inverted inputs, the OR operation is present when one or the other (or both) inputs are *not* present. In other words, *the negated input* OR *gate is activated when one or both inputs are not present.* Compare the negated input OR operation shown in Fig. 3-20a with the normal OR operation in Fig. 3-2 and determine the differences.

Next we will examine an AND operation with negated output. The inverter appears at the output rather than at the input of the AND gate, as in the OR gate example. Thus the AND operation is performed

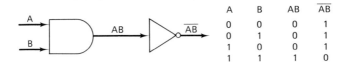

Figure 3-20 The NAND function

(a) Negated input OR operation

A	$\bar{A}$	B	$\bar{B}$	$\bar{A}+\bar{B}$
0	1	0	1	1
0	1	1	0	1
1	0	0	1	1
1	0	1	0	0

(b) Negated output AND operation

A	B	AB	$\overline{AB}$
0	0	0	1
0	1	0	1
1	0	0	1
1	1	1	0

before negation. The logic diagram and resulting truth table are shown in Fig. 3-20b.

At this point you will notice a rather startling resemblance between the truth tables resulting from the negated input OR operation and the negated output AND operation. In fact, they are *identical*; that is, *when two truth tables with identical inputs have the same outputs, the logic operations described by the truth tables are equal (identical).* This is one of the statements of *DeMorgan's Laws* and is also one of the most important statements of equality, $\bar{A} + \bar{B} = \overline{AB}$, in Boolean algebra. The conventional method (i.e., truth tables) of indicating equality of Boolean algebra functions is shown in Table 3-8.

The NOR Function

The second special case is one in which both inputs to an AND gate are fed through inverters. This configuration allows the AND gate to operate in the *complementary* (opposite) manner, just as the OR gate did in the first case. With inverted inputs, the AND operation is

Table 3-8 $\bar{A} + \bar{B} = \overline{AB}$

A	$\bar{A}$	B	$\bar{B}$	$\bar{A}+\bar{B}$	AB	$\overline{AB}$
L	H	L	H	H	L	H
L	H	H	L	H	L	H
H	L	L	H	H	L	H
H	L	H	L	L	H	L

 Equality

Derived Functions

present only when both inputs are not present. In other words, *the negated input* AND *gate is activated only when both inputs are not present.* Compare the negated input AND operation shown in Fig. 3-21a with the normal AND operation in Fig. 3-2 to determine the differences.

Now we will examine an OR operation with negated output. The inverter appears at the output rather than at the input of the OR gate, as in the AND gate example. Thus the OR operation is performed *before* negation. The logic diagram and resulting truth table are shown in Fig. 3-21b.

A	$\bar{A}$	B	$\bar{B}$	$\bar{A} \cdot \bar{B}$
0	1	0	1	1
0	1	1	0	0
1	0	0	1	0
1	0	1	0	0

(a) Negated input AND operation

A	B	A + B	$\overline{A + B}$
0	0	0	1
0	1	1	0
1	0	1	0
1	1	1	0

(b) Negated output OR operation

Figure 3-21 The NOR function

Once again, the resemblance between the negated-input AND operation truth table and the negated-output OR operation truth table is obvious. Since the truth tables are identical, the logic functions are identical. This is the second of DeMorgan's Laws, which in Boolean algebra becomes $\overline{A} \cdot \overline{B} = \overline{A + B}$. Comparative truth tables are shown in Table 3-9.

While we do not plan a detailed discussion of derived logic operations in this chapter, we would like to point out that the logic operations

Table 3–9 $\bar{A} \cdot \bar{B} = \overline{A + B}$

A	$\bar{A}$	B	$\bar{B}$	$\bar{A} \cdot \bar{B}$	A + B	$\overline{A + B}$
L	H	L	H	H	L	H
L	H	H	L	L	H	L
H	L	L	H	L	H	L
H	L	H	L	L	H	L

Equality

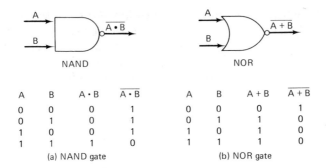

Figure 3-22 NAND and NOR gates

resulting from DeMorgan's Laws are commonly implemented in *hardware form.* A device called the NAND (NOT AND) gate (Fig. 3-22a) implements the first law, and the NOR (NOT OR) gate (Fig. 3-22b) implements the second law. Since both AND and OR operations may be derived from NAND and NOR gates, many manufacturers furnish only the latter types of logic elements. Economically, it may be more feasible for a logic designer to use a minimum number of different logic elements (that is, all NAND and all NOR) rather than mixed AND, OR, NAND, NOR, and inverter elements. Such designs may actually require more of one type of logic assembly in some cases, but the saving in quantity purchases and reduced inventory does reduce the overall system costs.

Using a Binary Counter To Show Identities

Our introduction of DeMorgan's Laws and an increasing use of truth-table identity proofs leads us to a more detailed discussion of the *logic signal source.* Since the identity of two truth tables is considered proof of logic identity, a device that would speed up such proofs by mechanizing the truth tables is highly desirable. Such a device, the logic signal source, has already been used to examine some of the Boolean algebra theorems. In Sec. 2-6, the FF was discussed in conjunction with counting. If we examine the counting table developed for the four-stage binary counter, we see that the entries match directly with the truth table developed for the logic signal source. Essentially, therefore the logic signal source is *a simple binary counter with outputs arranged and named so that they can be used to mechanize the standard truth table.* The two tables (3-10) are identical, thus establishing the identity of the counter and the logic signal source.

One additional fact: as previously stated, the total number of combinations to be evaluated when examining the operations of combina-

Derived Functions

Table 3–10 Binary counter versus logic-signal source

Counter table				Logic-signal source		
A	B	C	Decimal Value	A	B	C
0	0	0	0	0	0	0
0	0	1	1	0	0	1
0	1	0	2	0	1	0
0	1	1	3	0	1	1
1	0	0	4	1	0	0
1	0	1	5	1	0	1
1	1	0	6	1	1	0
1	1	1	7	1	1	1

Note: 0 = LOW and 1 = HIGH

tional logic circuits is 2^n, where n is the number of variables. Thus with two variables, 2^2 or four combinations must be evaluated. You might randomly determine the four possible combinations, but it is more efficient to organize the combinations into a usable list. By arranging the truth table entries in ascending order (starting with 0 and ending with $2^n - 1$), all possible combinations will occur in an easily remembered and verifiable order. In a two-variable truth table (Table 3-11), 2^2 entries (starting with 0 and ending with $2^2 - 1$ or decimal 3) are made. If three variables are to be examined, then 2^3 entries (starting with 0 and ending with $2^3 - 1$ or decimal 7) must be made. (See Tables 3-10 and 3-11.)

Table 3–11 2-variable truth table

Decimal value	A	B
0	0	0
1	0	1
2	1	0
3	1	1

Note: 0 = LOW and 1 = HIGH

The technique, mentioned in Chapter 2, that advocated visual representation of logic states is also extremely useful with mechanized truth tables. If the *light-on* is "true" (1 or HIGH) and the *light-off* is "false" (0 or LOW), then either a waveform display or any electrical indications are unnecessary in determining logic states. In actual practice, a special type of amplifier called a *lamp driver* (Fig. 3-23) provides sufficient power to operate the indicator lamp.

Each stage of the logic-signal source has an indicator, which will determine its state. In addition, indicators are attached to the output

Figure 3-23 Lamp Driver symbol

of each logic circuit being tested. The logic-signal source is then started through its sequence, observing the states of each stage and the outputs of the logic circuits. If the indicators attached to the outputs of the logic circuits being tested are in the same configuration for each input step, then those outputs are identical. Thus if the output indicators stay the same (i.e., changing together throughout all possible input conditions), the *identity* is established. Logic diagrams of typical Boolean algebra identity proofs, using both of DeMorgan's Laws, are shown in Fig. 3-24. Lamp drivers are connected so as to display the logic state of each circuit.

(a) $\overline{A} + \overline{B} = \overline{AB}$

(b) $\overline{A} \cdot \overline{B} = \overline{A + B}$

Figure 3-24 Use of Lamp Drivers in logic circuits to indicate proof of DeMorgan's Laws

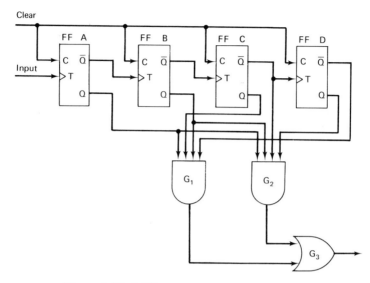

Figure 3-25 A "Count Decoding" logic circuit

3-7 A SIMPLE APPLICATION OF BASIC COMBINATIONAL LOGIC FUNCTIONS

The Logic Circuit

A practical application of combinational logic is the *count-decoding* method. Many logic applications require the counting of objects, events, timing pulses, and so forth. For example, consider the logic circuit in Fig. 3-25. Although the binary counter shown is sequential in nature (a concept that has not yet been discussed), the reader should be able to understand its operation from both the counting-situation and the logic-signal source discussions. Specifically, the number of input pulses counted may be determined by examining the states of the outputs of each flip-flop (FF). If we can assume that the FF interconnections provide the capability to count, then the connection of the FF outputs to G_1 and G_2 inputs *could* be used to decode specific counter configurations. G_3 should provide an output when either G_1, G_2, or both are activated. Now comes the problem: when will an active output appear at G_3, and what form will it take?

Waveform Analysis

One approach uses waveform analysis. The waveforms for the 4-FF counter, developed in Ch. 2, are reproduced in Fig. 3-26 along with the G_1, G_2 and G_3 outputs.

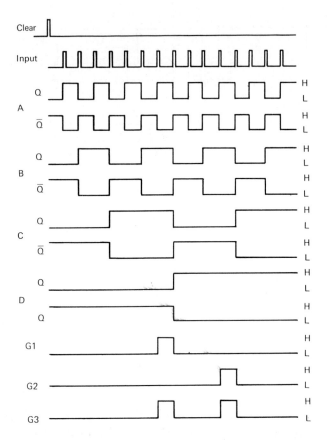

Figure 3-26 Counter and decoder waveforms

To activate the AND gate G_1 (i.e., cause its output to go HIGH), all inputs to G_1 must be HIGH. The inputs to G_1 are furnished by FFs A, B, C, and D, so each of these inputs must be HIGH to activate the gate. Thus the Q outputs of A, B, and C and the $\bar{Q}$ output of D must all be HIGH at the same time. The rules of operation for FFs state that the $\bar{Q}$ output is HIGH when the FF is in the CLEAR (RESET) state. G_1 will therefore be activated when A, B, and C are SET and D is RESET.

Waveform analysis* of this type of logic circuit usually begins at the most-significant-digit (MSD) FF and works backward. Investigation

---

*If any doubt exists as to how the waveforms are obtained, review the FF-counter section of Ch. 2.

of the $\overline{Q}$ output of FF D (one of the inputs to G_1) will show that it is HIGH from the time of the CLEAR pulse application through (and including) count 7. Thus G_1 *must* be decoding a count between 0 and 7. Observing the Q output of FF C (another input to G_1) indicates that it is HIGH only from count 4 through (and including) count 7. The count is now narrowed to the range between 4 and 7. The Q output of FF B is HIGH during counts 2 and 3, and counts 6 and 7. Since the Q output of C is LOW during counts 2 and 3, G_1 cannot be activated at that time. However the C FF Q output does provide an enabling input during counts 6 and 7. G_1 activation is now narrowed to either count 6 or 7. The A FF Q output is HIGH during count 7, so G_1 becomes enabled approximately from the time input pulse 7 arrives until the arrival of input pulse 8. In other words, G_1 is *decoding* count 7. (G_1 output is shown in the waveform diagram.) Of course, since any HIGH input to an OR gate results in a HIGH output, gate G_3 also provides an indication at count 7. Another output from G_3 occurs when G_2 is enabled. The same analytical approach may be applied to G_2 to show that it is decoding count 11. G_3 output then occurs at either count 7 *or* count 11. Figure 3-26 shows all the important waveforms.

Since waveform analysis provides a visual representation of logic-circuit functions, some readers may prefer to use this method. However as logic circuits become more complex, waveform drawings become more cumbersome, so Boolean algebra notations are preferred.

A few words must be said at this point to clarify the use of Boolean notation when discussing FFs. The standard FF has two outputs, commonly called the Q and the $\overline{Q}$ outputs. When the FF is in the SET state, the Q output is HIGH and the $\overline{Q}$ output is LOW. When FFs are named, (for example, A, B, C, D), the SET state is considered to be the *true* state, and the CLEAR (RESET) state becomes the *false* state. For example, if FF A is in the SET state, it is commonly denoted as A; if it is in the RESET state it is denoted as $\overline{A}$.

Evaluation of the decode 7 gate (G_1) shows that A, B, and C must be in the SET state and D must be in the RESET state (because the $\overline{Q}$ output rather than the Q output is used as an input to G_1). The Boolean expression for enabling G_1 then becomes $A \cdot B \cdot C \cdot \overline{D}$. G_2 is enabled by FFs A, B, and D in the SET state and C in the RESET state. The resulting expression is $A \cdot B \cdot \overline{C} \cdot D$. Remember, when the Boolean expression is written for an AND gate (under MIL-STD-806B rules), the variables shown must *all* be HIGH to activate the gate. Writing $A \cdot B \cdot C \cdot \overline{D}$ means that all four variables must be HIGH. Although $\overline{D}$ appears, this merely means that whatever the $\overline{D}$ input comes from must provide a HIGH output when D is in the *false* condition. This is accomplished in a FF by using the $\overline{Q}$ output.

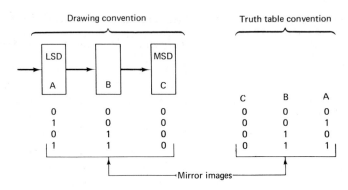

Figure 3-27 Drawing convention compared with truth table convention

Until now, we have written expressions describing gate outputs in the order of FF positions as they appear on the logic diagram. Actually, it makes no difference in what *order* the variables are ANDed. Thus it is more convenient to list the variables in alphabetical order (i.e., $A \cdot B \cdot C \cdot D$). Unfortunately, logic diagrams showing FFs are conventionally arranged with the LSD on the left, since information flows from LSD to MSD in such a circuit. The truth table, however, is arranged with the MSD on the left, which is convenient if all combinations of variables are placed in ascending decimal order (see Fig. 3-27).

Rather than break completely with these conventions, the reader should merely adapt a bit. Since much of our work is done by referring to logic diagrams, it is advisable to retain the diagram convention of data flow (from left to right) and to adapt the truth table to fit the situation. The table showing the sequential states of the three FFs is actually a "mirror image" of the truth table. Keeping this in mind, the advantages of both conventions can be retained.

When the outputs of G_1 and G_2 are combined in the OR gate (G_3), the resulting expression $A \cdot B \cdot C \cdot \overline{D} + A \cdot B \cdot \overline{C} \cdot D$ becomes the output of the multiple-count decoding circuit. Assignment of numerical values to each variable, based on the position of the FFs in the counter circuit, produces the actual decoded counts. Remember, A is the LSD FF and D is the MSD FF. The expression $A \cdot B \cdot C \cdot \overline{D}$ is equivalent to $\overline{8}$ (which is 0) plus 4 plus 2 plus 1 equals 7. $A \cdot B \cdot \overline{C} \cdot D$ means 8 plus $\overline{4}$ (0) plus 2 plus 1 equals 11. Study the Boolean expressions and the waveforms thoroughly to establish the correct correlations and relationships.

Since we had no intention of explaining all the complexities of binary counters—they were used here merely to develop input signals for the gates—please refer to Chapter 11 for further details about the counter's operation.

TABLE 3-12 4-variable truth table used to analyze count decoder

D	$\bar{D}$	C	$\bar{C}$	B	A	$AB C\bar{D}$	$AB\bar{C}D$	$AB C\bar{D} + AB\bar{C}D$
0	1	0	1	0	0	0	0	0
0	1	0	1	0	1	0	0	0
0	1	0	1	1	0	0	0	0
0	1	0	1	1	1	0	0	0
0	1	1	0	0	0	0	0	0
0	1	1	0	0	1	0	0	0
0	1	1	0	1	0	0	0	0
0	1	1	0	1	1	1	0	1
1	0	0	1	0	0	0	0	0
1	0	0	1	0	1	0	0	0
1	0	0	1	1	0	0	0	0
1	0	0	1	1	1	0	1	1
1	0	1	0	0	0	0	0	0
1	0	1	0	0	1	0	0	0
1	0	1	0	1	0	0	0	0
1	0	1	0	1	1	0	0	0

Truth-Table Analysis

Once the inputs to the gates have been defined, truth-table techniques can be used to analyze combinational logic circuits. In this particular case, four input variables are used. Applying Eq. (3-1), we see that a total of 16 possible combinations of the four variables must be examined. Arranging the 16 combinations in ascending decimal order results in Table 3-12. (Since negated values are used in both expressions, they are included in the table.) Now we merely examine each of the combinations for the *binary value* (truth value) of each expression. $A \cdot B \cdot C \cdot \bar{D}$ is true only at count 7, and $A \cdot B \cdot \bar{C} \cdot D$ is true only at count 11. The truth table thus proves the original analysis.

3-8 SUMMARY

The OR Operation

$$A + B$$

"If A is true, or B is true, or A and B are true together, then A OR B is true." From a more practical standpoint: "If A is HIGH, or B is HIGH, or A and B are HIGH together, then A OR B is HIGH." (See Figure at top left.)

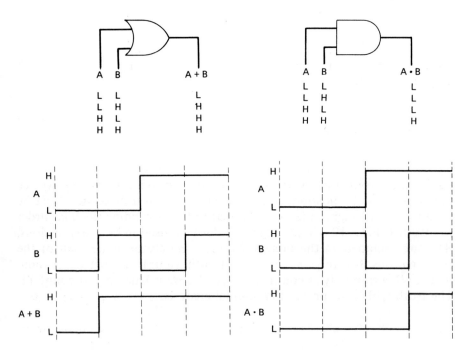

The AND Operation

$$A \cdot B$$

"If A and B are both true at the same time, then $A \cdot B$ is true." Stated in more common language: "If A and B are both HIGH at the same time, then $A \cdot B$ is HIGH." (See Figure above right.)

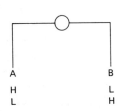

The NOT Operation

$$A = \bar{B}, \quad \bar{A} = B$$

"A is true if B is false; A is false if B is true." Stated differently: "A is HIGH when B is LOW; A is LOW when B is HIGH." (See Figure at right.)

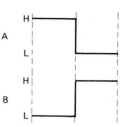

PROBLEM APPLICATIONS

3-1. What are the two general types of logic circuits? Define them.
3-2. What is a logic operation?
3-3. Define CONJUNCTION. AND
3-4. Define DISJUNCTION. OR
3-5. Define NEGATION. NOT
3-6. List all of the Boolean constants.
3-7. What is the maximum number of constant values a Boolean variable can assume?
3-8. Define truth table.
3-9. What is the purpose of a truth table?
3-10. Why is Boolean algebra less complex than conventional algebra?
3-11. Draw the MIL-STD806B symbols for devices that will perform the CONJUNCTION operation; the DISJUNCTION operation; the NEGATION operation.
3-12. In combinational logic terms, what is a gate?
3-13. Show the proper symbol, construct a truth table, and draw the waveforms for the Boolean expression $A + B + C$.
3-14. Explain at least three ways in which the OR operation may be put to practical use.
3-15. Show the proper symbol, construct a truth table, and draw the waveforms for the Boolean's expression ABC.
3-16. Explain at least three ways in which the AND operation may be put to practical use.
3-17. What are the restrictions on the use of the NOT symbol? amplifier
3-18. What is the most convenient method for examining multiple-combinational logic operations?
3-19. In the Boolean expression $A + B + C + D$, how many combinations of the four variables exist?
3-20. What is the order of performing operations in a complex Boolean expression?
3-21. What is a NAND gate?
3-22. What is a NOR gate?

4

ADVANCED COMBINATIONAL LOGIC

Our investigation of the postulates and theorems of Boolean algebra in Ch. 3 has prepared us for new territory—*the laws and identities of Boolean algebra.* In developing techniques for logic-circuit analysis, we must continually broaden our knowledge and understanding of the mathematics of logic as well as such other analytical "tools" as truth tables and waveform analysis.

Each logic diagram we encounter presents a new challenge. Some are relatively simple, so the use of logic symbols and the application of "logical-thought processes" will provide an adequate understanding of the diagram. Initially, due to inexperience, the reader will find few diagrams that can use this method of analysis. Instead, as increasingly complex logic circuits are encountered, truth tables must be compiled to decipher the intricacies of the logic functions. A reader with an extensive background in electronic troubleshooting, on the other hand, may find waveform analysis a more useful tool. For the mathematically oriented reader, the use of Boolean algebra is probably most appropriate. Undoubtedly, however, any combinational logic circuit will require the application of more than one (and perhaps even all) of these tools.

In this chapter, we will learn how to apply basic laws and identities of Boolean algebra to these various methods of combinational logic diagram analysis.

4-1 BASIC LAWS OF BOOLEAN ALGEBRA

In conventional mathematics, it is often difficult to provide sufficient nonmathematical proofs to help simplify the cornerstones—*the laws*—of the system. In Boolean algebra, however, truth tables and waveforms supply visual clues and explanations that fix the basic principles in the readers' minds. In addition, whenever it is appropriate, the actual mathematical manipulations are shown. Thus the reader can progress from simple to complex laws with a minimum of effort.

The Law of Identity

The *Law of Identity* states that *any letter, number, or expression is equal to itself*. Thus in Boolean algebra,

$$A = A \quad \text{and} \quad \overline{A} = \overline{A} \tag{4-1}$$

Some additional examples of this law are

$$B = B, \quad 0 = 0, \quad AB = AB, \quad \overline{X} = \overline{X}, \quad \text{and so on.}$$

It is unnecessary to prove the Laws of Identity by such methods as truth tables—the proof is implicit in the statement of the law.

The Commutative Law

The first part of the *Commutative Law* states that *when inputs to a logic symbol are* AND*ed, the order in which they are written does not affect the binary value of the output.* Stated algebraically,

$$AB = BA \tag{4-2}$$

The proof of this statement is easily established by truth tables. Furthermore, mechanization of the truth tables and the use of indicator lamps (discussed in Sec. 3-6) are employed in most of the proofs presented here. These logic circuits may be constructed by using any of the digital-logic training aids available. We strongly recommend that each of the logic circuits be investigated by using actual hardware.

When the output level is HIGH (1), the lamp is ON. If the lamp is not ON, the output level is LOW (0). As each input combination is established (each line on the truth table), note the condition of the indicator lamps. If both lamps are in the same condition (*either* OFF or ON), both circuits are reacting in the same manner to the specific input combination. The logic-signal source can then be set to the *next* input combination (the next line on the truth table) and the process repeated. If both lamps are in the same condition (*either* OFF or ON) for *all* possible input combinations, then the two circuits are identical

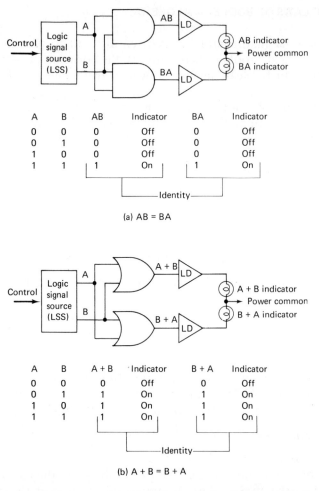

Figure 4-1 Logic diagram and truth table proofs for the Commutative Laws

in logic functions. Of course, the condition of the lamps determines whether the output is 1 or 0, and the truth table may be constructed directly by recording the lamp conditions for each input condition. Figure 4-1a shows the logic diagram and truth table proofs for the identity $AB = BA$.

In logic diagrams, the power connections are assumed to be made and are therefore *not* shown. In the case of the lamp drivers, when the input is LOW, the output is LOW (as defined by MIL-STD-806B), so the lamps are OFF. Conversely, HIGH input causes a HIGH output, so the lamps are ON. The lamp drivers act as *switches,* connecting power when the inputs are HIGH and removing power when the inputs are LOW.

Basic Laws of Boolean Algebra

The second part of the *Commutative Law* states that *when inputs to a logic symbol are ORed, the order in which they are written does not affect the binary value of the output.* In Boolean algebra terms

$$A + B = B + A \qquad (4-3)$$

A proof (Fig. 4-1b) similar to that used for the first part of the Commutative Law will suffice.

The Commutative Law is *not* limited to two variables. With three inputs (*A*, *C*, and *D*) provided to an AND gate, the output could be written as

$$A \cdot C \cdot D, \; A \cdot D \cdot C, \; C \cdot D \cdot A, \; C \cdot A \cdot D, \; D \cdot C \cdot A \quad \text{or} \quad D \cdot A \cdot C$$

Furthermore, the expression $R \cdot (S + T)$ may also be written as $(T + S) \cdot R$ —or in numerous other ways. $W + X + YZ$ and $ZY + X + W$ are the same, although the order of *both* the ANDed and ORed variables have been changed. The Commutative Law still applies, as proven in Fig. 4-2.

Figure 4-2 Logic diagram and truth table proofs $W + X + YZ = ZY + X + W$

Note that the 16 possible combinations of four variables are arranged in the truth table in ascending decimal order.

The procedure for developing a logic diagram from simple Boolean expressions consists of four steps.

1. Identify the general type of circuit.
2. Begin with the output expression, drawing the logic symbol that represents the last (final) operation.
3. Work toward the input, drawing logic symbols for each operation until all inputs are accounted for.
4. Label inputs and outputs of all required gates.

Naturally, as expressions become more complex, we will be forced to expand this basic procedure.

The identity just investigated may also be demonstrated by using waveforms. Figure 4-3, for example, is a waveform representation of the logic diagram and truth table of Fig. 4-2. Waveform diagrams show quite adequately the logic conditions defined by Boolean algebra, logic diagrams, and truth tables. HIGH and LOW conventions, discussed in previous chapters, are used in this chapter in all waveform diagrams. Many of the identities that will follow are explained by using waveform diagrams, in addition to the truth tables and algebraic proofs of Boolean algebra laws.

$(J + L)K$ is not the same as $(J + K)L$. The Commutative Law applies *only* to inputs to the *same* logic symbol. Note that the logic

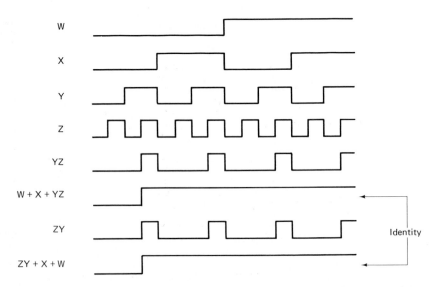

Figure 4-3 Waveform drawing for the identity $W + X + YZ = ZY + X + W$

Basic Laws of Boolean Algebra

values of the outputs do *not* match, thus showing that the two expressions are *not* identical. A logic diagram, truth table, and waveform proofs for $(J + L)K \neq (J + K)L$ are shown in Fig. 4-4.

A composite restatement of the Commutative Law might therefore say: "In a simple AND or OR relationship, the positions of the input signals may be interchanged with no change in the output signal."

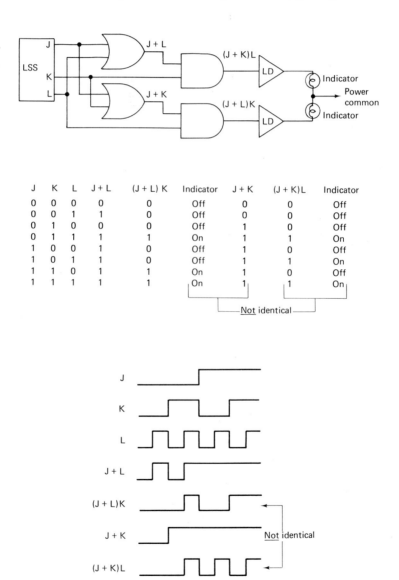

Figure 4-4 $(J + L)K \neq (J + K)L$ — truth table, logic diagram, and waveforms

The Associative Law

The *Associative Law* simply states that *it makes no difference in what order parts of an* AND *or an* OR *expression are combined.* Thus

$$A(BC) = ABC \qquad (4\text{-}4)$$

(See Fig. 4-5.)

and $\quad A + (B + C) = A + B + C \qquad (4\text{-}5)$

(See Fig. 4-6.)

Figure 4-5 Truth table, logic diagram, and waveform proofs for A(BC) = ABC (Associative Law)

Basic Laws of Boolean Algebra

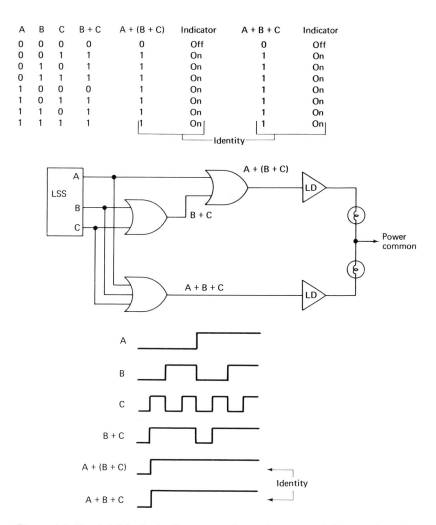

A	B	C	B + C	A + (B + C)	Indicator	A + B + C	Indicator
0	0	0	0	0	Off	0	Off
0	0	1	1	1	On	1	On
0	1	0	1	1	On	1	On
0	1	1	1	1	On	1	On
1	0	0	0	1	On	1	On
1	0	1	1	1	On	1	On
1	1	0	1	1	On	1	On
1	1	1	1	1	On	1	On

Figure 4-6 Truth table, logic diagram, and waveform proofs for A + (B + C) = A + B + C (Associative Law)

The reader may wonder why something as simple as the Associative Law gets so much attention. If so, consider the hardware implementation of $A + (B + C)$. Note that one 2-input OR gate is required to combine B with C, and another 2-input OR gate is needed to combine A with the results of ORing B with C. Implementing $A + B + C$ requires only one 3-input OR gate. Thus we have saved one gate! Even in the simpler digital devices, hundreds of such expressions may be required, so any simplification operations produce a great saving.

The Idempotent Laws

The *Idempotent Law* states that *if a binary variable is* ANDed *with itself or* ORed *with itself, the binary value of the output will be that of the input variable.*

$$AA = A \quad \text{and} \quad A + A = A \tag{4-6}$$

These expressions are merely restatements of Theorems 5 and 11, respectively, whose proofs were shown in Chapter 3.

The Distributive Law

The first identity of the Distributive Law is

$$A(B + C) = AB + AC \tag{4-7}$$

Replace our conventional methods of proof (i.e., logic diagram and the truth table) and try to "think this one out." Thus for the binary value of $A(B + C)$ to be 1, A must be 1 *and* either B or C must be 1. If B were 1, the AB combination would yield a 1. Likewise, if C were 1, the AC combination would also yield a 1. This type of approach —the *logical thought process*—is sometimes an effective method of simplification for Boolean expressions.

The truth table proof of $A(B + C) = AB + AC$ and the associated logic diagram and waveforms are shown in Fig. 4-7.

This first identity of the Distributive Law may be easily proven by applying the rules of *logical multiplication* (ANDing). When variables inside parentheses are logically multiplied by a variable outside the parentheses, the result is the logical product of the "outside" variable and each of the "inside" variables, joined by the operator connecting the "inside" variables. Thus by logical multiplication, $A(B + C) = AB + AC$. Note that logical multiplication in this case operates the same as algebraic multiplication.

The second identity of the Distributive Law is

$$A + BC = (A + B)(A + C) \tag{4-8}$$

This identity is valid in Boolean algebra but *not* in ordinary algebra.

The logic diagram, truth table, and waveform diagrams proving this identity are shown in Fig. 4-8.

This is only one form that a Boolean algebra proof can assume. Many other approaches (some with less steps) yield the same results. *But we have selected proofs for this and all the laws and identities that will demonstrate the most important uses of Boolean algebra.* The reader is encouraged to investigate other forms of algebraic proof.

Basic Laws of Boolean Algebra

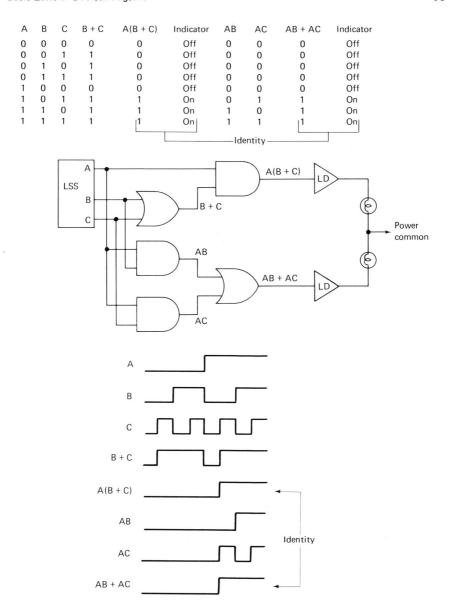

Figure 4-7 Truth table, logic diagram, and waveform proofs for A(B+C) = AB+AC (Distributive Law)

A	B	C	BC	A + BC	Indicator	A + B	A + C	(A + B)(A + C)	Indicator
0	0	0	0	0	Off	0	0	0	Off
0	0	1	0	0	Off	0	1	0	Off
0	1	0	0	0	Off	1	0	0	Off
0	1	1	1	1	On	1	1	1	On
1	0	0	0	1	On	1	1	1	On
1	0	1	0	1	On	1	1	1	On
1	1	0	0	1	On	1	1	1	On
1	1	1	1	1	On	1	1	1	On

———— Identity ————

Figure 4-8 Truth table, logic diagram, and waveform proofs for A+BC = (A+B)(A+C) (Distributive Law)

Application of the Distributive Law* to simplification operations significantly reduces hardware requirements and increases analytic capabilities.

DeMorgan's Law

DeMorgan's Law expresses a relationship between the OR and the AND operations in Boolean algebra. Stated in simple terms: *DeMorgan's Law says that inversion or negation of a Boolean expression may be accomplished by negating each variable/constant and by changing each AND to OR and OR to AND.* Equation (4-9) generalizes DeMorgan's Law.

$$\overline{A \cdot B} = \overline{A} + \overline{B} \qquad \text{and} \qquad \overline{A + B} = \overline{A} \cdot \overline{B} \qquad (4\text{-}9)$$

Be sure to note the *difference* between the expression $\overline{A + B}$ and the expression $\overline{A} + \overline{B}$. In the former, A and B are combined in an OR relationship; then the output is inverted. In the latter, both A and B are first inverted; then the inverted variables are combined in the OR relationship. Hence *the two expressions are not equal*, which is also evident from the truth tables and the waveforms in Fig. 4-9.

All these laws are powerful tools that can be used to simplify Boolean expressions and to minimize the number of logic devices necessary to perform a particular function, as the following examples illustrate.

Example 4-1. Simplify the expression $(\overline{A} + \overline{B} + \overline{C})D$. Then draw the logic diagrams for the original and the simplified expressions.

Solution. $(\overline{A} + \overline{B} + \overline{C})D = (\overline{ABC})D$ DeMorgan's Laws

See Fig. 4-10.

Example 4-2. Simplify the expression $R + S + (TV) + \overline{(RS)}$. Then draw the logic diagrams for the original and the simplified expressions and show truth-table proof of the equivalency.

Solution. $R + S + (TV) + \overline{(RS)} = R + S + (TV) + (\overline{R} + \overline{S})$ DeMorgan's Law
$= R + S + (\overline{R} + \overline{S}) + TV$ Commutative Law
$= R + S + \overline{R} + \overline{S} + TV$ Associative Law
$= R + \overline{R} + S + \overline{S} + TV$ Commutative Law
$= 1 + 1 + TV$ Theorem 12
$= 1$ Theorem 10, Postulate 8.

---

*The Distributive Law will be examined in greater detail in subsequent chapters.

Figure 4-9 DeMorgan's Laws—logic diagram, waveforms, and truth tables

See Fig. 4-11 for the logic diagrams and Table 4-1 for the truth-table proof.

DeMorgan's Law also states mathematically the *equivalence of certain gate forms*. Since $\overline{A \cdot B} = \overline{A} + \overline{B}$, the hardware implementation of each side of the equation will provide identical results (see Fig. 4-9). Thus the negated input OR gate is functionally equivalent to the negated output AND (NAND) gate. Similarly, since $\overline{A} \cdot \overline{B} = \overline{A + B}$, the negated input AND gate is equivalent to the negated output OR (NOR) gate.

Basic Laws of Boolean Algebra

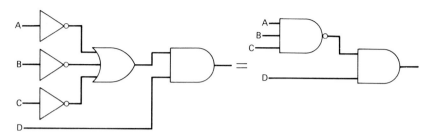

Figure 4-10 Logic diagram, Example 4-1

As we mentioned in Ch. 3, many hardware manufacturers supply only certain types of gates—in the interests of cost. By applying DeMorgan's Law and the Distributive Law, we can show how *all* combinational functions implemented by the AND, OR, and NOT method may *also* be implemented by the NOR method *alone* or by the NAND method *alone*.

Section 3-6 defines *a* NOR *gate as an* OR *gate followed by an inverter*. The truth table for the OR gate (see Fig. 3-2) reveals that when any one or more of the inputs to an OR gate is HIGH, the output is HIGH and that when all inputs are LOW, the output is LOW. Terminating all but one of the gate inputs—as recommended by the manufacturer's specification sheet—results in a 1-input OR gate. The 1-input OR gate still behaves according to Fig. 3-2—a HIGH input results in a HIGH output, a LOW input results in a LOW output. By using the 1-input OR gate followed by an inverter (a 1-input NOR gate), the "straight-through" operation of the 1-input OR gate is inverted, resulting in the logic operation of NOT. Thus it is possible to obtain the NOT operation by using only NOR gates.

Figure 4-12a shows how NOR gates can be combined to perform the AND operation. The truth table accompanying the diagram verifies both the equality and the NOR gate implementation of the 2-input AND gate.

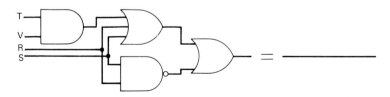

Figure 4-11 Logic diagram, Example 4-2

Table 4–1 Truth table $R + S + (TV) + (\overline{RS})$

R	$\overline{R}$	S	$\overline{S}$	T	$\overline{T}$	V	$\overline{V}$	$T \cdot V$	$R \cdot S$	$\overline{R \cdot S}$	$\overline{R}+\overline{S}$	$R + S + (TV) + (\overline{RS})$	$R + \overline{R} + S + \overline{S} + (TV)$
0	1	0	1	0	1	0	1	0	0	1	1	1	1
0	1	0	1	0	1	1	0	0	0	1	1	1	1
0	1	0	1	1	0	0	1	0	0	1	1	1	1
0	1	0	1	1	0	1	0	1	0	1	1	1	1
0	1	1	0	0	1	0	1	0	0	1	1	1	1
0	1	1	0	0	1	1	0	0	0	1	1	1	1
0	1	1	0	1	0	0	1	0	0	1	1	1	1
0	1	1	0	1	0	1	0	1	0	1	1	1	1
1	0	0	1	0	1	0	1	0	0	1	1	1	1
1	0	0	1	0	1	1	0	0	0	1	1	1	1
1	0	0	1	1	0	0	1	0	0	1	1	1	1
1	0	0	1	1	0	1	0	1	0	1	1	1	1
1	0	1	0	0	1	0	1	0	1	0	0	1	1
1	0	1	0	0	1	1	0	0	1	0	0	1	1
1	0	1	0	1	0	0	1	0	1	0	0	1	1
1	0	1	0	1	0	1	0	1	1	0	0	1	1

Basic Laws of Boolean Algebra

A	$\bar{A}$	B	$\bar{B}$	$\bar{A}+\bar{B}$	$\overline{\bar{A}+\bar{B}}$	AB
0	1	0	1	1	0	0
0	1	1	0	1	0	0
1	0	0	1	1	0	0
1	0	1	0	0	1	1

(a) AND operation using NOR gates—Logic diagram and truth tables

A	B	A+B	$\overline{A+B}$	$\overline{\overline{A+B}}$	A+B
0	0	0	1	0	0
0	1	1	0	1	1
1	0	1	0	1	1
1	1	1	0	1	1

(b) OR operation using NOR gates—Logic diagram and truth table

Figure 4-12 AND and OR operations using NOR gates—logic diagram and truth tables

The OR operation may also be implemented using only NOR gates. The logic diagram and substantiating truth table appear in Fig. 4-12b.

At this point, we should mention some alternate types of logic diagrams. In some cases a *block diagram* form is used, since it more closely resembles the physical configuration of the devices performing the logic functions. A typical illustration is the 7402 IC (integrated circuit) form, where four NOR gates are packaged in a *dual in-line IC package* (DIP). Physically, it is approximately 0.28 inch wide, 0.785 inch long, and 0.18 inch thick. Its diagram may assume the form of Fig. 4-13 and is equivalent to the logic diagram in Fig. 4-12b.

NAND gates may also be used to perform the AND, OR, and NOT operations. *A NAND gate* has been previously defined as *an* AND *gate followed by an inverter* (NOT). The NAND gate's output is LOW when all inputs are HIGH. If any one input is LOW, its output* is HIGH. A 1-input AND may be derived in a manner similar to the 1-input OR.

---

*An AND gate, in comparison, has a HIGH output when (and only when) all inputs are HIGH.

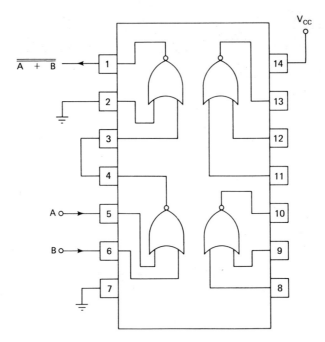

Figure 4-13 Physically oriented logic diagram

Termination of unused inputs must be in accordance with the manufacturer's specifications. The 1-input AND performs the same function as the 1-input OR; that is, HIGH in–HIGH out, and LOW in–LOW out. Combining the 1-input AND with the inverter (NOT) results in a 1-input NAND that performs the NOT operation.

The AND operation may be obtained from a NAND gate with two inputs A and B, followed by a 1-input NAND gate. Justification and diagrams appear in Fig. 4-14a. As the complexity of logic functions within a single IC package increases, it becomes less and less practical to use the form of diagram in Fig. 4-14a. It is quite common to provide the capability to perform complete arithmetic operations, such as addition, on a single IC "chip." Obviously the logic diagram of this type of operation would not fit inside the symbol provided and still be readable; hence, the form in Fig. 4-14b is used. Complete manufacturer's logic descriptions and diagrams must be available to perform logic analysis with this type of diagram. Figure 4-14b is logically equivalent to Fig. 4-14a.

If the output $A + B$ is desired, it can also be obtained from NAND gates. DeMorgan's Law states that $\overline{A + B}$ is equivalent to $\overline{A} \cdot \overline{B}$, which is

Basic Laws of Boolean Algebra

A	B	AB	$\overline{AB}$	$\overline{\overline{AB}}$	AB
0	0	0	1	0	0
0	1	0	1	0	0
1	0	0	1	0	0
1	1	1	0	1	1

(a) AND operation using NAND gates—Logic diagram and truth table

A	$\overline{A}$	B	$\overline{B}$	$\overline{A} \cdot \overline{B}$	$\overline{\overline{A} \cdot \overline{B}}$	A + B
0	1	0	1	1	0	0
0	1	1	0	0	1	1
1	0	0	1	0	1	1
1	0	1	0	0	1	1

(b) Physically oriented logic diagram (DIP case)

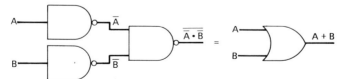

(c) OR operation using NAND gates—Logic diagram and truth table

Figure 4-14 AND and OR operations using NAND gates—logic diagram and truth tables

Table 4-2   AND-OR/NAND-NOR equivalences

A	$\bar{A}$	B	$\bar{B}$	$A \cdot B$	$A + B$	$\overline{A \cdot B}$	$\overline{A + B}$	$\bar{A} + \bar{B}$	$\bar{A} \cdot \bar{B}$
0	1	0	1	0	0	1	1	1	1
0	1	1	0	0	1	1	1	0	0
1	0	0	1	0	1	1	1	0	0
1	0	1	0	1	1	0	0	0	0

the output of a NAND gate with inputs $\bar{A}$ and $\bar{B}$. These inputs may be easily obtained by two 1-input NANDs being fed by A and B, respectively. Logic diagrams and the truth table are provided in Fig. 4-14c.

Thus any of the three basic operations of Boolean Algebra—*conjunction* (AND), *disjunction* (OR), and *negation* (NOT)—may be obtained either from NOR or NAND gates. As an aid to using and remembering the three basic logic operations and the two derived logic operations, Table 4-2 provides the logic characteristics of each in an easily comparable format. Innumerable combinations of AND, OR, NOT, NAND, and NOR operations are possible. Extensive coverage of all possible configurations is not intended. The basic fundamentals have been provided and will be expanded as examinations of more complex logic circuits are undertaken.

The Law of Absorption

The first identity of the *Law of Absorption* is

$$A + AB = A \qquad (4\text{-}10)$$

Actually, this identity is nothing more than common sense. Whenever A is 1, the value of $A + AB$ is 1. It is really unnecessary to consider what happens to the value of the expression when both A and B are 1. Regardless of the value of A and B together, the value of the expression is determined by the value of A alone. The truth table and diagrams in Fig. 4-15a verify this statement.

The second identity of the Law of Absorption

$$A(A + B) = A \qquad (4\text{-}11)$$

is demonstrated in Fig. 4-15b.

The Law of Expansion

The first identity of the *Law of Expansion* is

$$AB + A\bar{B} = A \qquad (4\text{-}12)$$

Basic Laws of Boolean Algebra

Figure 4-15 The Laws of Absorption

can be proven by any of the methods employed so far. For example, the following is an algebraic proof.

$$AB + A\bar{B} = A(B + \bar{B}) \quad \text{Factoring}$$
$$= A \qquad\qquad B + \bar{B} = 1$$
$$\therefore AB + A\bar{B} = A$$

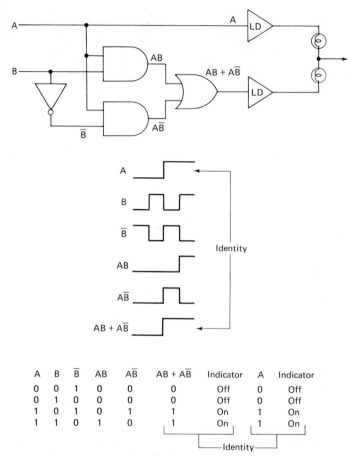

Figure 4-16(a) Truth table, logic diagram, and waveform proofs—Law of Expansion: (a) $AB + A\overline{B} = A$;

The logic diagram, waveforms, and accompanying truth table in Fig. 4-16a further demonstrate the Law of Expansion

Truth-table proof of the second identity of the Law of Expansion

$$(A + B)(A + \overline{B}) = A \qquad (4\text{-}13)$$

appears in Fig. 4-16b, along with the logic diagram, waveforms, and algebraic proofs.

Duality

Our investigation of the postulates, theorems, and laws of Boolean algebra points out an interesting and useful relationship. The Commutative Law, $AB = BA$ and $A + B = B + A$, may be used as an example.

Basic Laws of Boolean Algebra

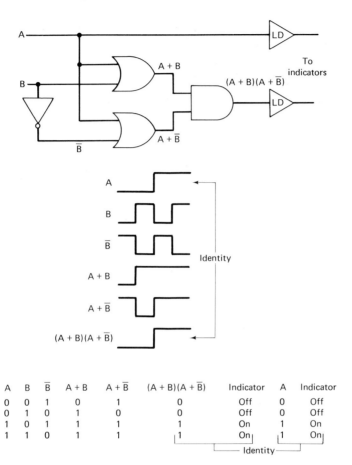

Figure 4-16(b) $(A + B)(A + \bar{B}) = A$

These identities as well as the other laws, theorems, and postulates, appear in *pairs*. One pair is usually in a *conjunctive form* (the ANDing of variables/constants), while the other is in a *disjunctive form* (the ORing of variables/constants). Both the conjunctive and the disjunctive versions of the Commutative Law have already been proven separately. In fact, perusal of the other laws, theorems, and postulates reveals that if the conjunctive version has been proven, the disjunctive version is also valid and vice versa. It can therefore be stated that *the conjunctive version of a postulate, theorem, or law is the dual of the disjunctive version* and that *the disjunctive version is the dual of the conjunctive version*.

The *symmetry* (duality) of Boolean algebra can be more formally stated.

Given any postulate, theorem, or law in Boolean algebra, if each 0 is replaced by 1, each 1 by 0, each + by ·, and each · by +, the result is the *dual* of the selected postulate, theorem, or law.

This effectively reduces by half the task of investigation of the theorems and laws of Boolean algebra. The mathematical derivation of the duality relationship is itself another subject and is not investigated in this text.

4-2 COMMON IDENTITIES OF BOOLEAN ALGEBRA

Considerable justification for the investigation of the identities of Boolean algebra exists from an economic viewpoint. In conventional mathematics, identities are used to facilitate the *mathematical simplification* of equations and expressions; in Boolean algebra such simplifications usually result not only in simplified forms of equations and expressions but also in *reduced requirements for hardware* to implement them. Only the most common identities encountered in everyday work are discussed here. Each of the identities is proven by at least two methods previously employed. The remaining proofs are left to the reader as exercises at the end of the chapter.

The Identity

$$A(\overline{A} + B) = AB \tag{4-14}$$

This identity may be analyzed by logical thinking. For $A(\overline{A} + B) = AB$ to be true (logical 1), both A and B must be 1. Confirmation is made in the following manner: If A is 0, it doesn't matter what value B possesses, since 0 ANDed with *anything* is 0. If A is 1 and B is 0, then the value of $A(\overline{A} + B)$ is 0, that is $1(0 + 0) = 0$. However, with $B = 1$ and $A = 1$, the value of $A(\overline{A} + B)$ becomes 1, since $1(0 + 1) = 1$. Note that the *logical thought* method of analysis closely resembles the truth-table method. If this "English-language" method is more easily understood by some readers, it should be applied wherever possible. Usually, however, the truth-table method provides the same information in a more organized manner. Figure 4-17 shows the proofs of this identity.

The Identity

$$A + \overline{A}B = A + B \tag{4-15}$$

Logically thinking, if $A = 1$, it doesn't matter whether $B = 0$ or $B = 1$, since 1 ORed with *anything* is 1. If $A = 0$, then $\overline{A} = 1$. B must equal 1 for $\overline{A}B = 1$. $\overline{A}B = 1$ so that $A + \overline{A}B = 1$, that is, $0 + (1 \cdot 1) = 1$. The logic diagram and algebraic proofs appear in Fig. 4-18.

Common Identities of Boolean Algebra

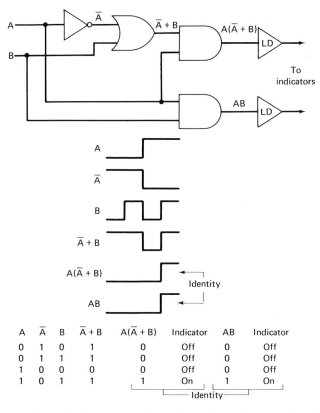

Figure 4-17 Logic diagram, truth table, and waveform proofs of the identity $A(\overline{A} + B) = AB$

Steps	Justification
$A + \overline{A}B = (A + AB) + \overline{A}B$	$A + AB = A$ (Absorption)
$= (AA + AB) + \overline{A}B$	$AA = A$
$= AA + AB + A\overline{A} + \overline{A}B$	$A\overline{A} = 0$
$= (A + \overline{A})(A + B)$	Factoring
$= 1 \cdot (A + B)$	$A + \overline{A} = 1$
$= A + B$	$1 \cdot A = A$

Figure 4-18 $A + \overline{A}B = A + B$: Logic diagram and algebraic proofs

Another Identity

$$AB(A + B) = AB \tag{4-16}$$

The identity in Eq. (4-16) is often encountered in digital logic circuits. Recognition of this identity allows the same operation to be performed with one AND gate instead of one OR gate *and* two AND gates. Proofs of the identity are shown in Fig. 4-19.

The EXCLUSIVE-OR Identity

$$\overline{AB}(A + B) = \overline{A}B + A\overline{B} \tag{4-17}$$

Investigation of the right-hand side of the equation shows that it can have a value of 1 only when A is 1 and B is 0 or when A is 0 and B is 1. This specific relationship is a special case of disjunction, called the EXCLUSIVE-OR *operation*.

Figure 4-19 AB(A+B) = AB

Common Identities of Boolean Algebra 111

It has a special symbol ⟫⟩— and a special sign ⊕.

This operation is commonly encountered in digital logic and can appear in many different forms. One of these forms is shown on the left-hand side of the equation.

As expressions and equations, such as the EXCLUSIVE-OR, become more complex, it becomes more difficult to use the logical-thought method of analysis. However increased experience with Boolean forms results in greater insight and capability. Although the approach may not appear immediately obvious, it will be shown here as an example of more advanced form of logical-thought.

For the value of the left-hand side of the equation to be 1, $\overline{AB}$ must be 1 and $(A + B)$ must be 1. For $\overline{AB}$ to be 1, AB must be 0. For AB to be 0, A must be 0, B must be 0, or both must be 0. For $(A + B)$ to be 1, A must be 1, B must be 1, or both must be 1. Note the discrepancy in the requirements for $\overline{AB}$ and $(A + B)$ to be 1.

If A and B are both 0, making $\overline{AB} = 1$, then $(A + B)$ is 0. $\overline{AB}(A + B)$ is then equal to 0, since $1 \cdot 0 = 0$. When A and B are both 1, $\overline{AB} = 0$, while $(A + B) = 1$. The combination is again equal to 0, since $0 \cdot 1 = 0$. The only condition under which the combination is equal to 1 is either when $A = 1$ and $B = 0$ or when $A = 0$ and $B = 1$. Once again, the use of *grouping marks* (vinculum and parentheses) aids in developing the logic diagram. The parentheses imply that $A + B$ is to be implemented by itself, and the vinculum implies that A is to be ANDed with B and negated as a single term. When these operations are completed, $\overline{AB}$ is to be ANDed with $(A + B)$. The truth table and logic diagram may be seen in Fig. 4-20.

The EXCLUSIVE-NOR Identity

$$A\overline{B} + \overline{A}B = A \cdot B + \overline{A} \cdot \overline{B} \tag{4-18}$$

This identity shows that the inverted form of the EXCLUSIVE-OR operation, at left side of the equal sign, is the same as the disjunction of the conjunction of the true form *and* the false form of the variables. In other words, by ORing A AND B with $\overline{A}$ AND $\overline{B}$, the inverted form of the EXCLUSIVE-OR operation is obtained. Although this identity may be examined using the logical-thought method, this approach becomes unwieldy when either many variables or numerous negations are encountered. It is recommended that truth-table or other forms of analysis be used in this and remaining identities. (See Fig. 4-21.)

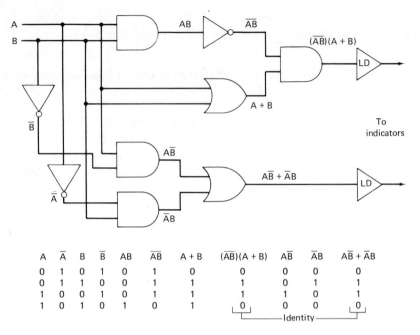

Figure 4-20 $(\overline{AB})(A + B) = A\overline{B} + \overline{A}B$

A Three-variable Identity

$$(A + B)(B + C)(A + C) = AB + BC + AC \qquad (4\text{-}19)$$

The commonly encountered identities are not limited to two variables. *Three variables,* arranged in the manner shown in Eq. (4-19), are often seen in digital logic circuits. Figure 4-22 shows the logic diagram and waveform proofs for this identity.

A Complex Three-variable Identity

$$(A + B)(\overline{A} + C) = AC + \overline{A}B \qquad (4\text{-}20)$$

For this identity to be true, at least one variable inside each set of parentheses on the left-hand side of the equal sign must be equivalent to 1. If $A = 1$, then $\overline{A} = 0$ and C must be equal to 1. The value of B is not important, since if $A = 1$, then $1 + B = 1$. Therefore AC is one of the combinations required. If $A = 0$, then $\overline{A} = 1$ and B must be equal to 1. The value of C is not important, since if $\overline{A} = 1$, then $1 + C = 1$. Therefore $\overline{A}B$ is the other combination required. The following is the algebraic proof.

A	$\bar{A}$	B	$\bar{B}$	$A\bar{B}$	$\bar{A}B$	$A\bar{B}+\bar{A}B$	$\overline{A\bar{B}+\bar{A}B}$	AB	$\bar{A}\bar{B}$	$AB+\bar{A}\bar{B}$
0	1	0	1	0	0	0	1	0	1	1
0	1	1	0	0	1	1	0	0	0	0
1	0	0	1	1	0	1	0	0	0	0
1	0	1	0	0	0	0	1	1	0	1

———— Identity ————

$$\overline{A\bar{B}+\bar{A}B} = \overline{(A\bar{B})}\,\overline{(\bar{A}B)} \qquad \text{DeMorgan's Law}$$
$$= (\bar{A}+B)(A+\bar{B}) \qquad \text{DeMorgan's Law}$$
$$= A\bar{A} + \bar{A}\bar{B} + AB + B\bar{B} \qquad \text{Multiplication}$$
$$= O + \bar{A}\bar{B} + AB + O \qquad A\bar{A} = O$$
$$= \bar{A}\bar{B} + AB \qquad A + O = A$$
$$= AB + \bar{A}\bar{B} \qquad \text{Commutation}$$

Figure 4-21 $\overline{A\bar{B}+\bar{A}B} = AB+\bar{A}\bar{B}$

$X = (A+B)(B+C)(A+C)$
$Y = AB+BC+AC$

Figure 4-22 $(A+B)(B+C)(A+C) = AB+BC+AC$

Common Identities of Boolean Algebra

$$(A + B)(\overline{A} + C) = A\overline{A} + AC + \overline{A}B + BC$$
$$= 0 + AC + \overline{A}B + BC$$
$$= AC + \overline{A}B + BC$$
$$= \overline{A}B + AC + (A + \overline{A})(BC)$$
$$= \overline{A}B + AC + ABC + \overline{A}BC$$
$$= \overline{A}B + \overline{A}BC + AC + ABC$$
$$= \overline{A}B(1 + C) + AC(1 + B)$$
$$= (\overline{A}B \cdot 1) + (AC \cdot 1)$$
$$= \overline{A}B + AC$$
$$= AC + \overline{A}B$$
$$\therefore (A + B)(\overline{A} + C) = AC + \overline{A}B$$

Justification for each step in this proof should be determined by the reader. The truth table for the identity $(A + B)(\overline{A} + C) = AC + \overline{A}B$ may be seen in Table 4-3.

Table 4-3 Truth Table $(A + B)(\overline{A} + C) = AC + A\overline{B}$

A	$\overline{A}$	B	$\overline{B}$	C	$\overline{C}$	$A + B$	$\overline{A} + C$	$(A + B)(\overline{A} + C)$	AC	$\overline{A}B$	$AC + \overline{A}B$
0	1	0	1	0	1	0	1	0	0	0	0
0	1	0	1	1	0	0	1	0	0	0	0
0	1	1	0	0	1	1	1	1	0	1	1
0	1	1	0	1	0	1	1	1	0	1	1
1	0	0	1	0	1	1	0	0	0	0	0
1	0	0	1	1	0	1	1	1	1	0	1
1	0	1	0	0	1	1	0	0	0	0	0
1	0	1	0	1	0	1	1	⌊1⌋	1	0	⌊1⌋

└─────── Identity ───────┘

The OR-AND Identity

$$AC + AB + B\overline{C} = AC + B\overline{C} \tag{4-21}$$

Logic diagrams and waveforms (Fig. 4-23) are used to prove this identity. Equation (4-21) may be considered to be an example of the Law of Absorption in more complex form. The term AB is effectively "absorbed" by the first term AC, which contains the variable A, and the third term $B\overline{C}$, which contains the variable B. Development of the algebraic proof (left to the reader as an exercise at the end of the chapter) makes this viewpoint more apparent.

The AND-OR Identity

$$(A + B)(B + C)(\overline{A} + C) = (A + B)(\overline{A} + C) \tag{4-22}$$

Truth-table proof of this identity is shown in Table 4-4, while the associated algebraic proof follows.

Figure 4-23 $AC+AB+B\overline{C} = AC+B\overline{C}$

Table 4–4 Truth table $(A+B)(B+C)(\bar{A}+C) = (A+B)(\bar{A}+C)$

A	$\bar{A}$	B	$\bar{B}$	C	$\bar{C}$	A+B	B+C	$\bar{A}$+C	(A+B)(B+C)($\bar{A}$+C)	A+B	$\bar{A}$+C	(A+B)($\bar{A}$+C)
0	1	0	1	0	1	0	0	1	0	0	1	0
0	1	0	1	1	0	0	1	1	0	0	1	0
0	1	1	0	0	1	1	1	1	1	1	1	1
0	1	1	0	1	0	1	1	1	1	1	1	1
1	0	0	1	0	1	1	0	0	0	1	0	0
1	0	0	1	1	0	1	1	1	1	1	1	1
1	0	1	0	0	1	1	1	0	0	1	0	0
1	0	1	0	1	0	1	1	1	1	1	1	1

Identity

$$(A + B)(B + C)(\overline{A} + C) = [(A + B)(B + C)][\overline{A} + C]$$
$$= [AB + AC + BB + BC][\overline{A} + C]$$
$$= \overline{A}AB + \overline{A}AC + \overline{A}BB + \overline{A}BC + ABC + ACC + BBC + BCC$$
$$= 0 + 0 + \overline{A}B + \overline{A}BC + ABC + AC + BC + BC$$
$$= \overline{A}B + \overline{A}BC + ABC + AC + BC$$
$$= \overline{A}B + BC(\overline{A} + A) + AC + BC$$
$$= \overline{A}B + BC + AC$$
$$\therefore (A + B)(B + C)(\overline{A} + C) = \overline{A}B + BC + AC = (A + B)(\overline{A} + C)$$

Once again the reader is encouraged to determine the justification for each step in this proof. The final step should be verified.

4-3 SUMMARY OF ALGEBRAIC TO LOGIC CONVERSION

Although they have not been specifically pointed out, the major techniques for converting algebraic expressions to equivalent logic diagrams have been discussed in this chapter. Collecting the various techniques and listing them in sequence of execution provides the reader with a valuable aid in the conversion of algebraic expressions to logic diagrams. A list of suggested procedures is provided in this section, along with explanatory notes.

1. *Identify the general type of circuit.* Except in very complex cases, a logic circuit may be categorized as either an AND-OR or an OR-AND type. The AND-OR type consists of two or more ANDed operations combined together in an OR operation. For example, $AB + CD$ is an AND-OR type of logic expression; it would be implemented with an AND gate to combine A and B, an AND gate to combine C and D, and an OR gate to combine the results of the two ANDed operations.

   The OR-AND logic circuit consists of two or more ORed operations combined together in an AND operation. For example $(A + B)(C + D)$ is a typical OR-AND logic expression; it would be implemented with an OR gate to combine A and B, an OR gate to combine C and D, and an AND gate to combine the results of the two ORing operations.

2. *Starting with the overall expression, draw the logic symbol that represents the last logic operation.* The final logic operation can be identified after deciding what the general type of circuit does. If it is an AND-OR circuit, the last operation is an OR operation; if it is an OR-AND circuit, the last operation is an AND operation. As logic

circuits become more complex, it may not be easy to determine the final operation. Experience with logic-circuit analysis provides the reader with the necessary skills to make this determination.

3. *Working toward the input, draw the logic symbol for each operation until all inputs are accounted for.* In the simple AND-OR example ($AB + CD$), it was shown that the last logic operation is an OR operation. Each of the inputs to the OR gate will have to come from AND gates, since the two terms being ORed are ANDed terms. Thus one of the inputs would come from an AND gate with A and B inputs; the other OR input would come from another AND gate with C and D inputs. The logic diagram is now complete, since all inputs have been accounted for.

Quite often, *apparent* simplifications to the logic diagram being constructed may be noted. Until added experience is gained, however, such random simplifications are discouraged.*

4-4 A COMBINATIONAL LOGIC CIRCUIT (BINARY HALF-ADDER)

Combinational logic problems appear in many different forms—most of which have been presented in this and preceding chapters. Now a partial combinational logic circuit, borrowed from a computer application, will be described using "English-language" statements. From this description, various other methods of presenting the same information will be shown.

Binary Addition

Digital computers use logic circuits called *adders* to obtain the *sum* of binary numbers. One of the added numbers is called the *augend*, which is the number to which the other number is to be added. The other number is called the *addend*. Conventionally, the augend is written first (on the top line), and the addend is written next (below the augend). The rules for arithmetic addition of positive binary integers follow.

0 *plus* 0 = 0,   0 *plus* 1 = 1,   1 *plus* 0 = 1,   1 *plus* 1 = 0 *carry* 1

A logic circuit that performs these operations is called a *half-adder*.

---

*Chapter 5 deals with algebraic methods of logic-circuit simplification where an organized method for reducing logic circuit requirements is discussed. Conversion techniques are also expanded in Ch. 5.

Although the derivation of the term "half-adder" is somewhat obscure, investigation shows that such a circuit must be capable of adding two single-digit binary numbers and providing the sum and carry outputs. An example of binary addition might prove helpful in clarifying the meaning of the term. When the binary addition to be performed concerns only two single-digit binary numbers, the half-adder quite adequately performs the function. Thus

	0	0	1	1
plus	0	1	0	1
sum	0	1	1	0
carry	0	0	0	1

It seems reasonable, therefore, that if the preceding addition is performed on the least significant digits of multidigit numbers, it is necessary to provide only one other half-adder for each order of the numbers. To make it simple: consider the addition of the 2-digit binary numbers 11 and 11. The carry generated in the least-significant order must be added to the next-most-significant order. But this results in a three input requirement, and the half-adder has only two inputs. Hence the name half-adder, since it can only do *half* the job. A *full-adder* is required in all but the least-significant positions to accommodate the carry inputs from previous orders. Full-adders are discussed in subsequent chapters.

Analysis of the Half-Adder

The operation of the half-adder logic circuit is described by the following definition.

> If the augend is zero and the addend is one or if the augend is one and the addend is zero, the sum is one. If the augend is one and the addend is one, the carry output is one.

This turns out to be a relatively complex statement, subject to misinterpretation and difficult to visualize. Perhaps substituting letter names for each of the variable names in the "English-language" statement would clarify the meaning. Thus the augend can be named A, the addend can be B, the sum can be S, and the carry can be C. Now the statement reads: "If A is zero and B is one or if A is one and B is zero, S is one. If A is one and B is one, C is one." By arbitrarily assigning the logical value 0 to the numerical value zero, and the logical value 1 to the numerical value one, another approach can be initiated. Consider logical 0 as a LOW and logical 1 as a HIGH. This statement is shown in truth-table form in Table 4-5a. Note that two variables are used in this

Common Identities of Boolean Algebra

Table 4–5 Half-adder truth tables

A	B	S	C
0	1	1	0
1	0	1	0
1	1	0	1

(a) Direct conversion of half-adder description to tabular form

A	B	S	C
0	0	0	0
0	1	1	0
1	0	1	0
1	1	0	1

(b) Conventional half-adder truth table

table. Only three of the four possible combinations of the two variables are shown. But we desire to expand the truth table to include all four combinations so that the table will more closely resemble standard truth tables (see Table 4-5b).

Remember that the designation of a variable as LOW (0) is usually considered to be the false (negated) form, and HIGH (1) is the true form. The two terms that result in a HIGH S output and the one term that results in a HIGH C output can be written in Boolean algebra equations. Thus

$$S = \overline{A}B + A\overline{B}$$
$$C = AB \tag{4-23}$$

Applying what was discussed earlier in this chapter allows this equation to be displayed in symbolic diagram form (Fig. 4-24a) and, finally, as a waveform (Fig. 4-24b).

Summary of Analysis Methods

The same combinational logic circuit has been described by five different methods. Descriptive methods to be used depend not only on the ultimate user of the material but also on the complexity of the description. The "English-language" description is commonly encountered in material written for nontechnical people, since the combinational logic circuit can be presented in a relatively nontechnical manner.

Algebraic and truth-table methods provide an excellent means of presenting circuit-function information to engineering groups, who may use these mathematical forms to develop actual logic circuits. These methods are also useful to technicians in analyzing and troubleshooting logic circuits. Symbolic diagrams are used by both engineers and technicians as a visual aid to understanding circuit operation. Waveform drawings and photographs assist troubleshooting during the detection

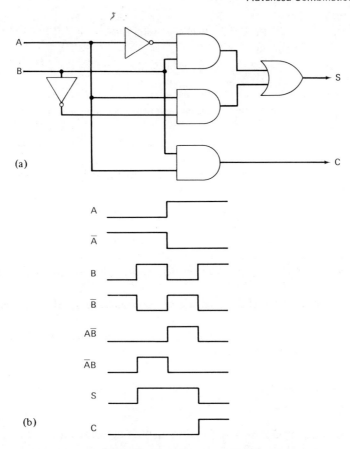

Figure 4-24 Half-adder logic diagram and waveforms

and repair of malfunctions in digital equipment. Of course, the use of any of these descriptive methods is not limited to a specific group. Comments made herein merely provide a general picture for those who want to find which method would be most useful.

PROBLEM APPLICATIONS

4-1. List four combinational logic circuit analysis methods.
4-2. Explain a common application of the Commutative Law in combinational logic.
4-3. Can the Associative Law be used to save hardware when implementing logic equations? Explain.
4-4. Why is the Distributive Law sometimes considered one of the most powerful of the laws of Boolean algebra?

A Combinational Logic Circuit (Binary Half-adder)

4-5. What is a NOR gate? Show how it can be developed from AND, OR, and NOT hardware elements.

4-6. What is a NAND gate? Show how it can be developed from NOR hardware elements.

4-7. Explain the second identity of the Law of Absorption, $A(A + B) = A$, using the "logical-thought" method.

4-8. What is duality? Why is it a valuable attribute of Boolean algebra?

4-9. Prove the identity $A(\bar{A} + B) = AB$ using the postulates, theorems, and laws of Boolean algebra.

4-10. Develop the truth table and waveform diagrams to prove the identity $A + \bar{A}B = A + B$.

4-11. Prove the identity $AB(A + B) = AB$ using the "logical-thought" method.

4-12. Using the postulates, theorems, and laws of Boolean algebra, prove the identity $\overline{AB}(A + B) = A\bar{B} + \bar{A}B$.

4-13. Show the waveform diagram that represents the identity in Problem 12.

4-14. Draw a logic diagram and show the waveforms that represent the identity $A\bar{B} + \bar{A}B = A \cdot B + \bar{A} \cdot \bar{B}$.

4-15. Develop the truth table and show the Boolean algebra proof for the identity $(A + B)(B + C)(A + C) = AB + BC + AC$.

4-16. Show the waveforms and logic diagram that represent the identity $(A + B)(\bar{A} + C) = AC + \bar{A}B$.

4-17. Develop the Boolean algebra proof for the identity $AC + AB + B\bar{C} = AC + B\bar{C}$.

4-18. Draw the waveforms that represent the identity $(A + B)(B + C)(\bar{A} + C) = (A + B)(\bar{A} + C)$.

Given the following logic circuit description, develop the

4-19. Boolean Algebra equation,

4-20. truth table,

4-21. logic diagram, and

4-22. waveform diagram.

> The light will illuminate if the switch at the top of the stairs is on and the switch at the bottom of the stairs is off, or if the switch at the top of the stairs is off and the switch at the bottom of the stairs is on.

Given the truth table on the right, develop the

4-23. Boolean algebra equation, $A\bar{B} + \bar{A}B$
$= \bar{A}\bar{B}(A+B)$
exclusive OR

4-24. logic diagram, and

4-25. waveform diagram.

A	B	C	X
0	0	0	0
0	0	1	0
0	1	0	1
0	1	1	1
1	0	0	1
1	0	1	1
1	1	0	0
1	1	1	0

5

ALGEBRAIC SIMPLIFICATION METHODS

5-1 WHAT IS LOGIC SIMPLIFICATION?

Logic simplification definitions are multitudinous—there are almost as many definitions as there are people to make them! In general, however, the word *simplification* implies *a reduction in complexity*. Since there are so many different ways to describe a logic circuit in terms of ANDs, ORs, NOTs, NANDs, and NORs, some arbitrary definition of which circuit is the simplest must be determined.

In some cases (e.g., many engineering situations), the basis for choosing a particular circuit is cost. But with minimum cost as a criterion, other parameters come into play. Can a relationship between the cost of simple diode gates and the number of inputs per gate be discovered? How much do NAND or NOR gates* cost in quantity compared to AND-OR-NOT circuits? Is it advantageous to purchase a "prepackaged" logic function in integrated circuit (IC) form? These are but a few of the many questions that must be considered when logic simplification is to result in minimum cost requirements. The increasing use of digital computers has allowed designers to consider these variables in the design phase of logic systems; hence cost-effective designs are resulting in a rapid expansion of digital applications to society's everyday needs.

---

*Economic reasons frequently limit the type(s) of gates available to implement the logic. Use of NAND gates or NOR gates may simplify the manufacturing process due to fewer *types* of modules in the system.

But cost-effective designs* are basically a problem for the design engineer. Since this text is using logic simplification merely as a way to develop skill in logic analysis and troubleshooting, a more appropriate definition must be developed. Sometimes logic expressions are simplified so that a *minimum number of logic elements* (gates, inverters, etc.) are required for implementation. Often this is equivalent to finding the "simplest" form for the logic function. Other simplification techniques result in Boolean expressions containing a minimum number of terms with a minimum number of variables in each term. As we mentioned earlier, it all depends on the definition of the "simplest form."

Two such definitions are used in this text. First, after applying the procedures for simplification that are detailed in this chapter, *the algebraic form that results in the smallest number of logic elements is considered the simplest form for the Boolean expression.* It may be necessary to examine several different alternatives to determine the final form.

Second, the Boolean expression should also be stated in the AND-OR form in such a way that *no variables are enclosed in parentheses and no vincula extend over more than one variable.* This second form may not necessarily be the form with the smallest number of logic elements, but we should be familiar with it in preparation for the chapter on graphic simplification.

The Study of Logic Simplification

Although some people find the study of a subject fascinating just for the "sheer enjoyment of learning something new," this may not be enough motivation for everyone. *From a strictly economic viewpoint, logic simplification is a useful undertaking.* We have already mentioned that considerable savings are realized if the capability to simplify logic is achieved. If the task of the reader is to design logic circuits, then there is no doubt that such a capability is valuable.

However many of our readers are involved in analysis and troubleshooting, both of which require *the recognition of many different logic operations and functions.* Besides performing the analysis and troubleshooting on actual equipment, perhaps one of the most satisfactory ways to recognize these logic circuits is to work with them on paper. And perhaps the best way to work with them on paper is to study the various methods of logic simplification.

As numerous logic circuits performing the same function are observed, *the many different ways of performing the same operations*

---

*Of course, cost cannot be disregarded completely. Implementing any logic function without regard to cost would be financial suicide.

become clearer. As a desirable by-product, the reader experiences actual logic simplification work and should therefore develop *a certain amount of capability in the area of simplification techniques.* Simplification is one of the steps toward a *design capability.* Such a capability, with the actual hardware experience that many technical personnel already possess, is and always will be in demand in the labor force.

Methods of Logic Simplification

Two general methods of logic simplification are in common use today. One method (the subject of this chapter) uses the postulates, theorems, laws, and identities of Boolean algebra. The other method uses graphic techniques and is detailed in Ch. 6. Each has its advantages and disadvantages—its easy roads and pitfalls. Certain similarities become obvious as the simplification methods are developed, and mastery of both methods is a big step toward a complete understanding of *combinational logic circuits.*

5-2 ALGEBRAIC SIMPLIFICATION PROCEDURES

A General Procedure

Every combinational logic circuit has its own peculiarities and requires its own analysis procedure. However enough similarities exist to develop a general procedure for all combinational logic circuits and then modify the procedure slightly for each new circuit encountered. Such a procedure is listed here.

1. Obtain the Boolean expression for the original circuit:
   (a) from the English language description (Sec. 3-1);
   (b) from a truth table (Sec. 4-1);
   (c) from a logic diagram (Sec. 4-3);
   (d) from the algebraic description (already in proper form).
2. Simplify algebraically:
   (a) apply postulates, theorems, laws, and identities as required (Ch. 3);
   (b) factor, substitute, expand, modify, and convert as required (Ch. 5);
   (c) try all over again to make sure the simplest form has been obtained.
3. Compare the new expression with the original expression by using truth tables (Sec. 4-1);
4. Draw the simplified logic diagram (Sec. 4-3);

Algebraic Simplification Procedures

As experience is gained, the number of steps required to reach a simplified version of an original expression diminishes. Once the factoring is accomplished (as in Ex. 5-1), the experienced analyst probably recognizes that both $(1 + B)$ and $(A + \bar{A})$ are equivalent to 1 and that any variable ANDed with 1 is equal to that variable. In other words, the step following the factoring step *could* have been $A + BC$ directly. But don't jump steps until it is obvious that such jumps can be taken. Many traps exist between steps, and a jump in the wrong direction may end up in a "deep hole!"

Example 5–1. Algebraically simplify $A + ABC + AB + \bar{A}BC$.
Solution.

$$
\begin{aligned}
A + ABC + AB + \bar{A}BC &= A + AB + ABC + \bar{A}BC & &\text{Commutative Law} \\
&= (A + AB) + (\bar{A}BC + ABC) & &\text{Associative Law} \\
&= (A \cdot 1 + AB) + (ABC + \bar{A}BC) & &\text{Theorem 3} \\
&= A(1 + B) + BC(A + \bar{A}) & &\text{Factoring} \\
&= A \cdot 1 + BC(A + \bar{A}) & &\text{Theorem 9} \\
&= A + BC(A + \bar{A}) & &\text{Theorem 3} \\
&= A + BC \cdot 1 & &\text{Theorem 12} \\
\therefore\ A + ABC + AB + \bar{A}BC &= A + BC & &\text{Theorem 3}
\end{aligned}
$$

Some New Terminology

Another word of caution is justified at this time. The use of the multiplication dot ($\cdot$) and the plus sign (+) in Boolean algebra is a carry-over from conventional arithmetic and algebra. Since the plus sign is used to designate disjunction in Boolean algebra, it is only reasonable that disjunction be called *logical addition.* Similarly, the dot's designation as a conjunctive operator results in conjunction being called *logical multiplication.* Thus sums and products exist in Boolean algebra. Of course, these are "logical sums" and "logical products," but the name "logical" is dropped by many users. It is not at all uncommon to see the result of ANDing called a "product" and the result of ORing called a "sum." But please remember that these terms are *logical operations— not* conventional arithmetic operations. When such terms are used in this text, they will be specifically expressed as *logical sum* and *logical product.*

In Boolean algebra, it is generally recognized that *any logic expression* can be stated as either *a conjunction of disjunctive terms* or *a disjunction of conjunctive terms.* In other words, any expression can be stated either as OR operations connected by AND operations or AND operations connected by OR operations. The expression $(A + B)(C + D)$ is an example of the first case and is said to be in the OR-AND form or the "logical-product-of-logical-sums" form. This same expression may be converted to the AND-OR form by "multiplying" the terms out to form $AC + AD + BC + BD$. Now the expression is said to be in the "logical-sum-of-logical-products" form. Be careful, though, because it is not uncommon to see these expressions called product of sums and sum of products. If logical operations are being performed at the time, be sure to recognize that these are logical products of logical sums and logical sums of logical products.

Procedural Summary

Once a simplified form is reached, there is no guarantee that it is the simplest form available. This is one of the main disadvantages of the algebraic-simplification method. Until considerable experience has been gained, much time may be wasted in trying different forms to determine which is the simplest. And there is still no guarantee that all of the forms have been tried!

Upon completion of the simplification steps, the new expression should be tested by using a truth table. Comparison of this new truth table with the truth table for the original expression assures us that both expressions are equivalent. It is not impossible to make a mistake while taking a short cut or to misinterpret a portion of the expression and end up with nonequivalence. Once equivalence is determined, the final step is to draw the simplified logic diagram and bask in the glory of a successful reduction in hardware requirements!

5-3 1-LEVEL LOGIC SIMPLIFICATION

In a 1-*level logic circuit*, a given logic variable goes through only *one logic element* between source and output. Such circuits (discussed in earlier chapters) have only one logic operation, either AND or OR. However, so that a starting point can be established in this chapter, some of the basic ideas of logic symbology and Boolean algebra notation are reviewed.

The common AND gate is an example of a 1-level logic circuit. Consider such a simple circuit with four inputs $A, B, C,$ and D and one output X. The definition of the AND operation states that if and only if inputs $A, B, C,$ and D are all HIGH *simultaneously*, the output X will be HIGH. This statement is written in Boolean notation as $A \cdot B \cdot C \cdot D = X$ or, more simply, as $ABCD = X$. The AND operator ($\cdot$) is conventionally omitted, unless it is needed to clarify the expression.

Now consider a logic circuit that is defined by the equation $ABAB = X$. The logic diagram is drawn in Fig. 5-1 along with the

1-Level Logic Simplification

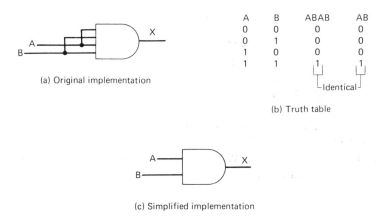

(a) Original implementation

(b) Truth table

(c) Simplified implementation

Figure 5-1 Logic diagram and truth table, ABAB = AB

equivalent truth table. Inspection of the truth table (Fig. 5-1b) should be sufficient to determine that the 4-input AND gate (Fig. 5-1a) required by the equation can be replaced by a 2-input AND gate with A and B each appearing once as inputs. However the steps in the algebraic simplification are shown to start developing the techniques to be used in more complex simplifications.

$$
\begin{aligned}
ABAB &= AABB & &\text{Commutative Law} \\
&= (AA)(BB) & &\text{Associative Law} \\
&= A(BB) & &\text{Idempotent Law} \\
&= AB & &\text{Idempotent Law}
\end{aligned}
$$
$$\therefore ABAB = AB$$

The truth table and simplified logic diagrams are drawn in Figs. 5-1b and c. Truth table comparison will reveal that $ABAB = AB$ is indeed valid and that the 4-input AND gate is replaceable by a 2-input AND gate. If the number of inputs to an AND gate is a direct cost factor, then the circuit has been simplified, and the cost cut in half.

Another example appears in Fig. 5-2. The logic diagram (Fig. 5-2a) shows that the variables A, B, $\overline{A}$, and $\overline{B}$ are being ORed together. This may be expressed as $A + B + \overline{A} + \overline{B} = X$. By developing the truth table (Fig. 5-2b), it is apparent that $X = 1$—no matter what value any of the variables take on. Therefore the 4-input OR gate can be replaced by a direct connection to a point that is always HIGH (Fig. 5-2c), eliminating a logic element—that is a *real* simplification! The algebraic simplification steps to justify this replacement are shown here.

$$\begin{aligned}
A + B + \overline{A} + \overline{B} &= A + \overline{A} + B + \overline{B} &&\text{Commutative Law}\\
&= (A + \overline{A}) + (\overline{B} + B) &&\text{Associative Law}\\
&= 1 + (\overline{B} + B) &&\text{Theorem 12}\\
&= 1 + 1 &&\text{Theorem 12}\\
&= 1 &&\text{Postulate 8}
\end{aligned}$$

$$\therefore\ A + B + \overline{A} + \overline{B} = 1$$

So although it might appear that very little simplification could take place in a 1-level logic circuit, we have shown that the recognition of certain relationships of variables has considerably reduced the requirements for hardware.

5-4 2-LEVEL LOGIC SIMPLIFICATION

In *2-level logic circuits,* a variable passes through *two cascaded logic elements* between input and output. It is assumed that both the true and the complemented values of variables are available as inputs to the circuits being simplified. However, when inverters are required to negate variables within the logic circuit, these inverters contribute to the overall level of the logic circuit. The most common configuration of 2-level logic circuits is AND circuits feeding OR circuits or vice versa.

A Simplification Starting with an English-Language Description

Our first example of a 2-level logic circuit is presented in terms of an English-language statement, and the procedural steps necessary to prepare the problem for simplification are demonstrated. This example is typical of industrial-control applications of logic. It is only a partial statement of an overall requirement, but it is sufficiently complete in itself to demonstrate how simplification is performed.

The operation under consideration deals with a conveyor belt in an industrial plant. The belt carries empty boxes that are to be filled with breakfast cereal. A portion of the operational requirement is to have an alarm sound if a dangerous situation or a situation requiring immediate attention should occur. The systems designer has provided the following information: An alarm should sound if a box is available and the cereal hopper is empty, or if no box is available and electric power is lost, or if the cereal hopper is empty and the electric power is lost. This description of circuit operation should be investigated to determine if any simplification can be made prior to system construction. Reference to the general procedure for logic simplification shows that the first thing to do is to develop the Boolean expression. Thus we assign variable names to each of the conditions mentioned in the English-language description so that a truth table, a logic diagram, and

2-Level Logic Simplification

(a) Original implementation

(b) Truth table

HIGH ─────────

(c) Simplified implementation
(a direct connection to a HIGH)

Figure 5-2 Logic diagram and truth table, $A + B + \bar{A} + \bar{B} = 1$

the Boolean expression can be developed. The following assignments are made.

- B = Box available
- $\bar{B}$ = No box available
- C = Cereal hopper empty
- $\bar{C}$ = Cereal hopper not empty
- P = Electric power available
- $\bar{P}$ = Electric power lost
- X = Alarm sounds

Substituting, the Boolean expression becomes

$$X = B \cdot C + \bar{B} \cdot \bar{P} + C \cdot \bar{P}$$

Note that the optional method of showing conjunctive terms is used in this equation to reduce the possibility of misinterpreting the term $\bar{B} \cdot \bar{P}$.

The truth table for $X = B \cdot C + \bar{B} \cdot \bar{P} + C \cdot \bar{P}$ appears in Fig. 5-3a and the equivalent logic diagram in Fig. 5-3b. Algebraic proof is shown in Example 5-2.

Example 5–2. Simplify the expression $B \cdot C + \bar{B} \cdot \bar{P} + C \cdot \bar{P}$.
Solution.

$$
\begin{aligned}
B \cdot C + \bar{B} \cdot \bar{P} + C \cdot \bar{P} &= B \cdot C + \bar{B} \cdot \bar{P} + C \cdot \bar{P}(B + \bar{B}) && \text{Theorem 3} \\
&= B \cdot C + \bar{B} \cdot \bar{P} + B \cdot C \cdot \bar{P} + \bar{B} \cdot C \cdot \bar{P} && \text{Distributive Law} \\
&= B \cdot C + B \cdot C \cdot \bar{P} + \bar{B} \cdot \bar{P} + \bar{B} \cdot C \cdot \bar{P} && \text{Commutative Law} \\
&= (B \cdot C + B \cdot C \cdot \bar{P}) + \bar{B} \cdot \bar{P} + \bar{B} \cdot C \cdot \bar{P} && \text{Associative Law} \\
&= BC(1 + \bar{P}) + \bar{B} \cdot \bar{P} + \bar{B} \cdot C \cdot \bar{P} && \text{Factoring} \\
&= BC + \bar{B} \cdot \bar{P} + \bar{B} \cdot C \cdot \bar{P} && \text{Theorems 10 \& 3} \\
&= BC + (\bar{B} \cdot \bar{P} + \bar{B} \cdot C \cdot \bar{P}) && \text{Associative Law} \\
&= BC + \bar{B} \cdot \bar{P}(1 + C) && \text{Factoring} \\
\therefore B \cdot C + \bar{B} \cdot \bar{P} + C \cdot \bar{P} &= BC + \bar{B} \cdot \bar{P} && \text{Theorems 10 \& 3}
\end{aligned}
$$

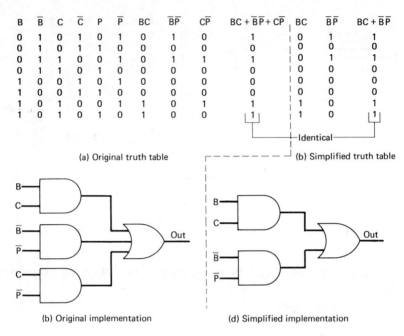

Figure 5-3 Logic diagram and truth tables $BC + \overline{B}\overline{P} + C\overline{P} = BC + \overline{B}\overline{P}$

Note that the third term of the original expression has been eliminated. Apparently it was not necessary to completely specify the expression. This may be verified by constructing the truth table for the expression (Fig. 5-3c) and comparing it with the truth table for the original expression (Fig. 5-3a). The truth tables are identical, thus verifying the equality of the original and the new expression. The logic diagram for the new expression (Fig. 5-3d) uses one less AND gate than the original logic diagram (Fig. 5-3b), so a reduction in number of gates has been accomplished. Additional attempts at simplification may be tried, but the form $B \cdot C + \overline{B} \cdot \overline{P}$ is probably the simplest. It meets the definition of "simplest" provided earlier: the smallest number of logic elements to perform the task apparently has resulted, and the expression is in the AND-OR form with no variables enclosed in parentheses and no vincula extending over more than one variable.

In the process of performing this simplification, English-language statements have been equated with variable names; truth tables and logic diagrams have been constructed; and theorems and laws of Boolean algebra have been applied. Although all simplification attempts may not require such a complete analysis, the steps performed in this example are typical, and the analyst should be prepared to apply any or all of these techniques.

2-Level Logic Simplification 133

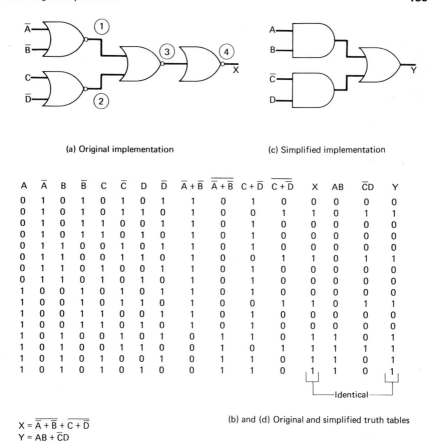

$X = \overline{\overline{A} + \overline{B}} + \overline{C + \overline{D}}$
$Y = AB + \overline{C}D$

Figure 5-4 Logic diagrams and truth tables $\overline{\overline{A} + \overline{B}} + \overline{C + \overline{D}} = AB + \overline{C}D$

A Simplification Starting with a Logic Diagram

Often it is necessary to convert a logic expression to a form that is readily implemented by the type of logic elements available. For example, a logic circuit may have been originally implemented strictly with NOR elements, due to a particular designer's choice or for economic reasons. As we mentioned before, when a logic system is designed exclusively with one type of logic element, cost reductions are realized in the design and fabrication processes. For the sake of discussion, however, assume that the logic diagram given in Fig. 5-4a must be implemented with AND-OR elements—not NOR elements as shown.

The first step is to determine the logic expression that this diagram represents. Derivation of the logic expression starts with inputs and moves toward the output—one gate or logic element at a time. The output at ① is $\bar{A} + \bar{B}$. At ② the output is $C + \bar{D}$. Combining these two expressions in the NOR gate results in the expression $\overline{\bar{A} + \bar{B} + C + \bar{D}}$ at ③. Inversion of the expression at ③ by the single input NOR gate provides the final output expression

$$\overline{\overline{\bar{A} + \bar{B} + C + \bar{D}}}$$

at ④. The double negation may be removed by applying Theorem 13, and the final expression is usually written as $\bar{A} + \bar{B} + C + \bar{D}$. Truth-table representation of this expression is provided in Fig. 5-4b and algebraic proof in Example 5-3.

Example 5–3. Convert the NOR form expression $\bar{A} + \bar{B} + C + \bar{D}$ to the AND–OR form.

Solution.
$\bar{A} + \bar{B} + \overline{C + \bar{D}} = \overline{\bar{A} \cdot \bar{B}} + \overline{C + \bar{D}}$ DeMorgan's Law
$\phantom{\bar{A} + \bar{B} + \overline{C + \bar{D}}} = AB + \overline{C + \bar{D}}$ Theorem 13
$\phantom{\bar{A} + \bar{B} + \overline{C + \bar{D}}} = AB + \bar{C} \cdot \bar{\bar{D}}$ DeMorgan's Law
$\phantom{\bar{A} + \bar{B} + \overline{C + \bar{D}}} = AB + \bar{C}D$ Theorem 13

$\therefore \bar{A} + \bar{B} + \overline{C + \bar{D}} = AB + \bar{C}D$

Comparison of the truth table for the simplified expression (Fig. 5-4d) demonstrates the equivalence of the NOR form and the AND-OR form in this example. Conversion from NOR form to AND-OR form is a commonly encountered situation in digital logic. The logic diagram of the AND-OR form in Fig. 5-4c shows a reduction in gate count from four NOR gates to two AND and one OR gate. More importantly, derivation of logic expressions from logic diagrams and application of DeMorgan's Law have been demonstrated.

A Simplification Starting with a Boolean Expression

Example 5–4. Simplify $(A + \bar{C})(A + D)(B + \bar{C})(B + D)$.
Solution.

$(A + \bar{C})(A + D)(B + \bar{C})(B + D) = (\bar{C} + A)(D + A)(\bar{C} + B)(D + B)$
 Commutative Law
$ = [(\bar{C} + A)(\bar{C} + B)] [(D + A)(D + B)]$
 Associative Law
$ = (\bar{C} \cdot \bar{C} + \bar{C}B + A\bar{C} + AB)(DD + DB + AD + AB)$
 Distributive Law
$ = [\bar{C}(1 + B + A) + AB][D(1 + B + A) + AB]$
 Factoring
$ = (\bar{C}1 + AB)(D1 + AB)$ Theorem 9
$ = (\bar{C} + AB)(D + AB)$ Theorem 3

2-Level Logic Simplification

$$= \overline{C}D + \overline{C}AB + ABD + ABAB$$
Distributive Law
$$= \overline{C}D + AB(\overline{C} + D + 1) \quad \text{Factoring}$$
$$= \overline{C}D + AB(1) \quad \text{Theorem 9}$$
$$= \overline{C}D + AB \quad \text{Theorem 3}$$
$$= AB + \overline{C}D \quad \text{Commutative Law}$$
$$\therefore (A + \overline{C})(A + D)(B + \overline{C})(B + D) = AB + \overline{C}D$$

Once again, no major steps are omitted. By now, it should be apparent that some steps may be combined, and as experience is gained, the reader may find it advantageous to do so. Comparative logic diagrams are shown in Figs. 5-5a and b, and a truth table showing both expressions is provided in Fig. 5-5c. This example demonstrates conversion from OR-AND to AND-OR form.

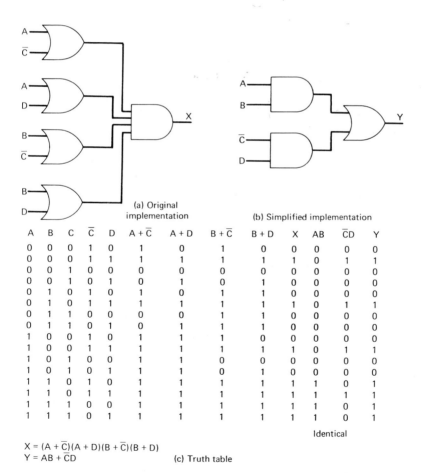

A	B	C	$\overline{C}$	D	$A+\overline{C}$	A+D	$B+\overline{C}$	B+D	X	AB	$\overline{C}D$	Y
0	0	0	1	0	1	0	1	0	0	0	0	0
0	0	0	1	1	1	1	1	1	1	0	1	1
0	0	1	0	0	0	0	0	0	0	0	0	0
0	0	1	0	1	0	1	0	1	0	0	0	0
0	1	0	1	0	1	0	1	1	0	0	0	0
0	1	0	1	1	1	1	1	1	1	0	1	1
0	1	1	0	0	0	0	1	1	0	0	0	0
0	1	1	0	1	0	1	1	1	0	0	0	0
1	0	0	1	0	1	1	1	0	0	0	0	0
1	0	0	1	1	1	1	1	1	1	0	1	1
1	0	1	0	0	1	1	0	0	0	0	0	0
1	0	1	0	1	1	1	0	1	0	0	0	0
1	1	0	1	0	1	1	1	1	1	1	0	1
1	1	0	1	1	1	1	1	1	1	1	1	1
1	1	1	0	0	1	1	1	1	1	1	0	1
1	1	1	0	1	1	1	1	1	1	1	0	1

Identical

$X = (A + \overline{C})(A + D)(B + \overline{C})(B + D)$
$Y = AB + \overline{C}D$

(c) Truth table

Figure 5-5 $(A + \overline{C})(A + D)(B + \overline{C})(B + D) = AB + \overline{C}D$

5-5 3-LEVEL LOGIC SIMPLIFICATION

A Simplification Starting with a Boolean Expression

When *any* variable in a Boolean expression must pass through *three cascaded logic elements* between input and output, the circuit is considered to be a *3-level logic circuit*. The Boolean expression $A(A + B) + B(\bar{A} + B)$ is a relatively simple example in which at least one of the variables in the expression passes through three logic elements.

Example 5–5. Simplify $A(A + B) + B(\bar{A} + B)$.
Solution.

$A(A + B) + B(\bar{A} + B) = AA + AB + \bar{A}B + BB$	Distributive Law
$= A + AB + \bar{A}B + BB$	Theorem 5
$= A + AB + \bar{A}B + B$	Theorem 5
$= A(1 + B) + \bar{A}B + B$	Factoring
$= A(1) + \bar{A}B + B$	Theorem 10
$= A + \bar{A}B + B$	Theorem 3
$= A + B(1 + \bar{A})$	Factor
$= A + B(1)$	Theorem 10
$= A + B$	Theorem 3

$\therefore A(A + B) + B(\bar{A} + B) = A + B$

The logic diagrams for both the original expression and the simplified expression appear in Fig. 5-6a, while the truth tables that are used to demonstrate equivalence are shown in Fig. 5-6b.

(a) Logic diagrams

A	$\bar{A}$	B	A + B	A(A + B)	$\bar{A}$ + B	B($\bar{A}$ + B)	A(A + B) + B($\bar{A}$ + B)
0	1	0	0	0	1	0	0
0	1	1	1	0	1	1	1
1	0	0	1	1	0	0	1
1	0	1	1	1	1	1	1

——————Identical——————

(b) Truth tables

Figure 5-6 Logic diagrams and truth tables $A(A + B) + B(\bar{A} + B) = A + B$

A Simplification Starting with a Logic Diagram

Another 3-level logic circuit is shown in Fig. 5-7a. The straightforward derivation of the Boolean expression is shown in the logic diagram. In the algebraic simplification that follows, note that as the *number of variables* in an expression increases, more and more of the theorems, laws, and identities come into play. Numerous exercises are provided at the end of this chapter to supply practice in the use of logic simplification techniques.

Example 5-6. Simplify $(A + \bar{B})C + (A\bar{B} + C)$.
Solution.

$(A + \bar{B})C + (A\bar{B} + C) = (AC + \bar{B}C) + (A\bar{B} + C)$	Distributive Law
$= AC + \bar{B}C + A\bar{B} + C$	Associative Law
$= AC + C + \bar{B}C + A\bar{B}$	Commutative Law
$= C(A + 1) + \bar{B}C + A\bar{B}$	Factoring
$= C(1) + \bar{B}C + A\bar{B}$	Theorem 9
$= C + \bar{B}C + A\bar{B}$	Theorem 3
$= C(1 + \bar{B}) + A\bar{B}$	Factoring
$= C(1) + A\bar{B}$	Theorem 10
$= C + A\bar{B}$	Theorem 3
$= A\bar{B} + C$	Commutative Law

$\therefore (A + \bar{B})C + (A\bar{B} + C) = A\bar{B} + C$

(a) and (b) Logic diagrams

A	B	$\bar{B}$	C	$A + \bar{B}$	$(A + \bar{B})C$	$A\bar{B}$	$(A\bar{B} + C)$	$(A + \bar{B})C + (A\bar{B} + C)$	$A\bar{B} + C$	
0	0	1	0	1	0	0	0	0	0	
0	0	1	1	1	1	0	1	1	1	1
0	1	0	0	0	0	0	0	0	0	
0	1	0	1	0	0	0	1	1	1	
1	0	1	0	1	0	1	1	1	1	
1	0	1	1	1	1	1	1	1	1	
1	1	0	0	1	0	0	0	0	0	
1	1	0	1	1	1	0	1	1	1	

(c) Truth tables — Identical

Figure 5-7 Logic diagrams and truth tables, $(A + \bar{B})C + (A\bar{B} + C) = A\bar{B} + C$

The simplified logic diagram is in Fig. 5-7b. Comparison of the original expression and the simplified expression is shown in the truth tables of Fig. 5-7c.

5-6 4-LEVEL LOGIC SIMPLIFICATION

Simplification Starting with a Logic Diagram

A path through *four cascaded logic elements* for any variable in a Boolean expression is the criterion for classifying a logic circuit as "4-level." Figure 5-8 is a logic diagram of a 4-level logic circuit. It represents a logic function that has been implemented strictly with NAND gates. But for the purpose of discussion, assume that it is necessary to implement this same function with AND-OR-INVERTER type logic elements. The first step is to determine the Boolean expression at each numbered output. Since two input variables A and B are present at the input to the first NAND gate, its output ① is $\overline{AB}$. The single-input NAND gate serves as an inverter, resulting in $\overline{\overline{AB}}$ at ② . $\overline{\overline{AB}}$ is combined with $\overline{C}$ in the third NAND gate, giving $\overline{(\overline{\overline{AB}})\overline{C}}$ at ③ . Finally, the last single input NAND gate, acting as an inverter, provides $\overline{\overline{(\overline{\overline{AB}})\overline{C}}}$ at ④ , the circuit output. Simplification of the output expression of this circuit is fairly easy.

Example 5-7. Simplify $\overline{\overline{(\overline{\overline{AB}})\overline{C}}}$.

Solution. $\overline{\overline{(\overline{\overline{AB}})\overline{C}}} = (\overline{\overline{AB}})\overline{C}$ Theorem 13
$\phantom{\overline{\overline{(\overline{\overline{AB}})\overline{C}}}} = (AB)\overline{C}$ Theorem 13
$\phantom{\overline{\overline{(\overline{\overline{AB}})\overline{C}}}} = AB\overline{C}$

Figure 5-8 Logic diagrams (AB)C = ABC

4-Level Logic Simplification

One simple 3-input AND gate performs the logic function previously requiring four NAND gates. From the viewpoint of the simplification criteria used in this text, a saving of three gates has been realized. Again, however, in extremely large and complex logic systems, a cost savings might still have been realized by implementing this logic function with NAND gates. It is not uncommon to purchase logic elements in quantities of 1000 or more units at half the single-unit price. This savings, coupled with simplifications in manufacturing processes when only one type of logic element is used, can quite often "swing the balance" toward single-element implementation of a logic system.

The reader should be prepared to encounter many NAND or NOR implemented systems. Recognition of the actual logic functions being performed may not be as easy as with AND-OR-INVERTER implemented systems. This situation calls for application of the techniques discussed in this chapter. Conversion of the logic expressions to straightforward AND-OR-INVERTER form can quite often make an apparently senseless logic circuit easily understandable.

Simplification Starting with a Boolean Expression

A Boolean expression that can be simplified by using one of the common identities of Boolean algebra is $[A\bar{B} + (\bar{A} + B)\bar{C}] + C$. Many of the theorems and laws of Boolean algebra are applied.

Example 5-8. Simplify $[A\bar{B} + (\bar{A} + B)\bar{C}] + C$.
Solution.

$[A\bar{B} + (\bar{A} + B)\bar{C}] + C$	Original problem
$[A\bar{B} + \bar{A} \cdot \bar{C} + B\bar{C}] + C$	Distributive Law
$A\bar{B} + (\bar{A} \cdot \bar{C} + B\bar{C}) + C$	Associative Law
$A\bar{B} + \bar{C}(\bar{A} + B) + C$	Factoring
$A\bar{B} + [C + \bar{C}(\bar{A} + B)]$	Commutative/Associative Laws
$A\bar{B} + [C + (\bar{A} + B)]$	Identity – Eq. (4–15)
$\bar{A} + A\bar{B} + B + C$	Associative/Commutative Laws
$\bar{A} + \bar{B} + B + C$	Identity – Eq. (4–15)
$\bar{A} + 1 + C$	Associative Law, Theorem 12
$1 + C$	Theorem 9
1	Theorem 10

Thus two AND and three OR gates are replaced by a *direct connection* to a HIGH. There should be no doubt that the simplified expression meets all criteria for the "least complex" implementation of the original expression: less logic elements are used; the cost of a direct connection is surely less than five logic gates; and no further algebraic simplification is possible. Comparative logic diagrams and truth tables appear in Fig. 5-9.

Recognition of Alternate Simplification Methods

The final example of 4-level logic circuit simplification demonstrates the use and application of many of the theorems and laws of Boolean algebra. In addition, the existence of alternate solution methods is mentioned, and the reader is provided with necessary clues to determine the best simplification approach.

Example 5-9.   Simplify $[AB(C + D + E) + DBA][\overline{A} + \overline{B} + \overline{E}]$.
Solution.

$[AB(C + D + E) + DBA][\overline{A} + \overline{B} + \overline{E}]$	Original problem
$[(ABC + ABD + ABE) + DBA][\overline{A} + \overline{B} + \overline{E}]$	Distributive Law
$[ABC + ABD + ABE + DBA][\overline{A} + \overline{B} + \overline{E}]$	Associative Law
$(ABC + ABD + ABE)(\overline{A} + \overline{B} + \overline{E})$	Commutative Law, Theorem 11
$(ABC + ABD + ABE)(\overline{ABE})$	DeMorgan's Law
$\overline{ABE}ABC + \overline{ABE}ABD + \overline{ABE}ABE$	Distributive Law
$\overline{ABE}ABC + \overline{ABE}ABD + 0$	Theorem 6
$\overline{ABE}ABC + \overline{ABE}ABD$	Theorem 7
$\overline{ABE}AB(C + D)$	Factoring
$AB(\overline{ABE})(C + D)$	Commutative/Associative Laws
$AB(\overline{AB} + \overline{E})(C + D)$	DeMorgan's Law
$(AB\overline{AB} + AB\overline{E})(C + D)$	Distributive Law
$(0 + AB\overline{E})(C + D)$	Theorem 6
$(AB\overline{E})(C + D)$	Theorem 8
$AB\overline{E}C + AB\overline{E}D$	Distributive Law
$ABC\overline{E} + ABD\overline{E}$	Commutative Law

An alternate method of solution to this simplification may be observed at the fourth and fifth steps in Example 5-9. The method selected in Example 5-9 involved recognizing the equivalency of the terms $(\overline{A} + \overline{B} + \overline{E})$ and $\overline{ABE}$ and using the $\overline{ABE}$ term to cancel the ABE term. However, at the fourth step, the expression could be changed to the form $(\overline{A} + \overline{B} + \overline{E})(ABC + ABD + ABE)$. By assigning single-variable names to replace certain multiple-variable names and by applying the Distributive Law, a means of expansion results that removes the parentheses and allows further simplification. Let

$$W = (\overline{A} + \overline{B} + \overline{E}), \quad X = ABC, \quad Y = ABD, \quad \text{and} \quad Z = ABE.$$

Algebraically Simplifying the Binary Adder

A	Ā	B	B̄	C	C̄	AB̄	Ā+B	C(Ā+B)	[AB̄ + (Ā+B)C̄]	[AB̄ + (Ā+B)C̄] + C
0	1	0	1	0	1	0	1	1	1	1
0	1	0	1	1	0	0	1	0	0	1
0	1	1	0	0	1	0	1	1	1	1
0	1	1	0	1	0	0	1	0	0	1
1	0	0	1	0	1	1	0	0	1	1
1	0	0	1	1	0	1	0	0	1	1
1	0	1	0	0	1	0	1	1	1	1
1	0	1	0	1	0	0	1	0	0	1

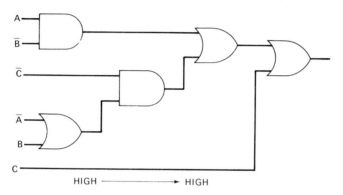

Figure 5-9 Logic diagram and truth table [AB̄ + (Ā + B)C̄] + C = 1

Expansion is accomplished by using the equality

$$W(X + Y + Z) = WX + WY + WZ$$

This approach is left to the reader as an exercise in simplification.

Either method of simplification reduces the requirement for three AND and three OR gates to two AND and one OR gate. Comparative logic diagrams appear in Figs. 5-10a and b. Truth-table proof of the equivalency of the original and final expression is left to the reader.

5-7 ALGEBRAICALLY SIMPLIFYING THE BINARY ADDER

When a digital computer must perform arithmetic addition, the half-adder (discussed in Ch. 4) is inadequate. A *full adder* is required in all but the least-significant positions to accommodate the carry input from previous orders. The full adder has two outputs, *SUM* and *CARRY*, just like the half-adder. However, it must handle not only the augend and the addend (as in the half-adder), but it must also be able to handle an input carry. The same basic rules for addition used in Ch. 4 apply to the full-adder, except the SUM and CARRY output expressions are more complex due to the additional inputs.

(a) Original implementation

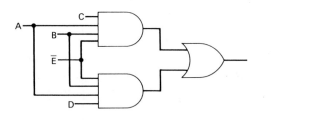

(b) Simplified implementation

Figure 5-10 Logic diagram $[AB(C + D + E) + DBA][\bar{A} + \bar{B} + \bar{E}]$ = $ABC\bar{E} + ABD\bar{E}$

Statement of the Problem

English-language statements for both outputs are given here.

> SUM *output.* If the augend is zero, the addend is zero, and the input carry is one; or if the augend is zero, the addend is one, and the input carry is zero; or if the augend is one, the addend is zero, and the input carry is zero; or if the augend, addend, and input carry are all one, the sum is one.

> OUTPUT CARRY *output.* If the augend is zero, the addend is one, and the input carry is one; or if the augend is one, the addend is zero, and the input carry is one; or if the augend is one, the addend is one, and the input carry is zero; or if the augend, addend, and input carry are all one, the output carry is one.

Substituting letter names for variable names and assigning the logic value of 0 to the numerical value of zero and the logic value of 1 to the

Algebraically Simplifying the Binary Adder

TABLE 5-1 Truth table, full-adder

A	B	C_i	S	C_o
0	0	0	0	0
0	0	1	1	0
0	1	0	1	0
0	1	1	0	1
1	0	0	1	0
1	0	1	0	1
1	1	0	0	1
1	1	1	1	1

numerical value of one, as was done in Ch. 4, puts this problem into a form that can be easily represented in truth tables. In fact, since both the SUM(S) and the OUTPUT CARRY (C_o) outputs are results of the same three inputs, both truth tables can be combined (see Table 5-1). Investigation of the output columns of the truth tables will show that there are four terms that give a 1 (true) output for the S output and four terms that give a 1 (true) output for the C_o output.

The resulting Boolean equations are

$$S = \overline{A} \cdot \overline{B} \cdot C + \overline{A} \cdot B \cdot \overline{C} + A \cdot \overline{B} \cdot \overline{C} + A \cdot B \cdot C$$
$$C_c = \overline{A} \cdot B \cdot C + A \cdot \overline{B} \cdot C + A \cdot B \cdot \overline{C} + A \cdot B \cdot C$$

Note that both equations are in the logical sum of logical products form, which is the form that always results when the *true outputs* are obtained from the truth table. All that remains is to submit each of these equations to algebraic simplification methods to determine if less complex equations may be obtained. The logic diagram for each of the outputs is shown in Fig. 5-11 and is used comparatively when the simplified logic diagram is developed.

Simplification

As with most simplification efforts, numerous methods and approaches are available. In the case of the full adder, an almost *optimum simplification* exists! Due to the widespread use of the full adder in digital equipment many component manufacturers produce integrated circuit (IC) assemblies that are complete full adders (all in one package). In fact, some IC packages contain more than one full adder. But such an approach is almost "cheating," and it doesn't provide much experience in the manipulation of Boolean algebra expressions.

Another possibility might be to use two half-adders to make a full adder—if such a method could be devised. Since half-adders are also in common usage in digital equipment, they are available in prepackaged form (two or more to an IC chip). An investigation of the "two half-adder" approach will now follow.

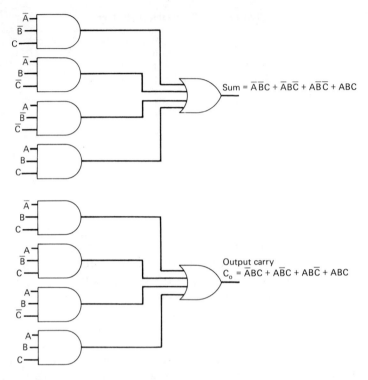

Figure 5-11 Logic diagram and equations for a full-adder

Some additional background concerning the half-adder will be useful while investigating various approaches to the full-adder scheme. The sum equation for the half-adder is commonly found in logic equations and circuit implementation, as pointed out in Sec. 4-3. It may be directly implemented, as shown in Fig. 5-12a, or indirectly implemented, as shown in Figs. 5-12b, c, and d. Derivation of the indirectly implemented versions of the EXCLUSIVE-OR, keyed to Fig. 5-12, follows.

$$S = A\bar{B} + \bar{A}B$$
$$= A\bar{A} + A\bar{B} + \bar{A}B + B\bar{B} \quad \text{Theorems 6, 7, and 8}$$
$$= (A\bar{A} + A\bar{B}) + (\bar{A}B + B\bar{B}) \quad \text{Associative Law}$$
$$= A(\bar{A} + \bar{B}) + B(\bar{A} + \bar{B}) \quad \text{Factoring}$$
$$= (A + B)(\bar{A} + \bar{B}) \quad \text{Distributive Law (see Fig. 5-12b)}$$
$$= (A + B)(\overline{AB}) \quad \text{DeMorgan's Law (see Fig. 5-12c)}$$
$$= \overline{\overline{(A + B)(\overline{AB})}} \quad \text{Theorem 13}$$
$$= \overline{\overline{(A + B)} + \overline{\overline{AB}}} \quad \text{DeMorgan's Law}$$
$$= \overline{\overline{(A + B)} + AB} \quad \text{Theorem 13 (see Fig. 5-12d)}$$

Algebraically Simplifying the Binary Adder

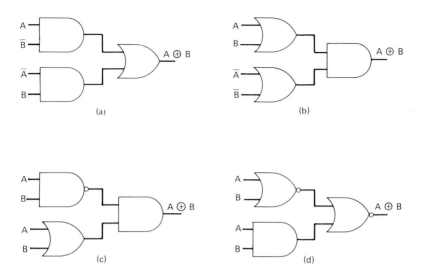

Figure 5-12 Half-adder logic implementations

The truth table for the sum output of the half-adder describes the EXCLUSIVE-OR operation. The output is 1 *only* when either A is 1 or B is 1—but not when *both* are 1. Thus the EXCLUSIVE-OR operation excludes the combination of both $A = 1$ and $B = 1$ at the same time, while the normal (inclusive) OR includes that combination. Many other uses will be discovered for the EXCLUSIVE-OR operation. The EXCLUSIVE-OR may be constructed with separate AND, OR, NAND, or NOR gates, or it may be obtained in an IC package, just like the other logic functions.

The SUM equation for the full adder can be manipulated into many different forms: like terms may be factored; theorems and laws of Boolean algebra applied; and identities recognized and used. These actions may be to no avail if the criteria of "logical sum of logical products" form is adhered to. In fact, the SUM equation as read directly from the truth table is in its simplest form within the criteria established in this text. Certain deviations from these criteria will be investigated later, but for our present discussion, the SUM equation is used "as is." Justification for this assumption is provided in the following chapter, where graphic simplification methods are demonstrated.

The CARRY equation, however, may be simplified algebraically with a resultant decrease in not only number of logic elements but also in the number of inputs per logic element. *One* method of simplification is shown.

$$\begin{aligned}
C_o &= \bar{A}BC + A\bar{B}C + AB\bar{C} + ABC \\
&= \bar{A}BC + ABC + A\bar{B}C + AB\bar{C} & \text{Commutative Law} \\
&= BC(A + \bar{A}) + A(\bar{B}C + B\bar{C}) & \text{Factoring} \\
&= BC + A(\bar{B}C + B\bar{C}) & \text{Theorems 12 and 13} \\
&= BC + A(B + C)(\overline{BC}) & \text{See EXCLUSIVE–OR derivation} \\
&= BC + AB\overline{BC} + AC\overline{BC} & \text{Distributive Law} \\
&= \underbrace{BC}_{X} + \underbrace{\overline{BC}}_{\bar{X}} \underbrace{(AB + AC)}_{Y} & \text{Factoring} \\
& & \text{Recognize the form } X + \bar{X}Y \\
C_o &= BC + AB + AC & \text{Proven identity}
\end{aligned}$$

Thus the original output carry function of the full adder may be implemented with three 2-input AND gates feeding a 3-input OR gate. This represents a reduction of one gate, and the gates used in the simplified form require less inputs per gate.

Now that the groundwork has been laid, the actual investigation of using two half-adders to form one full adder can begin. For reference purposes, the half-adder and the full adder equations and truth tables are reproduced in Fig. 5-13.

Earlier it was stated that the only reason for the full adder configuration was *to include the possible input carry from an addition operation in a previous order of operation.* The half-adder quite adequately performs addition of two single-digit binary numbers. Logically thinking, then, the second half-adder must add the SUM output of the first half-adder to the input CARRY—if any. Neglecting the output carries momentarily and concentrating on the SUM output, the equation for adding the SUM output of the first half-adder (S_1) to the input carry (C_i) is $S_1 \cdot \bar{C}_i + \bar{S}_1 \cdot C_i$.

If A represents the augend and B represents the addend, then $S_1 = A\bar{B} + \bar{A}B$. Substituting, the equation becomes

$$S_2 = (A\bar{B} + \bar{A}B)(\bar{C}_i) + \overline{(A\bar{B} + \bar{A}B)}(C_i).$$

Conversion of this to a logical sum of logical products results in

$$\begin{aligned}
S_2 &= (A\bar{B} + \bar{A}B)(\bar{C}_i) + \overline{(A\bar{B} + \bar{A}B)}(C_i) \\
&= A\bar{B}\bar{C}_i + \bar{A}B\bar{C}_i + \overline{(A\bar{B} + \bar{A}B)}(C_i) & \text{Distributive Law} \\
&= A\bar{B}\bar{C}_i + \bar{A}B\bar{C}_i + (\bar{A} \cdot \bar{B} + AB)(C_i) & \text{Identity} \\
&= A\bar{B}\bar{C}_i + \bar{A}B\bar{C}_i + \bar{A}\bar{B}C_i + ABC_i & \text{Distributive Law}
\end{aligned}$$

Note that the equation for S_2 is identical to the equation for the SUM output of the full adder, thus establishing the validity of using two half-adders to obtain the SUM function of the full-adder.

Algebraically Simplifying the Binary Adder

A	B	S	C		A	B	C_1	S	C_0
0	0	0	0		0	0	0	0	0
0	1	1	0		0	0	1	1	0
1	0	1	0		0	1	0	1	0
1	1	0	1		0	1	1	0	1
					1	0	0	1	0
					1	0	1	0	1
					1	1	0	0	1
					1	1	1	1	1

$S = \bar{A}B + A\bar{B}$
$C = AB$

$S = \bar{A} \cdot \bar{B} \cdot C + \bar{A} \cdot B \cdot \bar{C} + A \cdot \bar{B} \cdot \bar{C} + A \cdot B \cdot C$
$C_o = \bar{A} \cdot B \cdot C + A \cdot \bar{B} \cdot C + A \cdot B \cdot \bar{C} + A \cdot B \cdot C$

Figure 5-13 Half-adder and full-adder truth tables and equations

The CARRY output of the full-adder is not as immediately obvious. But a clue or two is available. The CARRY output of the first adder (AB) is also a term in the CARRY output equation of the full adder. The other two terms are to be ORed with AB according to the CARRY output equation. Assuming that the CARRY output is $AB + AC + BC$ implies that an OR gate performs the function of the terms $AC + BC$ or the equivalent can be provided as additional inputs. In other words, the CARRY output of the full adder (C_o) is $AB + (X)$, where X is the equivalent to $AC + BC$. Another way to look at it is that when some input X is ORed with AB, the overall expression is equivalent to $AB + AC + BC$. The only outputs where C is available in combination with A and B is the SUM output of the second half-adder (S_2) and the CARRY output of the second half-adder (C_2). Perhaps C_2 should be checked, since this output has not yet been used.

With the output of the first half-adder ($A\bar{B} + \bar{A}B$) as one input and C_i as the other input to the second half-adder, $C_2 = (A\bar{B} + \bar{A}B)(C_i)$. Combining C_2 with C_1 in an OR gate yields $C_o = AB + C_i(A\bar{B} + \bar{A}B)$. Expanding this expression to $AB + A\bar{B}C_i + \bar{A}BC_i$ yields very little, as far as the required output is concerned. Recognizing that the term ($A\bar{B} + \bar{A}B$) is the EXCLUSIVE-OR function and that sometimes equivalent Boolean expressions work better algebraically than the originals leads to substitution of $(A + B)(\bar{A} + \bar{B})$ for $A\bar{B} + \bar{A}B$. Expansion of this equivalent expression with C_i gives the same answer as the original.

$(A + B)(\overline{AB})$ appears to be more promising. As we mentioned in Sec. 5-5, it is advantageous to be on the lookout for terms in the expression being simplified that are *complements* of each other. This allows the use of the theorems, laws, and identities of Boolean algebra in the simplification process. Expansion and simplification of the CARRY output equation are shown.

$$C_o = AB + [(A+B)(\overline{AB})(C_i)]$$
$$= AB + [(AC_i + BC_i)(\overline{AB})] \quad \text{Distributive Law}$$
$$= AB + AC_i\overline{AB} + BC_i\overline{AB} \quad \text{Distributive Law}$$
$$= \underbrace{AB}_{} + \overline{AB}(AC_i + BC_i) \quad \text{Factoring}$$
$$= A + \overline{A} \cdot B \quad \text{Recognition of form } A + \overline{A}B$$
$$C_o = AB + AC_i + BC_i$$

It has been demonstrated that two half-adders and an OR gate provide the required outputs for the full adder function. The block/logic diagram with Boolean expressions appears in Fig. 5-14. Although very little has been shown in this discussion as far as actual logic elements are concerned, an increasingly popular method of implementing logic functions was used. "Prepackaged" IC logic functions are extremely popular, and where they are cost effective, they are replacing multiple-element implementation of the same function.

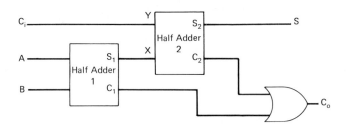

Figure 5-14 Full-adder from two half-adders

PROBLEM APPLICATIONS

5-1. Define logic simplification.
5-2. What is another name for the sum of products form of a Boolean expression?
5-3. What is another name for the product of sums form of a Boolean expression?
5-4. List the steps necessary to algebraically simplify a logically implemented form of an industrial operation expressed in English-language form.
5-5. A logic diagram of a Boolean function is supplied. What steps are necessary to algebraically simplify this logic diagram?
5-6. What algebraic form results from reading a Boolean expression from a truth table?
5-7. Define single-level, two-level, three-level, and four-level logic circuits.

Algebraically simplify:
5-8. $\overline{A} \cdot B + (A\overline{B} + \overline{A}B)$
5-9. $(A + \overline{B})(A\overline{B} + C)(C)$

Algebraically Simplifying the Binary Adder

5-10. $(A + \bar{B})(C) + (A\bar{B} + C)$ $C + A\bar{B}$

5-11. $(AB + C) + (\overline{AB + C})(A + CD)$ $A + C$

5-12. $[(A + C)(A + D)] [(B + C)(B + D)]$ $AB + CD$

5-13. $\bar{A} \cdot \bar{C} + \bar{A}B + BC + \bar{B} \cdot \bar{C} + AB$

5-14. $AB + A\bar{D} + BC + A\bar{B} + BD + \bar{A}B$

5-15. $\bar{A} \cdot \bar{B} + A\bar{D} + B\bar{D} + \bar{B}D + \bar{C} \cdot \bar{D}$

5-16. $[(AB + AC) + (BC + A\bar{B})] + \bar{A}B$

5-17. $A\bar{D} + \bar{A}D + CD + BC + \bar{A}C$

5-18. $AB + BC + \bar{A}C + A\bar{C} + \bar{B}C$

5-19. $(AB)(CD) + \overline{\bar{A} + \bar{B} + \bar{C} + \bar{D}} + A$

5-20. $AB\bar{C} + \bar{A} \cdot \bar{B} \cdot \bar{C} + \overline{\bar{A} + B}$

5-21. $\overline{M \cdot L} \cdot H \cdot \bar{J} \cdot K \cdot H + \bar{J}(H + K)$

5-22. $\overline{AB} + \bar{C} + \bar{D}A + ABAC$

5-23. $A\bar{B} + C + DE + \overline{\bar{A} + \bar{B} + F} + (\overline{\bar{A} + D})(\overline{\bar{A} + B + E})\bar{\bar{D}}$

5-24. $DE(\bar{A} \cdot \bar{C} + A + \bar{B} + C) + \bar{C} + \bar{D} + E$

5-25-34. Algebraically simplify Probs. 16-24 in Chapter 6.

6

GRAPHIC SIMPLIFICATION METHODS

In Chapter 5 we showed how logic expressions* may be simplified by using the postulates, theorems, laws, and identities of Boolean Algebra. It requires considerable ingenuity, judgment, and experience to perform such logic simplification consistently and satisfactorily. This chapter shows another method of logic simplification that requires a minimum of Boolean algebra background.

Just as graphic representation of a conventional algebraic expression makes that expression much easier to visualize, the graphic portrayal of logic expressions reduces the trauma of logic simplification. A minimized expression, requiring the *fewest possible logic elements,* is the goal of both graphic and algebraic simplification. For those who have studied Ch. 5, it will soon be apparent that in cases where graphic methods are employed, the simplification solution is more obvious and easier to obtain. But it may be apparent also that graphic solutions by themselves may not always work. Combinations of algebraic and graphic solutions are often required.

Numerous nonalgebraic methods of logic simplification employing charts, graphs, and maps have been and are being used. The graphic methods of John Venn, a nineteenth century British mathematician, are still used to demonstrate simple theorems and laws of Boolean algebra visually.

*Justifications and ground rules for simplification of logic expressions appear in Sec. 5-1, which should be examined at this time if Ch. 5 was not read previously.

Types of Graphs   151

The staff of the Harvard Computation Laboratory developed a chart method of simplification in the 1950s. During the same period, W. V. Quine proposed another chart method. Both the Harvard chart and the Quine method are somewhat cumbersome and time consuming. However the concepts behind both are currently being used in computer-aided logic-simplification operations despite the relative difficulty of their "paper-and-pencil" methods.

In 1952 E. W. Veitch proposed a diagrammatic means for simplifying logic functions. His method attempted to represent logic functions in a manner that would make algebraic factors more evident. The following year, M. Karnaugh suggested a reorganization of the *Veitch diagram.* Identification of factorable terms was more obvious in *Karnaugh's map,* and variations and simplifications have seen widespread use. The similarities of the Veitch diagram and the Karnaugh map have generated much confusion. So as not to perpetuate this confusion, only the term "map" is used in this text.

The purpose of this chapter is to develop the use of the *map method of simplification* as a tool for digital logic analysis. Investigations of computer-based methods are not covered but are left to the individual reader.

6-1 TYPES OF GRAPHS

Venn Diagrams

One graphic method used to display logical expressions of Boolean algebra is the *Venn diagram.* Venn's geometric method of representing logic relationships consist of (1) a rectangle that represents a class of elements with certain characteristics and (2) circles that represent subclasses or more sharply defined groups within an overall class. Since Boolean algebra is considered an algebra of classes, and since a *class* is a collection of *all objects having some specific characteristics in common,* Venn diagrams lend themselves well to an investigation of the logic expressions of Boolean algebra.

For example: let A equal all engineering technicians who work on digital equipment and B equal all engineering technicians who are reading this book. The Boolean expression for the combination of both is $A \cdot B$. The Venn diagram representing this condition appears in Fig. 6-1a. Shading is used to indicate the area within the rectangle that represents the special group of engineering technicians who work on digital equipment *and* who are reading this book, i.e., $A \cdot B$. Correlation with the truth table for the AND operation reveals that the shaded area represents the "true" (1) condition for ANDing two variables. The unshaded area

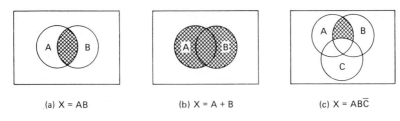

Figure 6-1 Venn diagrams

is the "false" (0) condition, which includes $\bar{A} \cdot \bar{B}$, $\bar{A} \cdot B$, and $A \cdot \bar{B}$. Note the ease with which the expression can be identified and visualized.

Representation of all engineering technicians who work with digital equipment *or* who are reading this book $(A + B)$ is provided in Fig. 6-1b. The shaded area includes all of A, all of B, and the intersection of A and B. Verification of the validity of this graphic representation can be made by noting that the truth table for an OR operation shows a true condition when A is true, B is true, or both A and B are true.

The Venn diagram for the 3-variable term $AB\bar{C}$ is shown in Fig. 6-1c. This diagram shows that the only condition under which the ANDed combination of A, B, and $\bar{C}$ is true is when both A and B are true and C is not true. Although Venn diagrams are used with four variables occasionally, we recommend that no more than three variables be used. The most common application of such diagrams is to demonstrate graphically the simple theorems of Boolean algebra, although they may also be used to simplify simple laws and identities.

Veitch Diagrams

Veitch diagrams also use rectangles to represent logic relationships. Instead of employing circles to represent variables, however, Veitch divided the rectangle into *squares*. Sufficient squares are provided so that every possible combination of all of the variables in the logic expression is represented. Thus the *Veitch diagram* is just another way of displaying *the same information that the truth table provides*. A 3-variable diagram, as originally presented by Veitch, is shown in conjunction with a 3-variable truth table in Fig. 6-2a. Each square on the diagram represents one line of the truth table. Actually, each square represents the conjunctive combination of the three variables and/or their complements. The row labeled 0 could be written as $\bar{A} \cdot \bar{B} \cdot \bar{C}$; the row labeled 1 as $\bar{A} \cdot \bar{B} \cdot C$; and so on. This labeling scheme, previously used in the manipulation of truth tables, clarifies the configuration of the Veitch diagram. Figure 6-2b shows correlation of truth table entries and Veitch diagram positions.

Types of Graphs

Label	A	B	C
0	0	0	0
1	0	0	1
2	0	1	0
3	0	1	1
4	1	0	0
5	1	0	1
6	1	1	0
7	1	1	1

		0	1	A
BC	00	0	4	
	01	1	5	
	10	2	6	
	11	3	7	

		0	1	A
BC	00	$\overline{A}\overline{B}\overline{C}$	$A\overline{B}\overline{C}$	
	01	$\overline{A}\overline{B}C$	$A\overline{B}C$	
	10	$\overline{A}B\overline{C}$	$AB\overline{C}$	
	11	$\overline{A}BC$	ABC	

(a) Truth table and original map (b) Map representation of terms

Figure 6-2 Veitch diagrams

The intent of the Veitch diagram is to present truth table information in a form that more clearly displays the factorable terms in a Boolean expression. Algebraically factoring a simple Boolean expression such as $\overline{A}BC + ABC$ demonstrates what may be done with the Veitch diagram. By inspection, we see that $\overline{A}BC + ABC = (\overline{A} + A)BC$. Further simplification yields (1) BC, or BC. If the two terms are arranged so as to easily identify the existence of common factors such as BC, the algebraic expression is simplified by inspection. (This is what Veitch did when he developed his diagram.)

$\overline{A}BC$ is represented on the diagram by square 3, and ABC by square 7. Marks are placed in each of the identified squares to define the fact that those terms are included in the expression being simplified. All of the squares are then investigated, and when marked squares having a difference of only one variable are recognized, those squares are grouped for future reference. In this case, squares 3 and 7 can be grouped together, since B and C remain the same for both squares and since A is 0 for square 3 and 1 for square 7. When a variable and its complement are both present in an AND-OR expression, that variable may be discarded, as demonstrated in the algebraic simplification of the expression being used. Thus in a 3-variable Veitch diagram, when two squares are properly grouped, the new term described by the grouped squares has one less variable than the original.

A bit of difficulty exists, however, in identifying those squares that can be grouped together. To take advantage of the factoring idea, recall that squares may differ by only *one* variable. No difficulty exists in terms of the squares making up the $A = 0$ column and the $A = 1$ column. Obviously, squares 0, 1, 2, and 3 are opposite in A value from squares 4, 5, 6, and 7. But what about squares 0 and 2 in terms of the variable B? Or squares 1 and 3 in terms of B? Or 4 and 6, or 5 and 7?

These combinations all meet the criterion of squares differing by only one variable, although from the circling or grouping operation it may not be apparent immediately. Considerable familiarity with the use of Veitch diagrams is required before such relationships are recognized automatically. Since other methods of graphic simplification not requiring the added effort of mentally grouping squares are available, the *original* Veitch diagram is rarely used.

Many additional capabilities of the Veitch diagram (along with the intricacies of its use) have not been covered here. But the contributions of Veitch should not be overlooked. Without his early efforts in portraying logic simplification graphically, the revised diagrams and maps in use now might have taken much longer to develop.

Karnaugh Maps

The same general concepts exist for both Veitch diagrams and Karnaugh maps—*rectangles* represent *logic relationships* and *squares* represent *conjunctive combinations* of all variables in the expression being evaluated. The basic difference between the two types of graphic representation lies in the *arrangement of the squares* inside the rectangle. By rearranging the squares, it is possible to have any square that is adjacent to any other square differ by only one variable. Figure 6-3a shows the 3-variable map as originally presented by Karnaugh. The commonly used simplified version is reproduced in Fig. 6-3b. By filling in the actual terms for each square, it is obvious that no square differs from any adjacent square by more than one variable, thereby facilitating the factoring process (see Fig. 6-3c).

Placing *identification numbers* in each square reveals that there may not be an easy method for determining how Karnaugh maps are constructed (Fig. 6-3d). So it may be interesting to consider some of the ideas behind the construction. Random tries could be initiated—terms tried in various locations until the proper combination is found. But trial and error is not really efficient! What really needs to be done is to find a way of identifying the squares so that there is no doubt of the outcome and so that a minimum of time is spent in determining how to construct a map for a situation not previously encountered. A recognizable means of identifying the squares is also desirable. It is apparent that sequentially using the binary-coded-decimal (BCD) equivalent of each row on the truth table is of no avail. Perhaps some version of the BCD code other than the 8-4-2-1 version could be used.

One such version of the BCD code, called the *Gray code,* has only a one-variable difference as counting progresses from one number to the next. Straight BCD code (4-2-1 BCD) and the Gray code are compared

Types of Graphs

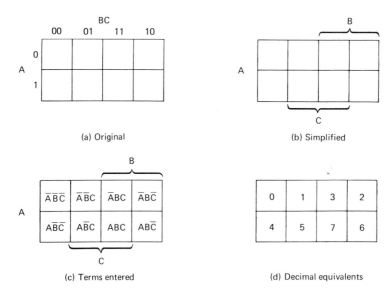

Figure 6-3 Karnaugh maps

in Table 6-1. But can this code help build a Karnaugh map? Taking the original arrangement that Karnaugh proposed for the physical layout of his map and placing the appropriate Boolean terms in each square (Fig. 6-3c) again leads to an interesting observation. Note that each square differs from each adjacent square by *only one* variable. If the decimal equivalent of the Gray code version of each term is entered (Fig. 6-4), a numbering pattern emerges. Thus the Karnaugh map is *not* a random development. It resulted from a recognition of the basic characteristics of Boolean expressions, number systems, and truth tables.

TABLE 6-1 4-2-1 versus Gray code

4–2–1				Gray		
A	B	C	Decimal	A	B	C
0	0	0	0	0	0	0
0	0	1	1	0	0	1
0	1	0	2	0	1	1
0	1	1	3	0	1	0
1	0	0	4	1	1	0
1	0	1	5	1	1	1
1	1	0	6	1	0	1
1	1	1	7	1	0	0

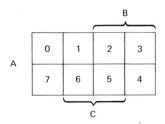

Figure 6-4 Three-variable Karnaugh map using Gray Code numbering

6-2 PREPARATION FOR THE USE OF MAPS

Map Construction

Maps used in graphic simplification of Boolean expressions are actually truth tables in another form. As such, the same number of possible combinations presented in the truth table must be accounted for in the map. Thus for n variables, 2^n combinations exist, so a 2-variable truth table must have 2^2 or 4 entries, a 3-variable truth table must have 2^3 or 8 entries, and so on. Two-variable maps must then have 4 squares, 3-variable maps 8 squares, and so on. The actual physical placement of the squares on the map depends on individual variations of the common map methods. As long as the criterion of each square varying from each adjacent square by only one variable is met, the physical layout is unimportant. Figure 6-5 shows the map layout used in this text for 2, 3, and 4-variable expressions.

In actual use, maps do not have the number identification (terms) indicated within each square. These are provided initially so that the pattern of square identification can be established. Careful study of the map forms shows that (1) each square varies from each adjacent square by only one variable and (2) an identification pattern does exist. Note that the identification numbers are based on the Gray code* representation of decimal numbers as are the Karnaugh maps.

Forms of Boolean Expressions

To use map methods of simplification, the Boolean expression must be in a form that can be represented on the map. Two basic forms of Boolean expressions exist (neglecting negated expressions).

    1. The logical sum of logical products form, which has been previously mentioned, contains terms ORed together, and

---

*See Chapter 13 for the conversion table from BCD to Gray to decimal codes.

Preparation for the Use of Maps

each term consists of one or more variables ANDed together.
2. The logical product of logical sums form contains factors ANDed together, and each factor consists of one or more variables ORed with other variables.

Each of these is discussed below.

Logical sum of logical products expressions are known by many different names: *AND-OR, sum-of-products, standard sum,* and *minterm form.* Actually, the last two names require a definition more explicit than merely "logical sum of logical products." Not only must the expression be in the logical sum of logical products form, but each term must contain all variables in either the normal or complemented (negated) form. Rather than confuse the situation with numerous definitions, *minterm* is used in this text to define a Boolean expression in which *each term contains all of the variables* (in either the normal or complemented form) AND*ed together and all of the terms* OR*ed together.* Minterm comes from a mathematical definition of a particular type of expression called a "minimal polynomial term."

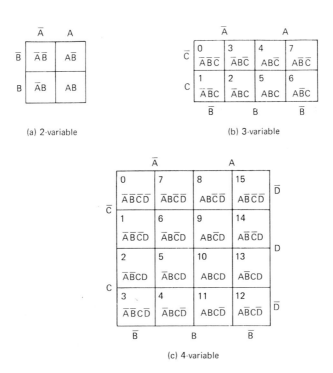

Figure 6-5 Construction of maps

The logical product of logical sums form is also known as the *OR-AND,* the *products of sums,* the *standard product,* and the *maxterm form.* Again, the standard product and the maxterm forms have more explicit definitions than the product of sums form definition. In the standard product and the maxterm forms, *all variables* (in either the normal or complemented form) *are ORed together in each factor and all of the factors are* ANDed. *Maxterm* is used exclusively in this text.

Conversion of Boolean expressions to minterm form (the form required for entry of the expression into the logic map) requires use of the theorems, laws, and common identities of Boolean algebra. For example, $\overline{A + B}$ is not in minterm form. However the vincula may be split, using DeMorgan's Law, and $\overline{A + B}$ becomes $\overline{A} \cdot \overline{B}$. Individual terms should also be simplified, like a variable that appears more than once within the term ($ABCA = ABC$). When a variable and its complement appear within the same ANDed term, simplification should also be accomplished ($ABC\overline{A} = 0$). Where appropriate, the Distributive Law should be applied to remove the parentheses $AB(C + D) = ABC + ABD$. The reader will be provided with considerable practice examples and problems.

Another way to obtain the minterm form for *any* expression is to use truth tables. Construct a truth table for the expression being used; then examine the output column for combinations of inputs that provide an output equal to 1. Each of these terms may then be combined disjunctively to give the minterm form. Thus when a Boolean expression is read from a truth table, it is in the logical sum of logical products or minterm form.

Truth Table Forms

Since the map is merely another way of presenting truth table information, the actual *form of the information* in the truth table must be realized. As originally defined, a truth table is a tabular listing of all possible combinations of input conditions to a logic device with a resultant output for each case. This definition remains valid in studying map methods of logic simplification. Right now, the important consideration is the form in which the *output conditions* are read from the truth table.

To take advantage of the *factorability* of logical expressions, the expressions should be in the logical sum of logical products form. When in this form, factorable terms are more easily identified, as demonstrated during the Veitch diagram discussion. *The major premise of the map method of logic simplification is based on having the Boolean expression in the logical sum of logical products form.* If the expression is not in this form, algebraic methods must be used to convert it. In

Preparation for the Use of Maps

X	$\bar{X}$	Y	$\bar{Y}$	Z	$\bar{Z}$	$X\bar{Y}Z$	XY	$\bar{X}Y\bar{Z}$	$X\bar{Y}Z + XY + \bar{X}Y\bar{Z}$	
0	1	0	1	0	1	0	0	0	0	
0	1	0	1	1	0	0	0	0	0	
0	1	1	0	0	1	0	0	1	1	$\bar{X}Y\bar{Z}$
0	1	1	0	1	0	0	0	0	0	
1	0	0	1	0	1	0	0	0	0	
1	0	0	1	1	0	1	0	0	1	$X\bar{Y}Z$
1	0	1	0	0	1	0	1	0	1	$XY\bar{Z}$
1	0	1	0	1	0	0	1	0	1	XYZ

As shown in the truth table $X\bar{Y}Z + XY + \bar{X}Y\bar{Z} = \bar{X}Y\bar{Z} + X\bar{Y}Z + XY\bar{Z} + XYZ$

(a)

$$X\bar{Y}Z + XY + \bar{X}Y\bar{Z} = X\bar{Y}Z + XY(1) + \bar{X}Y\bar{Z}$$
$$= X\bar{Y}Z + XY(Z+\bar{Z}) + \bar{X}Y\bar{Z}$$
$$= X\bar{Y}Z + XYZ + XY\bar{Z} + \bar{X}Y\bar{Z}$$

$XY(1) = XY$
$Z + \bar{Z} = 1$
Distributive laws

(b)

Figure 6-6 Expansion of an expression to standard sum form; (a) using the truth table, (b) algebraic expansion

many cases, the algebraic conversion results in the simplest form, but the converted expression should be examined with maps to assure that such is the case. When the expression is converted to the logical sum of logical products form, the truth table is constructed, and the output is determined in this form. Once again, the Boolean expression must be in the logical sum of logical products form to apply map simplification operations. This is why one of the criteria of "simplest circuit" in Chapter 5 was that the final expression be in logical sum of logical products form.

Each term contains *all* of the variables in either the inverted or noninverted form when the expression is read from the truth table. Thus it may be entered directly into the simplification map. An expression that is not in the standard sum form must either be subjected to truth table manipulations or be converted to the proper form. The Boolean expression $X\bar{Y}Z + XY + \bar{X}Y\bar{Z}$ is not in standard sum form because it does not contain all of the variables in the second term. Expansion to standard sum form by both truth table and algebraic methods is shown in Fig. 6-6.

Maxterms are less commonly used than minterms. The maxterm form of a given expression can be obtained (1) by writing the inverse of the desired expression in minterm form and (2) by inverting the expression obtained in step (1). For example, the maxterm is derived for the expression described by the truth table, Table 6-2.

Table 6-2 Truth table $W = \bar{X}\bar{Y}\bar{Z} + \bar{X}\bar{Y}Z + \bar{X}Y\bar{Z} + X\bar{Y}\bar{Z} + X\bar{Y}Z$

X	Y	Z	W	$\bar{W}$
0	0	0	1	0
0	0	1	1	0
0	1	0	1	0
0	1	1	0	1
1	0	0	1	0
1	0	1	1	0
1	1	0	0	1
1	1	1	0	1

The minterm form is read directly from the truth table as

$$W = \bar{X}\cdot\bar{Y}\cdot\bar{Z} + \bar{X}\cdot\bar{Y}\cdot Z + \bar{X}\cdot Y\cdot\bar{Z} + X\cdot\bar{Y}\cdot\bar{Z} + X\cdot\bar{Y}\cdot Z$$

Writing the inverse of W in minterm form from the truth table yields

$$\bar{W} = \bar{X}\cdot Y\cdot Z + X\cdot Y\cdot\bar{Z} + X\cdot Y\cdot Z$$

Inverting $\bar{W}$ provides the required maxterm form

$$W = (X + \bar{Y} + \bar{Z})(\bar{X} + \bar{Y} + Z)(\bar{X} + \bar{Y} + \bar{Z})$$

The use of maxterm forms in map simplification will be demonstrated later.

Entering Information on Maps

Each square on a map represents only *one unique combination* of the variables in the expression being entered. It is assumed that the expression has been converted or expanded, as necessary, to the form required for entry. Each term of the expression is examined, and a 1 is placed in the square representing the true condition for that term. Zeros may be inserted in all of the remaining squares to represent the false terms, but common practice dictates omitting the zeros for clarity. 2-variable through 4-variable maps are shown in Fig. 6-7. Each map has a Boolean expression, demonstrating the information entry process.

The meaning of the words minterm and maxterm may become more apparent by examining the 3-variable map. Construction of the map is such that eight separate squares are contained within the rectangle defining the 3-variable expression. The *minimum* distinguishable area capable of being individually identified is one of the squares. Identification of a single square is made by conjunctively combining all three of the variables in either their normal or complemented form. A term composed of all of the variables conjunctively combined in either their normal or complemented form has been defined as a minterm. Thus *a minterm is characterized by the minimum distinguishable*

Preparation for the Use of Maps

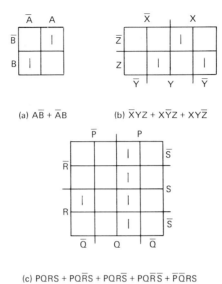

Figure 6-7 Completed maps

area on a logic map. Since the maxterm is the complement of the minterm, all of the rest of the squares on the map represent the maxterm. *The maxterm, then, is characterized by the maximum distinguishable area on a logic map.* This relationship may assist the reader in remembering and identifying the minterm and maxterm forms when they are encountered during map simplification exercises.

6-3 2-VARIABLE MAP SIMPLIFICATION

A Simple Two-variable Expression $AB + \bar{A}B$

An easy way to get started with the map method of simplification is to compare *map simplification* (of a simple 2-variable expression) with *algebraic simplification* (of the same expression). The many advantages of map simplification may not be immediately obvious with simple 2-variable expressions, but the use of maps can be demonstrated without too much difficulty. Consider the expression $AB + \bar{A}B$. Since the variable B is common to both terms, it may be factored, resulting in $B(A + \bar{A})$. $(A + \bar{A}) = 1$ by Theorem 12, thus simplifying the expression to $B(1)$. Theorem 3 states that $B(1) = B$, so the original expression reduces simply to B. The truth table and comparative logic diagram appear in Fig. 6-8.

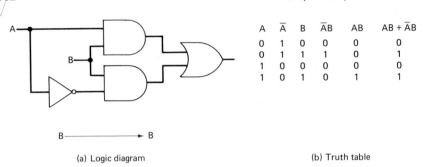

A	$\bar{A}$	B	$\bar{A}B$	AB	$AB + \bar{A}B$
0	1	0	0	0	0
0	1	1	1	0	1
1	0	0	0	0	0
1	0	1	0	1	1

(a) Logic diagram (b) Truth table

Figure 6-8 Logic diagram and truth table, $AB + \bar{A}B$

Figure 6-5 showed the physical layout of a map for simplification of 2-variable expressions. Section 6-2 defined the form of the Boolean expression required for entry on the map and demonstrated such entry. Examination of the expression to be simplified discloses that it is in the required minterm form—all we must do is insert 1s into the squares corresponding to each of the minterms. The completely filled-in map appears in Fig. 6-9a.

The next step is to learn how to use the map. Since the physical layout of the map is such that each square varies from each adjacent square by only one variable, examination of adjacent squares can yield valuable information. For example, the 1s in Fig. 6-9a are adjacent. While each 1 occupies a square that is equivalent to the variable B, 1 of the squares shares B with $\bar{A}$ and the other square shares B with A. It should be evident that the map is displaying graphically the same thing as the algebraic factoring $B(A + \bar{A})$. When a variable and its complement are ORed together, the result is logic 1. It is also known that a variable ANDed with 1 is equivalent to that variable. Thus the map performs factoring in much the same manner as algebraic factoring.

Identification of patterns of filled squares is usually made by circling groups of squares. When using the 2-variable map, the following patterns (shown in order of preferences) are allowed.

1. All four squares filled represent a constant value of logic 1 for the expression entered.
2. Any two adjacent squares filled represents a single-variable term.
3. A single square filled represents a 2-variable term.

In the example being simplified, only the two adjacent squares shown in Fig. 6-9b can be grouped. By the given rule, this must represent a single-variable term. When the grouping operation encompasses both a variable and its complement, that variable and its complement

Preparation for the Use of Maps

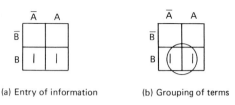

(a) Entry of information   (b) Grouping of terms

Figure 6-9 Map simplification of AB + $\bar{A}$B

are cancelled. Only the variable that is included in the grouping operation by itself (i.e., with no complement) is left. In this case, $A + \bar{A}$ cancels, leaving only B. As stated earlier, when two squares are properly grouped, the new term described by the grouped squares has *one less* variable than the original.

A Complex Two-variable Expression $A(A + B) + B(\bar{A} + B)$

A more complex 2-variable logic simplification problem is now demonstrated. To enter this expression in the map, it must be in minterm form. The steps required to perform the conversion to this standard form are presented here.

$A(A + B) + B(\bar{A} + B) = AA + AB + \bar{A}B + BB$ Distributive Law
$= A + B + AB + \bar{A}B$ Theorem 5, Commutative Law

$= A(1) + B(1) + AB + \bar{A}B$ Theorem 3
$= A(B + \bar{B}) + B(A + \bar{A}) + AB + \bar{A}B$ Theorem 12
$= AB + A\bar{B} + AB + \bar{A}B + AB + \bar{A}B$ Distributive Law, Theorem 5

∴ $A(A + B) + B(\bar{A} + B) = AB + A\bar{B} + \bar{A}B$ Theorem 11

Truth tables and logic diagrams for this expression are shown in Fig. 5-6. The minterm form entered in a 2-variable map is shown in Fig. 6-10a.

The same types of patterns used in the previous example are used in this map. Considering the map as two separate maps, we see that two 1s can be grouped (*looped*) at the bottom of the map (Fig. 6-10b) and that two 1s can also be looped at the right-hand side of the map (Fig. 6-10c). The two 1s at the bottom of the map represent the variable B. Actually, one square represents $\bar{A}B$, and the other square represents AB. It has been shown previously, however, that the map performs the task of algebraically simplifying expressions of the form $AB + \bar{A}B$. Therefore A and $\bar{A}$ cancel, allowing the two looped 1s at the bottom of the map to represent the single variable B. Similar reasoning can be applied to the two 1s at the right-hand side of the map to obtain the

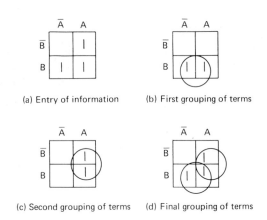

(a) Entry of information
(b) First grouping of terms
(c) Second grouping of terms
(d) Final grouping of terms

Figure 6-10 Map simplification of $AB + A\overline{B} + \overline{A}B$

single variable A. Placing the two single maps together (Fig. 6-10d) discloses that the square equivalent to AB is shared by both maps. When a 1 is placed in a square, that square is considered filled. If mapping another term requires a 1 in a filled square, no additional action is necessary. When such a situation occurs, it merely means that a redundancy exists for the term indicated by the filled square. All additional terms falling within the filled square will be ignored during the map-reading operations. Sharing squares is equivalent to Postulate 8 or $(1 + 1 = 1)$.

When more than one term is required to describe the patterns of 1s on the map, the terms are joined by the OR operator. Since the map is another way of presenting truth table information, the *joining of terms by the* OR *operator* is analogous to reading the output column of the truth table in logical sum of logical products form. Interpretation of map simplification of the Boolean expression $A(A + B) + B(\overline{A} + B)$, using the above information, results in the simplified expression $A + B$.

6-4 3-VARIABLE MAPS

A Complex Three-variable Expression

$$XY\overline{Z} + \overline{X}Y\overline{Z} + \overline{X} + \overline{Y} + \overline{Z} + X + Y + Z + Z(\overline{X} + Y)$$

The real advantages of map simplification begin to appear when three or more variables are encountered. Entry of the expression into the map is not feasible in the original form. Algebraic manipulation is necessary to obtain the minterm form. Thus

3-Variable Maps

$X Y \bar{Z} + \bar{X} Y \bar{Z} + \overline{X + \bar{Y} + \bar{Z}} + \overline{X + Y + Z} + Z(\bar{X} + Y)$ Original expression
$X Y \bar{Z} + \bar{X} Y \bar{Z} + \bar{X} Y Z + \bar{X} \cdot \bar{Y} \cdot \bar{Z} + Z(\bar{X} + Y)$ DeMorgan's Law
$X Y \bar{Z} + \bar{X} Y \bar{Z} + \bar{X} Y Z + \bar{X} \cdot \bar{Y} \cdot \bar{Z} + \bar{X} Z + Y Z$ Distributive Law
$X Y \bar{Z} + \bar{X} Y \bar{Z} + \bar{X} Y Z + \bar{X} \cdot \bar{Y} \cdot \bar{Z} + \bar{X} Y Z + \bar{X} \cdot \bar{Y} \cdot Z + X Y Z + \bar{X} Y Z$ Expansion
$X Y \bar{Z} + \bar{X} Y \bar{Z} + \bar{X} Y Z + \bar{X} Y Z + \bar{X} Y Z + \bar{X} \cdot \bar{Y} \cdot \bar{Z} + \bar{X} \cdot \bar{Y} \cdot Z + X Y Z$ Commutative Law
$X Y \bar{Z} + \bar{X} Y \bar{Z} + \bar{X} Y Z + \bar{X} \cdot \bar{Y} \cdot \bar{Z} + \bar{X} \cdot \bar{Y} \cdot Z + X Y Z$ $(A + A = A)$

Since the expression contains three variables (either in normal or complemented form), a total of eight possible combinations exists; hence a map with eight squares is required. The squares are arranged as we discussed in Sec. 6-2 so that each square differs from each adjacent square by only one variable. But now a problem arises. What about the squares on the left side and on the right side of the rectangle that have no adjacent squares on one side? Reference to Fig. 6-5 may help solve the dilemma. Note that the square numbered 0 ($\bar{A} \cdot \bar{B} \cdot \bar{C}$) differs from the square numbered 7 ($A \cdot \bar{B} \cdot \bar{C}$) and that the square numbered 1 ($\bar{A} \cdot \bar{B} \cdot C$) differs from the square numbered 6 ($A \bar{B} C$) by only one variable. Thus square 0 is adjacent to square 7, and square 1 is adjacent to square 6, as though the map were inscribed on a *cylinder*. If the 3-variable map is visualized in this manner, no difficulty should be encountered in simplifying 3-variable Boolean expressions.

As in the 2-variable map, a 1 is placed in the square representing each of the terms in the expression. Figure 6-11a shows the completed map. Rules for simplifying the expression by use of the map are similar to the 2-variable map rules. Certain expansions are necessary, however, to compensate for the additional squares. The following patterns of 1s (once again shown in order of preference) are allowed.

1. All eight squares filled represent a constant value of logic 1 for the expression entered.
2. Any four adjacent squares filled represents a single-variable term. Remember, squares at opposite ends of rows are adjacent.
3. Any two adjacent squares filled represents a 2-variable term.
4. A single square filled represents a 3-variable term.

The preference in all cases is to detect *patterns having the maximum number of 1s*, as long as the number of 1s is a power of 2 (i.e., 1, 2, 4, 8, 16, etc.). In the 3-variable map, the first order of preference is to have *all* eight squares filled. This would represent a logic value of 1 for the complete expression, and all of the gates required to implement the expression could be replaced with a direct connection. In this example, squares without 1s exist, so the first order of preference cannot

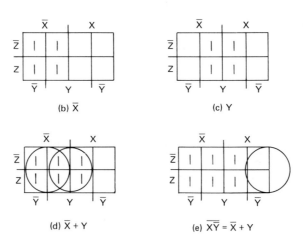

Figure 6-11 Map simplification of $XY\bar{Z} + \bar{X}Y\bar{Z} + \overline{X + \bar{Y} + \bar{Z}} + \overline{X + Y + Z} + Z(\bar{X} + Y)$

be met. The second order of preference is a grouping of four 1s in adjacent squares. Figure 6-11b shows one possible grouping, and Fig. 6-11c shows another possible grouping. As in the 2-variable map, 1s may be shared in more than one loop. Thus two loops of four 1s exist, and all 1s are within loops (Fig. 6-11d). The resulting equation should be a 2-term equation, consisting of one variable or its complement in each term. The grouping in Fig. 6-11b represents $\bar{X}$. Actually, it represents $\bar{X}YZ + \bar{X}\cdot\bar{Y}\cdot Z + \bar{X}Y\bar{Z} + \bar{X}\cdot\bar{Y}\cdot\bar{Z}$, but the factoring properties of the map allow *direct* interpretation of the display without any algebraic manipulation. Similarly, the grouping in Fig. 6-11c represents Y. Thus the simplified version of the complex expression is $\bar{X} + Y$.

Truth-table proof is shown in Table 6-3 to verify this simplification.

Another very important characteristic of map simplification is demonstrated by the Boolean expression currently being examined. If the expression is viewed as an equation such that $F = XY\bar{Z} + \bar{X}Y\bar{Z} + \overline{X + \bar{Y} + \bar{Z}} + \overline{X + Y + Z} + Z(\bar{X} + Y)$, then the completed map can be considered to represent F. The portion of the map that is *not* marked represents all that is not F, or $\bar{F}$. Circling of the unmarked squares (Fig.

Table 6–3 Truth table proof $XY\bar{Z} + \bar{X}Y\bar{Z} + \overline{X+\bar{Y}+\bar{Z}} + \overline{X+Y+Z} + Z(\bar{X}+Y) = \bar{X}+Y$

X	$\bar{X}$	Y	$\bar{Y}$	Z	$\bar{Z}$	$XY\bar{Z}$	$\bar{X}Y\bar{Z}$	$X+\bar{Y}+\bar{Z}$	$\overline{X+\bar{Y}+\bar{Z}}$	$X+Y+Z$	$\overline{X+Y+Z}$	$(\bar{X}+Y)$	$Z(\bar{X}+Y)$	W^*	$\bar{X}+Y$
0	1	0	1	0	1	0	0	1	0	0	1	1	0	1	1
0	1	0	1	1	0	0	0	1	0	1	0	1	1	1	1
0	1	1	0	0	1	0	1	1	0	1	0	1	0	1	1
0	1	1	0	1	0	0	0	0	1	1	0	1	1	1	1
1	0	0	1	0	1	0	0	1	0	1	0	0	0	0	0
1	0	0	1	1	0	0	0	1	0	1	0	0	0	0	0
1	0	1	0	0	1	1	0	1	0	1	0	1	0	1	1
1	0	1	0	1	0	0	0	1	0	1	0	1	1	1	1

* $W = XY\bar{Z} + \bar{X}Y\bar{Z} + \overline{X+\bar{Y}+\bar{Z}} + \overline{X+Y+Z} + Z(\bar{X}+Y)$

Equal

6-11e) and interpreting their meaning results in the Boolean equation $F = X\overline{Y}$. Negating both sides of the equation and applying DeMorgan's Law provide the equation $F = \overline{X} + Y$, which is identical to the result obtained by interpreting the marked squares. In cases where there are relatively few unmarked squares, this approach may prove effective in quickly recognizing the simplified expression.

Graphic Simplification Starting with a Logic Diagram

Often, the only information available from which to perform logic simplification is *a portion of a logic diagram.* It then becomes necessary to first develop the basic algebraic expression that describes the diagram's operation. Starting at the input side of the diagram, each logic element is examined and its output recorded. The logic diagram in Fig. 6-12 is evaluated in this manner.

OR 1	$A + \overline{C}$
OR 2	$\overline{A} + C$
AND 1	$(A + \overline{C})(\overline{A} + C)$
NOR 1	$\overline{B + (A + \overline{C})(\overline{A} + C)}$
NOR 2	$\overline{A + C}$
AND 2	$AB\overline{C}$
OR 3	$\overline{B + (A + \overline{C})(\overline{A} + C)} + \overline{A + C} + AB\overline{C}$

Conversion to sum of products form follows.

$\overline{B + (A + \overline{C})(\overline{A} + C)} + \overline{A + C} + AB\overline{C}$	
$\overline{B}\overline{(A + \overline{C})} + \overline{(\overline{A} + C)} + \overline{A} \cdot \overline{C} + AB\overline{C}$	DeMorgan's Law
$\overline{B}(\overline{A}C) + A\overline{C} + \overline{A} \cdot \overline{C} + AB\overline{C}$	DeMorgan's Law
$\overline{B} \cdot \overline{A} \cdot C + A\overline{C} + \overline{A} \cdot \overline{C} + AB\overline{C}$	Distributive Law

Then to the minterm form.

$\overline{B} \cdot \overline{A} \cdot C + A\overline{C}(B + \overline{B}) + \overline{A} \cdot \overline{C}(B + \overline{B}) + AB\overline{C}$	Theorem 3
$\overline{B} \cdot \overline{A} \cdot C + AB\overline{C} + A \cdot \overline{B} \cdot \overline{C} + \overline{A}B\overline{C} + \overline{A} \cdot \overline{B} \cdot \overline{C} + AB\overline{C}$	Distributive Law
$\overline{A} \cdot \overline{B} \cdot C + A \cdot \overline{B} \cdot \overline{C} + \overline{A}B\overline{C} + \overline{A} \cdot \overline{B} \cdot \overline{C} + AB\overline{C}$	Theorem 11, Commutative Law

The minterm form expression is entered on the 3-variable map, and the grouping of 1s performed (Fig. 6-13b) according to rules previously defined. $\overline{C}$ is interpreted from the "4-loop," and $\overline{A} \cdot \overline{B}$ is interpreted from the "2-loop." Both terms are joined by the OR operator to form the expression $\overline{C} + \overline{A} \cdot \overline{B}$. The simplified logic diagram appears in Fig. 6-13a.

3-Variable Maps

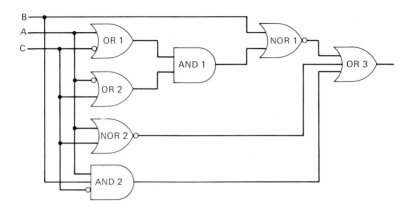

Figure 6-12 Sample logic diagram

Using Logical Sum of Logical Products Expressions

Now that a number of simplifications have been accomplished using the map method, it is time to investigate some "shortcuts." Note that much of the effort expended with the map method is in *preparing the expression* so that it can be entered on the map. Getting from the original expression to the minterm expression may be difficult at times, to say the least. The minterm form, of course, is a special case of the logical sum of logical products form, where all variables in the expression are represented in each term in either their normal or complemented form. As you may have noted, many terms in AND-OR form expressions do not contain all of the variables. It has been necessary to expand such terms to include the missing variable(s) to reach the minterm form. In each case, additional terms have been generated, and the complexity of the expression has grown.

Recognition of the operations performed allows these operations to be performed mentally without the attendant increase in complexity. An example may prove helpful. Consider the expression just simplified. Application of DeMorgan's Law and the Distributive Law to the expression $B + (A + \overline{C})(\overline{A} + C) + A + C + AB\overline{C}$ resulted in the 4-term AND-OR

(a) Logic diagram   (b) Map

Figure 6-13 Simplified logic diagram and map for Figure 6-12 expression $\bar{A} \cdot \bar{B} \cdot C + A\bar{C} + \bar{A} \cdot \bar{C} + AB\bar{C}$. Expansion to minterm form resulted in the 6-term expression $\bar{A} \cdot \bar{B} \cdot C + AB\bar{C} + A \cdot \bar{B} \cdot \bar{C} + \bar{A}B\bar{C} + \bar{A} \cdot \bar{B} \cdot \bar{C} + AB\bar{C}$. The 6-term expression fulfilled the requirements for the minterm form.

The operation that generated the additional terms in the minterm form was the expansion of the 2-variable terms $A\bar{C}$ and $\bar{A} \cdot \bar{C}$ to include both the normal and the complemented form of the missing variable in each case. Obviously, a relationship exists between the *number of variables* in a term and the *number of squares* occupied by that term on the map. In a 3-variable map, for example, a term containing all three variables occupies only one square. That one square is completely defined by the intersection of the rows and columns on the map used to describe each of the variables. When the number of variables in the term of a 3-variable map drops to only two, then two squares must be used to describe that term. This occurs because the 2-variable term may occupy the square that it shares with the missing variable and also the square that it shares with the complement of the missing variable. Reduction of the number of variables in a term to one results in four squares being required to describe that term. Now the variable in the term must share squares with the normal and complemented forms of *both* of the missing variables. The implied relationship between number of variables in a term and the size of the map can be expressed in the following manner.

1-variable term	1/2 of the squares
2-variable term	1/4 of the squares
3-variable term	1/8 of the squares
4-variable term	1/16 of the squares

Note that as the number of variables in the term increases, the number of squares occupied decreases by a power of 2 (or is halved). In the 3-variable map, a 1-variable term occupies half (four) of the squares. A 2-variable term occupies one-fourth (two) of the squares. Note the reduction by a power of 2. A 3-variable term occupies one-eighth (one) of the squares. Again, note the reduction by a power of 2. Thus an n-variable term occupies $1/2^n$ squares. This relationship holds for all maps to be discussed in this chapter.

Using this relationship, it can be seen that the two terms $A\bar{C}$ and $\bar{A} \cdot \bar{C}$ each occupy two squares, since a 2-variable term must occupy one-fourth of the total squares on the map. The algebraic expansion to minterm form has shown that in each case, the two squares are those squares shared with the missing variable and its complement. With a little practice, placement of logical sum of logical product form expressions in less than full minterm form on maps becomes a real timesaver.

Some of the examples that follow show entry of logical sum of logical products form expressions directly on the map without con-

4-Variable Maps

version to minterm form so that the techniques of such operations can be demonstrated. When in doubt, however, always expand to the full minterm form. A little time may be lost, but it ensures that the complete expression is entered.

6-5 4-VARIABLE MAPS

Advancing to 4-variable maps does not appreciably complicate matters. With four variables, sixteen possible combinations exist, so the map must have sixteen squares on which to plot variable combinations. The map is constructed (Fig. 6-5c) so that all adjacent squares differ by only one variable to take advantage of factoring possibilities. The techniques already discussed also apply to 4-variable maps. Following entry of the algebraic expressions on the map, the only difference in map analysis is the *meaning of identified adjacencies.* The general rule in Sec. 6-4 can be followed, and very little difficulty should be encountered. As stated, a 1-variable term is identified by one-half of the squares being marked. In the case of the 4-variable map, any allowable grouping of eight squares represents a 1-variable term. A 2-variable term is represented by a grouping of four squares, a 3-variable term by two squares, and a 4-variable term by one square. However, identification of allowable groups varies somewhat from that of the 2-variable and 3-variable configuration.

Adjacency Identification

Adjacency identification can be demonstrated by map simplification of the expression

$$AB + \overline{A} \cdot \overline{B} \cdot \overline{C} \cdot \overline{D} + \overline{A}BCD + \overline{BC}\overline{D} + \overline{B} \cdot \overline{C} \cdot D + \overline{A} \cdot \overline{D} + A\overline{B} + \overline{A}B\overline{C}D.$$

Once again, the logical sum of logical products form of the expression is entered on the map rather than conversion to full minterm form. The reader should verify each of the entries for the expression on the map (Fig. 6-14e). Should any difficulty arise in verifying the entries, the expression should be expanded to minterm form and separate maps used for each term.

Reading the map progresses in almost a straightforward manner. It is apparent that not all squares are filled, so the value of the expression is not logic 1. Since sixteen squares cannot be circled, eight (or one-half of the squares) should be examined. Figure 6-14a shows an identifiable group (A), Fig. 6-14b another (B), and Fig. 6-14c a third ($\overline{C}$). But another group of eight squares can be circled. Just as the 3-variable map was visualized as being inscribed on the surface of a cylinder, so the 4-variable map should be visualized as being inscribed on the surface of a *sphere.* Under these circumstances, the left-hand column

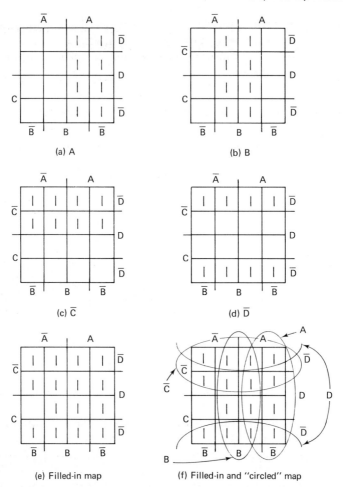

Figure 6-14 Map entries—$AB + \overline{A}\overline{B}\overline{C}\overline{D} + \overline{A}BCD + \overline{B}C\overline{D} + \overline{B}\overline{C}D + A\overline{D} + A\overline{B} + \overline{A}B\overline{C}\overline{D} = (A + B + \overline{C} + \overline{D})$

of squares is adjacent to the right-hand column of squares, and the top row of squares is adjacent to the bottom row of squares. Reference to Fig. 6-5c verifies that each square in the left-hand column differs from its adjacent square in the right-hand column by only one variable. The same is true for the top and bottom rows. Thus the eight squares comprising the top and bottom rows can be circled (Fig. 6-14d). The simplified expression then becomes A OR B OR $\overline{C}$ OR $\overline{D}$ and is shown in complete map form in Figs. 6-14e and f.

This expression also lends itself to simplification by using the *unplotted squares method* discussed in Sec. 6-4. The unplotted square is equivalent to $\overline{A} \cdot \overline{B} \cdot C \cdot D$ and is the *complement* of the mapped expression. If the mapped expression is equated to $\overline{X}$, then the unplotted square is equivalent to X. Complementing both sides to obtain X and applying DeMorgan's Law result in $X = \overline{\overline{A} \cdot \overline{B} \cdot C \cdot D}$ or $A + B + \overline{C} + \overline{D}$, which was the simplified expression obtained by circling all of the marked squares. Either method of map reading is equally valid. Just as in the 3-variable example, this operation is the same as taking the maxterm form directly from the map.

Simplification Starting with a Truth Table

Consider now a situation where the only information given is a truth table of an unknown logic function. As demonstrated in the previous chapter, the Boolean algebra expression may be read directly from the output column. For example, Table 6-4 shows all possible combinations of four variables with an output called E, which has some logical relationship to those four variables. Remember, the truth table output column is read to show all of the conditions of the variables that provide a true (1) output. Note that the expression is recovered from the truth table in logical sum of logical products form.

With the algebraic expression available, simplification can proceed in the usual manner, using either algebraic simplification or map simplification. This expression yields readily to map simplification. The 4-variable map is used to plot the minterms of the derived algebraic expression. Since minterms have been used, each term is plotted in only one square on the map. The seven terms in the expression thus occupy seven squares, as shown in Fig. 6-15a. Applying the concepts discussed earlier results in identification of two groups of four filled squares: one group represents the term $B\overline{C}$ (Fig. 6-15b), and the other group represents $\overline{A} \cdot \overline{D}$ (Fig. 6-15c). Note that the identification of the latter group depends on the visualization of the map inscribed on a sphere to obtain adjacent squares. The completed map appears in Fig. 6-15d.

6-6 SIMPLIFYING THE BINARY ADDER GRAPHICALLY

One practical application of simplification by map methods may be illustrated by simplifying the problem used in Ch. 5 to demonstrate arithmetic addition in the binary number system. Most of the concepts of map usage developed in this chapter will be included in this demonstration. The English-language statements for both outputs of the binary adder, plus the descriptive truth table and Boolean equations, were developed in Sec. 5-7. Final simplification is compared with Fig. 5-11 to determine the actual reduction in logic component requirements.

Table 6-4 Truth table of a 4-variable complex function

A	Ā	B	B̄	C	C̄	D	D̄	ĀB̄C̄D̄	ĀB̄CD̄	ĀBC̄D̄	ĀBC̄D	ABC̄D̄	ABC̄D	E*
0	1	0	1	0	1	0	1	1	0	0	0	0	0	1
0	1	0	1	0	1	1	0	0	0	0	0	0	0	0
0	1	0	1	1	0	0	1	0	1	0	0	0	0	1
0	1	0	1	1	0	1	0	0	0	0	0	0	0	0
0	1	1	0	0	1	0	1	0	0	1	0	0	0	1
0	1	1	0	0	1	1	0	0	0	0	1	0	0	1
0	1	1	0	1	0	0	1	0	0	0	0	0	0	0
0	1	1	0	1	0	1	0	0	0	0	0	0	0	0
1	0	0	1	0	1	0	1	0	0	0	0	0	0	0
1	0	0	1	0	1	1	0	0	0	0	0	0	0	0
1	0	0	1	1	0	0	1	0	0	0	0	0	0	0
1	0	0	1	1	0	1	0	0	0	0	0	0	0	0
1	0	1	0	0	1	0	1	0	0	0	0	1	0	1
1	0	1	0	0	1	1	0	0	0	0	0	0	1	1
1	0	1	0	1	0	0	1	0	0	0	0	0	0	0
1	0	1	0	1	0	1	0	0	0	0	0	0	0	0

* $E = \bar{A}\bar{B}\bar{C}\bar{D} + \bar{A}\bar{B}C\bar{D} + \bar{A}B\bar{C}\bar{D} + \bar{A}B\bar{C}D + AB\bar{C}\bar{D} + AB\bar{C}D$

Simplifying the Binary Adder Graphically

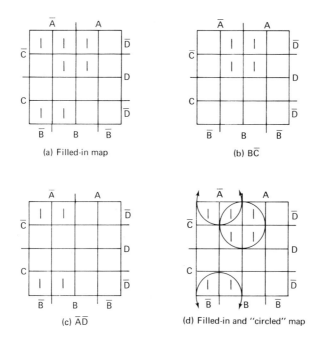

Figure 6-15 Map simplification of a 4-variable complex function (Table 6-4)

Simplification

The conventional approach to binary-adder map simplification is to treat each of the output expressions *individually,* i.e., develop two separate simplified logic circuits for each expression. Both the SUM and the OUTPUT CARRY expressions are in minterm form, as read from the truth table. Entering the SUM expression into the required 3-variable map (Fig. 6-16a) and applying the rules for looping adjacent squares reveals that no real simplification is attained by using the map method. No two marked squares are adjacent, which means that none of the terms may be simplified. Thus the original expression for the SUM output of the binary adder does not yield to simplification using the map method in the conventional manner.

Entering the OUTPUT CARRY expression into the 3-variable map, however, shows some promise. The four marked squares in Fig. 6-16b can be looped (Fig. 6-16c), resulting in three 2-variable terms instead of four 3-variable terms. The two equations for the full adder are now

$$S = \overline{A} \cdot \overline{B} \cdot C + \overline{A} \cdot B \cdot \overline{C} + A \cdot \overline{B} \cdot \overline{C} + A \cdot B \cdot C$$
$$C_o = A \cdot B + A \cdot C + B \cdot C$$

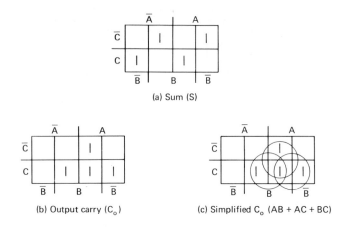

Figure 6-16 Full adder-map simplification

The logic diagram resulting from direct map simplification shows a reduction of one 3-input AND gate and the replacement of three 3-input AND gates with three 2-input AND gates. From a practical standpoint, the change from 3-input AND gates to 2-input AND gates would probably be disregarded (unless 2-input AND gates were commonly available in the logic system being used), since 3-input AND gates are fairly standard.

Another approach that sometimes yields a reduction in logic-element count is to look for *common squares* on the maps of the separate outputs. The binary adder is typical of many combinational logic circuits—it has common inputs but separate outputs. By placing the maps of the separate outputs in proximity to each other, common squares may be identified. If the same square is marked on each map, it means that the term represented by those squares is required in both output expressions. One gate can be used to form the common term, and the rest of the terms developed by additional gates can form the separate outputs. Although significant simplification cannot be accomplished this way by the binary adder, the small amount of simplification that does result is a good demonstration of the technique. Comparison of the SUM expression map of Fig. 6-17a and the CARRY OUTPUT expression map of Fig. 6-17b reveals that one square is common to both maps. The term ABC is represented by the common square, and this is required in both of the output expressions. Thus logic-element count is reduced by one AND gate by this method.

From a strict logic-element count, hardware implementation of this simplified binary adder does not appear any more advantageous than the previously discussed simplification. Where many binary adders

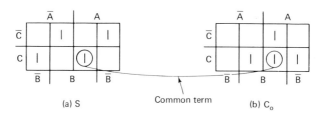

Figure 6-17 Common square in the SUM and CARRY outputs

are used, however, some reduction in physical wiring and interconnection requirements may be realized. This is really a design problem, but it does point out one reason why there may be different implementations of the simple binary adder function.

Numerous other configurations of binary adders exist. Both algebraic and map methods of simplification are used to determine these configurations, but the techniques are generally beyond the scope of this text. Further discussion tends to complicate and detract from the real purpose of discussing map methods of simplification—applying the concepts of maps to analysis of digital logic circuits. As more complex logic circuits are encountered, the fundamentals discussed in this chapter will become increasingly important. The ability to manipulate Boolean algebra expressions into different forms either algebraically or by the use of maps is of great value in determining operation of logic circuits and circuit equivalency.

6-7 COMPARISON OF SIMPLIFICATION METHODS

Many of the commonly encountered combinational logic simplification situations have been discussed in this chapter. Of course, variations of all these methods, plus some expansion of most of the methods to more complex problems, will be encountered in logic engineering tasks. For the digital logic analyst, though, a sufficient variety has been presented. The map methods, in conjunction with the algebraic simplification methods of earlier chapters, should handle practically all the logic analysis problems that will be encountered.

The reader has undoubtedly noted that work in both map and algebraic simplification has much in common. A good percentage of the expressions requiring simplification will not be in the required logical sum of logical products form needed for map simplification. Algebraic methods are required to obtain the required form. Until considerable experience is gained, it is good practice to convert all expressions to be simplified to the logical sum of logical products form before making the decision about which simplification method to employ. If the logical sum of logical products form of the expression contains many more than four terms and two variables, the map method will usually yield a faster answer. (We should also mention that the map method may yield more than one simplified answer.) At this point, the logic analyst will have to depend on ingenuity and insight to determine which of the multiple answers (if any) is the one that will best simplify in accordance with the definitions used for the system being analyzed.

If more than four variables are contained in the expression, both the map method and the algebraic method become exceedingly complex. The expression should be separated into two or more smaller expressions; each of those expressions should be simplified; and the results should then be combined either algebraically or with maps to yield an overall simplification.

The logic analyst should also be on the lookout for expressions that do not have to be converted to logical sum of logical product form to provide a simplified expression. This type of expression will usually be encountered when working with multiple applications of DeMorgan's Law, as demonstrated in Sec. 5-4.

Devising a fixed set of rules to cover all logic analysis cases is out of the question. The general guidelines and examples shown in Chapters 5 and 6 merely prepare the reader for the more complex work to follow. Each analyst must choose the method that best suits application needs. Hopefully, sufficient information has now been provided to enable the reader to make such a decision.

6-8 SUMMARY

Entering Information on Logic Maps*

1. Convert the Boolean algebra expression to AND-OR (logical sum of logical products) or minterm form.
2. Plot one term at a time, according to the following rules.
    (a) Determine the column(s) according to the column headings.
    (b) Determine the row(s) according to the row headings.
    (c) Place 1s in all squares covered by the intersection of the selected row(s) and column(s). When more than

*Figure 6-5 shows how to construct 2 through 4-variable maps.

one term generates a 1 in the same square, a single 1 is sufficient. This is equivalent to the logic operation $1 + 1 = 1$.
3. The number of squares to be marked for any term is determined by the relationship $X = 1/2^n$, where X is the number of squares to be marked and n is the number of variables in the term. For example,

1-variable term	1/2 of the squares
2-variable term	1/4 of the squares
3-variable term	1/8 of the squares

Extracting Information from Logic Maps

1. Look for groups of squares that are adjacent to each other.
   (a) An adjacent square is one which differs from another square by only one variable.
   (b) Squares may be considered adjacent under the following conditions.
      (1) They are physically next to each other in the same row or column.
      (2) They are at opposite ends of rows or columns.
      (3) They occupy identical positions in adjacent maps.
2. Adjacent squares may be grouped in integral powers of 2; that is, 1, 2, 4, 8 . . . . Groups should have as many squares included together as possible, as long as they are in integral powers of 2. Circles or loops are usually drawn around the group for identification purposes.
3. The number of variables in each term can be determined by the relationship shown in the rules for entering information on logic maps.
4. A marked square may be used in as many loops as needed.
5. Combine all of the extracted terms disjunctively (logic OR) to form the simplified expression.
6. When unmarked squares consist of only one term, those squares may be grouped and considered as the complement of the expression.
7. Logic maps can be used to simplify multioutput logic expressions where more than one output is developed from common inputs. The maps for each of the output expressions are compared. If the same squares are marked, the terms are common and can be used by all output expressions.

PROBLEM APPLICATIONS

6-1. What is the basic reason for performing logic simplification?
6-2. How does a Venn diagram differ from a Veitch diagram?
6-3. What is the major application of Venn diagrams?
6-4. Construct a Venn diagram representing $\overline{A}B$.
6-5. What was the original intent of the Veitch diagram?
6-6. Construct a Veitch diagram representing $\overline{A} + BC$.
6-7. How does a Veitch diagram differ from a Karnaugh map?
6-8. What was the original intent of the Karnaugh map?
6-9. Construct a Karnaugh map representing $\overline{A} + BC$.
6-10. What number system code is used to label divisions of a Karnaugh map? Discuss the characteristics of the code.
6-11. Define "sum of products" form as applied to Boolean expressions.
6-12. Define "minterm" form as applied to Boolean expressions.
6-13. Define "product of sums" form as applied to Boolean expressions.
6-14. Define "maxterm" form as applied to Boolean expressions.
6-15. What form must a Boolean expression possess to be entered on a logic map?

Graphically simplify the following Boolean expressions.

6-16. $B\overline{C} + ABC$
6-17. $AB + \overline{AB} + B\overline{C} + \overline{B}C$
6-18. $ABC + \overline{A}BC + \overline{B} \cdot \overline{C}$
6-19. $\overline{A}C + B\overline{C} + \overline{B} \cdot \overline{C}$
6-20. $AB + \overline{A} \cdot \overline{B} + B\overline{C} + AC + \overline{B} \cdot \overline{C}$
6-21. $ABCD + ACD + AD$
6-22. $\overline{A}B\overline{C}D + \overline{A}BCD + AB\overline{C}D + ABCD$
6-23. $W\overline{Z} + W\overline{Y}Z + WXY + \overline{W} \cdot \overline{X} \cdot \overline{Y} \cdot Z$
6-24. $[W(X\overline{Z} + YZ) + (\overline{W} + \overline{X})YZ][X + Y + WZ + \overline{Z}(\overline{X}Y + \overline{W}X) + WX\overline{Y}Z]$

7

ANALYSIS AND TROUBLESHOOTING OF COMBINATIONAL LOGIC

Analysis and troubleshooting of combinational logic circuits follows a relatively *standard sequence of events.* This chapter provides a proposed procedure, listing and explaining each preparatory and procedural step that leads to a complete understanding of combinational logic-circuit operation.

Troubleshooting is also organized into a procedural operation. Following discussion of the systems concept and troubleshooting philosophy, the flowchart concept is used to develop a step-by-step method of troubleshooting combinational logic circuits.

7-1 A PROPOSED ANALYSIS PROCEDURE

Preparatory Steps

Any investigation of combinational logic-circuit operation must begin with an overall description of the logic function(s) of the circuit being analyzed. The description may be in any of the commonly encountered forms: English-language statements, Boolean algebra, or waveforms. From a practical standpoint, either the Boolean algebra or the waveform description is the most desirable, since analysis is usually performed either algebraically or with waveforms. If English-language operational descriptions are provided, conversion to the Boolean algebra form *must* be performed.

A normal "fallout" of the logic description of a combinational logic circuit is *a definition of the inputs and outputs.* In other words, if the operation of a logic circuit has been described from an overall viewpoint, it will usually have been described in terms of outputs for various inputs. If such is not the case, all inputs should be provided with arbitrary names so that the rules of Boolean algebra can be applied during investigation. In any case, it should be possible with either the information at hand or that arbitrarily provided to describe all outputs in terms of inputs following logic-circuit analysis. Some logic circuits also operate with *timed inputs,* i.e., they are enabled to perform their function for a given period of time and then disabled until needed again. If such timing is used in the circuit being evaluated, it should be described either in terms of Boolean algebra or waveforms, depending on the method(s) of analysis being used.

Finally, any nonstandard logic symbols or operations being performed should be carefully defined.* Most manufacturers provide Boolean algebra descriptions of their logic packages; so as new logic operations become available, it will be necessary to apply only the provided descriptions to understand the operation of the logic element.

In summary: *certain basic steps must be performed prior to the start of any combinational logic-circuit investigation.* The logic description of the overall function should be obtained, along with definitions of inputs and outputs if possible. Where inputs and outputs are not adequately defined, arbitrary naming is allowed. Any necessary timing should be established, even if it is necessary to once again arbitrarily assign values for the purpose of analysis. Finally, any nonstandard logic symbols must be defined in terms of MIL-STD-806B symbols to facilitate understanding.

Procedural Steps

Combinational logic analysis usually begins at the *input portion of the logic circuit* and proceeds to the output(s). If the preparatory steps have been adequately performed, all inputs and timing will have been identified or synthesized, and investigation can begin with the selection of *one arbitrary path of signal flow* from input toward the output. The selected path should be followed until a level of logic is reached that requires other inputs. All of the tools for analysis that have been discussed in previous chapters should be applied where needed during the evaluation of the combinational logic circuit. Boolean algebra descriptions are commonly used in the case of simple combinational logic

---

*Unless otherwise stated, it will be assumed that all symbols used in logic diagrams are from MIL-STD-806B. Nonstandard symbols are described in Appendix C, where they are correlated with MIL-STD-806B.

circuits, but other methods are equally applicable. Waveforms often prove invaluable by making the actual circuit operation more "visible." Truth tables and maps may also be used to help understand circuit operation.

When the level of logic is reached where additional inputs are needed, we must return to the beginning and develop the required inputs. This process continues—new inputs being developed as required—until *all* inputs and outputs are accounted for. As might be imagined, some combinational logic-circuit analysis problems can become quite complex. It is therefore quite necessary to adequately document each step being performed to reduce errors caused by complex circuit configurations.

The complete procedure is now outlined.

A Proposed Analysis Procedure
I. *Preparatory steps*
   A. Obtain description of overall logic function.
   B. Define or synthesize inputs and outputs.
   C. Establish or synthesize timing if necessary.
   D. Define nonstandard rules of logic.
II. *Procedural steps*
   A. Perform preparatory steps.
   B. Select an arbitrary path from input to output.
   C. Using the tools of logic analysis, proceed from the input toward the outputs until a level of logic is reached that requires other inputs.
   D. Return to the beginning and develop those inputs.
   E. Continue, developing new inputs as required until all outputs are obtained.
   F. Document results of each step as developed.

7-2 A TYPICAL ANALYSIS

Our proposed procedure can be demonstrated by a typical problem. Figure 7-1a is the logic diagram of a device that is to be analyzed. Note that the diagram consists of standard logic symbols and is made up completely of NAND gates. The only other information available is that the logic circuit performs the full-adder function.

Preparatory Steps

With the logic diagram and its functional description available, the preparatory steps of the procedure may be accomplished. The knowledge that the logic circuit is a *full adder* is of considerable aid in developing the complete description of the overall logic function. A small

Figure 7-1 Boolean algebra description and truth table (full-adder)

amount of analysis has already been performed on the full adder, and advantage should be taken of the information gained in earlier chapters. Both the Boolean algebra description and the truth table form of the full adder are reproduced in Fig. 7-1b. Derivation of the truth table and Boolean expressions from the basic English-language description were previously discussed.

Waveform analysis is attempted following algebraic and operational analysis using the binary inputs. Logic circuit inputs and outputs are defined by the Boolean algebra description. However it may not be immediately obvious which output is the SUM and which is the CARRY

A Typical Analysis   **185**

OUT. Additionally, the proper labels to be applied to each input may be a mystery. Temporarily, the inputs are arbitrarily named X, Y, and Z, and the outputs are named U and V. Upon completion of the procedural steps, the proper assignment of variables will be apparent. No timing requirements are indicated on the logic diagram, since the required number of inputs for a full adder (3) are the only inputs shown. Therefore it is not necessary to establish or synthesize circuit timing.

Since standard logic symbology prevails in this logic diagram, no additional rules of logic are required. The logic circuit is now ready for the procedural steps.

Procedural Steps

Starting at the left side of the diagram (where the inputs are usually located), the first gate to be encountered is G_1. The inputs to G_1 are X and Y, and since G_1 is a NAND gate, the output is $\overline{XY}$. The output of G_1 is routed to both G_2 and G_3, so a purely arbitrary direction of analysis is assumed. If the output U is to be developed first, then G_2 may be examined next. With inputs X, Z, and $\overline{XY}$ applied, the output of NAND gate G_2 is $\overline{(X)(Z)(\overline{XY})}$. G_8 is the next gate encountered. Two inputs for G_8 have already been determined (from G_1 and G_2), so it is necessary to determine only the third input, furnished by G_3.

A level of logic has now been reached that requires returning to a lower (previous) level to obtain other inputs. Evaluation of G_3 shows that the three required inputs are Y, Z, and the output of G_1, which is $\overline{XY}$. G_3's output, then, is $\overline{(Y)(Z)(\overline{XY})}$. The three inputs to G_8 are therefore $\overline{XY}$, $\overline{(X)(Z)(\overline{XY})}$, and $\overline{(Y)(Z)(\overline{XY})}$. Combining these three inputs in a NAND operation results in $\overline{[\overline{XY}][\overline{(X)(Z)(\overline{XY})}][\overline{(Y)(Z)(\overline{XY})}]}$, which is the output of G_8 (labeled U in the logic diagram). But the expression for U does not look anything like *either* of the full-adder outputs shown in Fig. 7-1. The time has come to apply some of the tools of logic analysis to determine which of the full-adder outputs is being represented by U.

The expression for the output of G_8 is by no means in its simplest form. In fact, without considerable experience, the real meaning of the expression may not be apparent. To determine which (if either) of the full-adder outputs is represented by the output of G_8, we must convert the NAND form to another form, perhaps the AND-OR (logical sum of logical products) form.

One of the first things that must be done when converting from the NAND to the AND-OR form is to remove the *vincula* (using DeMorgan's Law) *from all but single variable terms*. Expansion to the AND-OR form is then accomplished. *One* of the many methods of algebraic simplification is shown here.

$$[\overline{XY}][\overline{(X)(Z)(\overline{XY})}][\overline{(Y)(Z)(\overline{XY})}] = \overline{[\overline{XY}]} + \overline{[(X)(Z)(\overline{XY})]} + \overline{[(Y)(Z)(\overline{XY})]}$$
$$= XY + [(X)(Z)(\overline{XY})] + [(Y)(Z)(\overline{XY})]$$
$$= XY + [(X)(Z)(\overline{X}+\overline{Y})] + [(Y)(Z)(\overline{X}+\overline{Y})]$$
$$= XY + XZ\overline{X} + XZ\overline{Y} + YZ\overline{X} + YZ\overline{Y}$$
$$= XY + X\overline{Y}Z + \overline{X}YZ$$

The expression is now in the AND-OR form, but it still doesn't seem to resemble either of the full-adder outputs. Further algebraic simplification could be accomplished, but this is a good place to show a practical application of logic maps. The three terms of the simplified expression are shown entered and grouped in the 3-variable map in Fig. 7-2. Evaluation of the three loops results in the expression $XY + YZ + XZ$, which is the *exact* equivalent of the CARRY OUT output of the full adder. Thus the orderly application of the proposed procedure and use of the common tools available for logic-circuit analysis has resulted in identification of the function of a circuit that showed no resemblance to previously encountered circuits. Documentation should include the *outputs of all gates encountered* in addition to the simplification effort.

Having gained the knowledge that U is the CARRY OUT output removes some of the mystery about the logic circuit. Output V must be the SUM output, so all that remains is to check whether the logic circuit provides the correct output. If the Boolean algebra expression for the logic circuit matches the Boolean expression for the full adder function, then the complete circuit will have been analyzed.

Three inputs are furnished to the output gate (G_7): G_4 provides one input; G_5 provides the second input; and G_6 furnishes the final input. Examining G_4 first shows that the three inputs are X, $[\overline{(X)(Z)(\overline{XY})}]$, and $\overline{XY}$. Combining these inputs in the NAND gate results in $\overline{[X] [\overline{(X)(Z)(\overline{XY})}] [\overline{XY}]}$ as the output expression for G_4. Gates G_5 and G_6 are approached similarly, providing a G_5 output of $\overline{[Z] [\overline{(X)(Z)(\overline{XY})}] [\overline{(Y)(Z)(\overline{XY})}]}$
and a G_6 output of $\overline{[Y] [\overline{(Y)(Z)(\overline{XY})}] [\overline{XY}]}$. The result of combining the G_4, G_5, and G_6 outputs in G_7 is the somewhat awesome expression

$$\overline{\{\overline{[X][\overline{(X)(Z)(\overline{XY})}][\overline{XY}]}\}\{\overline{[Z][\overline{(X)(Z)(\overline{XY})}][\overline{(Y)(Z)(\overline{XY})}]}\}\{\overline{[Y][\overline{(Y)(Z)(\overline{XY})}][\overline{XY}]}\}}$$

that once again in no way resembles what is known to be the SUM output of a full adder! Both algebraic and map methods are used to determine the equivalent logic expression that will be compared with the full-adder expression.

Since any expression that is in simplified form has all vincula removed, except those over single variables, we may appropriately apply

A Typical Analysis

Figure 7-2 Map simplification of $X Y + X \bar{Y} Z + \bar{X} Y Z = X Y + Y Z + X Z$

DeMorgan's Law until this criterion is met. The first step splits the full-length vinculum at major divisions in the G_7 output expression, resulting in

$$\{\overline{[X][\overline{(X)(Z)(\overline{XY})}][\overline{XY}]}\} + \{\overline{[Z][\overline{(X)(Z)(\overline{XY})}][\overline{(Y)(Z)(\overline{XY})}]}\} + \{\overline{[Y][\overline{(Y)(Z)(\overline{XY})}][\overline{XY}]}\}.$$

Note that the expression is now composed of three separate subexpressions that are ORed together—each subexpression topped by double vincula. It is known that double negation (defined by the double vincula) is equivalent to the original expression, so all double vincula are removed. The expression simplifies to

$$\{[X][\overline{(X)(Z)(\overline{XY})}][\overline{XY}]\} + \{[Z][\overline{(X)(Z)(\overline{XY})}][\overline{(Y)(Z)(\overline{XY})}]\} + \{[Y][\overline{(Y)(Z)(\overline{XY})}][\overline{XY}]\}.$$

Due to the complexity of the total expression, each of the three terms is simplified separately, and the simplified versions are recombined to form the new expression.

The left subexpression is attacked first. DeMorgan's Law removes the vincula from the center and last factor, resulting in simplification to $[[X][\bar{X} + \bar{Z} + \overline{(\overline{XY})}][\bar{X} + \bar{Y}]$. Of course, the double vincula over the term XY may be removed to reach our first goal of having no vincula extending over more than a single variable. Perhaps, however, rearrangement of this expression and regrouping to $[X][\bar{X} + \bar{Y}][\bar{X} + \bar{Z} + XY]$ will make it easier to visualize the expansion steps. The first two factors (X and $\bar{X} + \bar{Y}$) are expanded to $X\bar{X} + X\bar{Y}$ and are then further expanded using the $[\bar{X} + \bar{Z} + XY]$ factor. $X\bar{X}$ cancels, and combining $X\bar{Y}$ with $\bar{X} + \bar{Z} + XY$ makes the AND-OR form ($X\bar{Y}\bar{X} + X\bar{Y}\bar{Z} + X\bar{Y}XY$) available for any further simplification.

Examination of the AND-OR form shows that, in this case, algebraic simplification is relatively simple. The first term is rewritten in the form $\bar{X}X\bar{Y} = 0$, simplifying the term to $0 \cdot \bar{Y}$. The theorems of Boolean algebra state that conjunctively combining any variable with 0 results in a logic value of 0. Therefore the first term simplifies to 0. The last term in the AND-OR form of the expression is $X\bar{Y}XY$. This term is rearranged to the form $XX Y\bar{Y}$, which may be easier to understand.

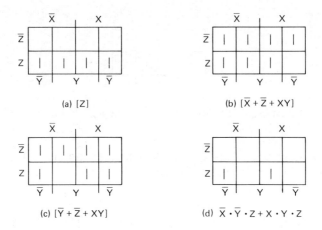

Figure 7-3 Logic maps for $[Z][\overline{X} + \overline{Z} + XY][\overline{Y} + \overline{Z} + XY]$

$XX = X$, of course, and $Y\overline{Y} = 0$. $X \cdot 0 = 0$, so the last term in the AND-OR form is 0. Now the AND-OR form of the expression is $0 + X\overline{Y}\overline{Z} + 0$ or, simply, $X\overline{Y}\overline{Z}$. Quite a reduction in complexity from the original $[X][\overline{(X)(Z)(\overline{XY})}][\overline{XY}]$ isn't it?

Close persual of the right-hand subexpression in the original expression reveals that it is of the same form as the left-hand subexpression. It will also simplify to a single 3-variable term $(\overline{X}YZ)$.

The center subexpression does not lend itself quite as well to algebraic manipulation. Two of the three terms are complex, and expansion can become troublesome. Following removal of the vincula by DeMorgan's Law, the center subexpression is further simplified by the use of maps. The map method is not an absolute requirement in this case, but it does serve as an excellent example of map ANDing. The center term $([Z][\overline{(X)(Z)(\overline{XY})}][\overline{(Y)(Z)(\overline{XY})}])$ becomes $[Z][\overline{X} + \overline{Z} + (XY)]$ $[\overline{Y} + \overline{Z} + (XY)]$, after application of DeMorgan's Law and the double negation theorem. Each of the three terms is mapped separately, and the maps are ANDed. Z is shown mapped in Fig. 7-3a; $\overline{X} + \overline{Z} + XY$ in Fig. 7-3b; and $\overline{Y} + \overline{Z} + XY$ in Fig. 7-3c. Superimposing the maps reveals that only two common squares are marked on all three maps. These two squares, which are identified as $\overline{X}\overline{Y}Z$ and XYZ, represent the ANDing of the three terms in the center subexpression (Fig. 7-3d). Therefore

$$[Z][\overline{(X)(Z)(\overline{XY})}] \, [\overline{(Y)(Z)(\overline{XY})}] = \overline{X}\overline{Y}Z + XYZ.$$

All of the subexpressions have now been examined, and simplified forms have been developed. Substitution of the simplified forms for

A Typical Analysis

the original subexpressions results in $X\overline{Y}\overline{Z} + \overline{X}YZ + XYZ + \overline{X}Y\overline{Z}$ as the equivalent expression for the output of G_7. Comparison of this expression with the SUM output expression for the full adder in Fig. 7-1 shows that the two expressions are the same *except* for the arrangement of terms. Both the SUM and the CARRY OUT outputs of the NAND gate full adder have been shown to be equivalent to the outputs developed from the standard truth table for the full adder. Thus the circuit being studied is indeed capable of performing the functions of a full adder, and except for documentation of the results, the analysis is complete. Figure 7-4 shows typical documentation of a logic-analysis problem using the full-adder circuit and methods of analysis just completed.

Another possible method of investigating a logic circuit, such as the full-adder, is to try all possible combinations of the inputs and then record the output for each of the conditions tried. This has effectively been done by constructing the truth table, but it may be enlightening to actually try the eight combinations that can exist for the 3-input full-adder and determine if the truth table is really a valid presentation of full-adder operation. The use of a logic-signal source to furnish the input signals is one way to implement such a test. However the logic levels defined by the truth table are assumed to be present, and the logic levels are followed throughout the circuit to determine the effect of each combination. Certain input combinations will produce an almost obvious output. Other combinations will require tracing logic levels through the complete circuit to establish output conditions.

Prior to beginning this investigation, therefore, a review of the operation of the logic element concerned would be helpful. Remember that the NAND gate functions as an AND gate followed by an inverter. Thus when all inputs to the NAND gate are HIGH, the output is LOW. Another viewpoint is that if any input is LOW, the output is HIGH. Both viewpoints will prove valuable while tracing logic levels through the full-adder. It should also be recalled that logic 1 is often used in place of the term HIGH, and logic 0 is used in place of the term LOW. In a device such as the full-adder, it is advantageous to use logic 1s and 0s in place of HIGHs and LOWs, due to the correlation with actual binary numbers that are processed in the adder. So 1s and 0s are used here, although the reader may substitute HIGHs and LOWs if it makes the circuit more understandable.

As with any task, it is best to organize and plan what is to be done prior to actually starting. So much is already known about the full adder that little preparation (other than determining how to record the data) is necessary. Since the results of this analysis will be closely allied with the truth table for the full-adder function, we can record data in tabular form. Therefore the usual physical layout of the truth table is

FULL ADDER ANALYSIS

1. Logic diagram (see Figure 7–1a)
2. Gate output equations:

   $G_1 = \overline{XY}$
   $G_2 = \overline{(X)(Z)(\overline{XY})}$
   $G_3 = \overline{(Y)(Z)(\overline{XY})}$
   $G_4 = \overline{[X]\;[\overline{(X)(Z)(\overline{XY})}]\;[\overline{XY}]}$
   $G_5 = \overline{[Z]\;[\overline{(X)(Z)(\overline{XY})}]\;[\overline{(Y)(Z)(\overline{XY})}]}$
   $G_6 = \overline{[Y]\;[\overline{(Y)(Z)(\overline{XY})}]\;[\overline{XY}]}$
   $G_7 = \overline{\left\{\overline{[X]\;[\overline{(X)(Z)(\overline{XY})}]\;[\overline{XY}]}\right\}\;\left\{\overline{[Z]\;[\overline{(X)(Z)(\overline{XY})}]\;[\overline{(Y)(Z)(\overline{XY})}]}\right\}\;\left\{\overline{[Y]\;[\overline{(Y)(Z)(\overline{XY})}]\;[\overline{XY}]}\right\}}$
   $G_8 = \overline{[\overline{XY}]\;[\overline{(X)(Z)(\overline{XY})}]\;[\overline{(Y)(Z)(\overline{XY})}]}$

3. Algebraic simplification:

   $G_8 = \overline{[\overline{XY}]\;[\overline{(X)(Z)(\overline{XY})}]\;[\overline{(Y)(Z)(\overline{XY})}]}$
   $ = [\overline{\overline{XY}}] + [(X)(Z)(\overline{XY})] + [(Y)(Z)(\overline{XY})]$
   $ = [XY] + [(X)(Z)(\overline{XY})] + [(Y)(Z)(\overline{XY})]$
   $ = [XY] + [(XZ)(\overline{X}+\overline{Y})] + [(YZ)(\overline{X}+\overline{Y})]$
   $ = [XY] + [X\overline{X}Z + X\overline{Y}Z] + [\overline{X}YZ + Y\overline{Y}Z]$
   $ = XY + X\overline{X}Z + X\overline{Y}Z + \overline{X}YZ + Y\overline{Y}Z$
   $ = XY + X\overline{Y}Z + \overline{X}YZ$

   $G_7 = \overline{\left\{\overline{[X][\overline{(X)(Z)(\overline{XY})}][\overline{XY}]}\right\}\left\{\overline{[Z][\overline{(X)(Z)(\overline{XY})}][\overline{(Y)(Z)(\overline{XY})}]}\right\}\left\{\overline{[Y][\overline{(Y)(Z)(\overline{XY})}][\overline{XY}]}\right\}}$
   $ = \left\{[X][\overline{(X)(Z)(\overline{XY})}][\overline{XY}]\right\} + \left\{[Z][\overline{(X)(Z)(\overline{XY})}][\overline{(Y)(Z)(\overline{XY})}]\right\} + \left\{[Y][\overline{(Y)(Z)(\overline{XY})}][\overline{XY}]\right\}$
   $ = \left\{[X][\overline{(X)(Z)(\overline{XY})}][\overline{XY}]\right\} + \left\{[Z][\overline{(X)(Z)(\overline{XY})}][\overline{(Y)(Z)(\overline{XY})}]\right\} + \left\{[Y][\overline{(Y)(Z)(\overline{XY})}][\overline{XY}]\right\}$

   $ = \{[X][\overline{X}+\overline{Z}+\overline{\overline{XY}}][\overline{XY}]\} + \{[Z][\overline{X}+\overline{Z}+XY][\overline{Y}+\overline{Z}+\overline{\overline{XY}}]\} + \{[Y][\overline{Y}+\overline{Z}+\overline{\overline{XY}}][\overline{XY}]\}$
   $ = \{[X][\overline{X}+\overline{Z}+XY][\overline{XY}]\} + \{[Z][\overline{X}+\overline{Z}+XY][\overline{Y}+\overline{Z}+XY]\} + \{[Y][\overline{Y}+\overline{Z}+XY][\overline{XY}]\}$
   $ = \{[X][\overline{X}+\overline{Y}][\overline{X}+\overline{Z}+XY]\} + \{[\overline{X}Z+Z\overline{Z}+XYZ][\overline{Y}+\overline{Z}+XY]\} + \{[Y][\overline{X}+\overline{Y}][\overline{Y}+\overline{Z}+XY]\}$
   $ = \{[X\overline{X}+X\overline{Y}][\overline{X}+\overline{Z}+XY]\} + \{[\overline{X}Z+XYZ][\overline{Y}+\overline{Z}+XY]\} + \{[\overline{X}Y+Y\overline{Y}][\overline{Y}+\overline{Z}+XY]\}$
   $ = \{[X\overline{Y}][\overline{X}+\overline{Z}+XY]\} + \{\overline{X}\overline{Y}Z + \overline{X}Z\overline{Z} + \overline{X}ZXY + XY\overline{Y}Z + XYZ\overline{Z} + XYZXY\} + \{[\overline{X}Y][\overline{Y}+\overline{Z}+XY]\}$
   $ = \{[X\overline{X}Y + X\overline{Y}\overline{Z} + X\overline{Y}XY]\} + \{[\overline{X}\overline{Y}Z + XYZ]\} + \{[\overline{X}Y\overline{Y} + \overline{X}Y\overline{Z} + \overline{X}YXY]\}$
   $ = X\overline{Y}Z + \overline{X}\overline{Y}Z + XYZ + \overline{X}Y\overline{Z}$

   $G_7 = X\overline{Y}Z + \overline{X}\overline{Y}Z + XYZ + \overline{X}Y\overline{Z}$

4. Map Simplification

   For the G_7 output there is no need to map the first and last terms.
   Simplifications are immediately apparent as soon as the logical sum of logical products form is obtained.
   The G_7 middle term is mapped in Figure 7–3.
   Figure 7–2 maps the G_8 expression.

Figure 7-4 Documentation of a logic circuit analysis

used, and the same sequence of investigating inputs is employed. Instead of providing output columns for only the SUM and CARRY OUT outputs, *each* gate is shown. As each combination of inputs is investigated, it is necessary to place only the proper 1 or 0 in the column for each gate. When the analysis is completed, the table contains the output of each gate for all possible combinations of the three input vari-

TABLE 7-1 Complete truth table for full adder of Figure 7-1a

Row	Z	Y	X	G_1	G_2	G_3	G_8	G_4	G_5	G_6	G_7
0	0	0	0	1	1	1	0	1	1	1	0
1	0	0	1	1	1	1	0	0	1	1	1
2	0	1	0	1	1	1	0	1	1	0	1
3	0	1	1	0	1	1	1	1	1	1	0
4	1	0	0	1	1	1	0	1	0	1	1
5	1	0	1	1	0	1	1	1	1	1	0
6	1	1	0	1	1	0	1	1	1	1	0
7	1	1	1	0	1	1	1	1	0	1	1

ables. The completed truth table for "1 and 0" analysis of the full adder is shown in Table 7-1.

Just as with the algebraic/graphic investigation, the starting point is the inputs, and each gate is discussed as encountered. By examining the circuit and from our previous analysis, we see that both outputs require signals from G_1, G_2, and G_3. Although it does not result in *immediate* insight into the circuit operation, this slight deviation from the procedure of analyzing gates that are not in a specific path from input to output will speed the analysis process. G_1 is examined first—not only because it is the first gate with inputs applied but also because it is probably the simplest to understand. Only two inputs, X and Y, are provided. By definition, only when all inputs are 1s (HIGH) will the output be 0 (LOW). When either input is 0, the output is 1. Using the X and Y input columns in the table, we see that both X and Y are 1 at the same time only in rows 3 and 7—all other combinations of X and Y have at least one 0 present. The output of G_1 is 0 at rows 3 and 7; all other rows result in 1 outputs.

G_2 is studied next: it is a 3-input NAND gate, and it requires X to be 1, Z to be 1, and the output of G_1 to be 1 *simultaneously* to be activated. Z is 0 in rows 0 through 3, so obviously the output of G_2 will be 1 for at least these four combinations—no matter what the other variables are. G_1 is 0 at row 7, so G_2 is marked 1 for that row. Rows 4, 5, and 6 remain to be checked. X is 0 in rows 4 and 6, leaving only row 5. Careful perusal of row 5 shows that X, Z, and G_1 outputs are all 1 in this case. The output of G_2 will then be 0, which is desired.

Analysis of G_3 proceeds in a similar manner. It also is a 3-input NAND gate requiring the output of G_1 to be 1, Y to be 1, and Z to be 1 simultaneously to activate the gate. Rows 0 through 3 provide a 0 input from Z with the resultant 1 output from G_3. Y is 0 in rows 4 and 5, while still keeping the output at 1. In row 6, Y is 1, Z is 1, and according to the truth table, the output of G_1 is also 1. G_3 is activated under these circumstances, and its output becomes 0. Row 7 shows that the output of G_1 is 0, so the G_3 output becomes 1 again. All three gates common to both the SUM and the CARRY OUT outputs have now been checked, and their logic values determined for all possible combinations of inputs. G_1 through G_8 may now be evaluated in a relatively simple manner without numerous trips back to develop new inputs.

The CARRY OUT output from G_8 is checked now that G_1, G_2, and G_3 outputs have been determined. Evaluation of the G_1, G_2, and G_3 columns (one row at a time) allows selection of the proper entry in the G_8 column. Perhaps an easier method is to look for 0s in either the G_1, G_2, or G_3 columns for each row. Any 0 input to a NAND gate results in a 1 output, and the 0s tend to "stand out" from the 1s, due to their relative scarcity in the table. Just locate a 0 in any of the three columns representing inputs to G_8, and the row that contains the 0 has a 1 placed in the G_8 column. This effectively recognizes that DeMorgan's Law works, since $\overline{ABC} = \overline{A} + \overline{B} + \overline{C}$. After all eight rows have been examined and the G_8 outputs recorded, the full-adder truth table of Fig. 7-1b should be consulted and the CARRY OUT column compared with the G_8 column. One-to-one correspondence between the two tables should be observed, indicating once again that the NAND version of the full adder performs the same CARRY OUT function as other versions previously used.

The SUM output is developed from G_7. With inputs supplied from G_4, G_5, and G_6, we must now develop the outputs of each of these gates. There is no particular advantage to selecting a specific gate to be examined, so G_4 is arbitrarily evaluated first. G_4's inputs are X, G_1, and G_2. (Remember, any 0 input results in a 1 output.) Rows 0, 2, 4, and 6 contain 0s in the X column, so the corresponding locations in the G_4 column contain 1s. The output of G_1 is 0 in rows 3 and 7, and the output of G_2 is 0 in row 5. Ones are entered in the same locations in the G_4 column. Row 1 is the only location left without an entry, and evaluation of the X, G_1, and G_2 columns indicates that they all have a value of 1. Therefore the row 1 location in column G_4 has a 0 inserted.

G_6 is approached in a similar manner. In this case, Y is combined with G_1 and G_3. The result is but one case of the eight possible input combinations where the output is 0 (row 2). All other combinations of inputs to G_6 produce a 1 output, due to at least one 0 input for each row on the table. G_5 combines the G_2 and G_3 outputs with Z, and a different form of output appears. Investigation of the columns for the three inputs reveals that at least one 0 exists for all rows *except* rows 4 and 7. Ones appear then in all of the rows that provide at least *one* 0 input, and 0s appear in rows 4 and 7 because *all* inputs are 1 in these two rows.

Final evaluation of the full-adder circuit is now in sight. The SUM output is formed by combining the G_4, G_5, and G_6 outputs in G_7. Using the three input columns, once again scan from row 0 to row 7, placing a 1 in the G_7 column for any row that has at least one 0 in the

A Typical Analysis

input columns. Rows 1, 2, 4, and 7 have at least one 0 in the G_4, G_5, or G_6 column. The other four rows have 1s as inputs, and 0s are placed in the G_7 column. Comparison of the G_7 column with the SUM output column in the truth table for the full adder (Fig. 7-1b) once again shows a one-to-one correspondence. Hence the NAND gate full adder has been proven functionally identical to other versions of the full adder. If truth table analysis has been used, Table 7-1 should be included in the documentation of the logic-diagram evaluation.

Still another approach using 1s and 0s can be shown. This method is not as rigorous in its requirements to *record the outputs of all gates.* The basic concept that any 0 input to a NAND gate results in a 1 output is used to trace data flow through the NAND gate full adder. Each of the input combinations is examined, and based on the value of the inputs, a decision about the value of the outputs is made *directly* —without recording the intermediate values.

For example, the first combination to be evaluated is row 0 on the table: $X = 0$, $Y = 0$, and $Z = 0$. (Remember, any 0 input results in a 1 output.) Investigating X first reveals that it is fed directly to G_1, G_2, and G_4. Regardless of any other inputs, G_1, G_2, and G_4 will have 1 outputs. Thus the input from G_2 to G_8 is 1, and the input from G_4 to G_7 is 1. Now we must determine the other inputs to G_8 and G_7 to decide what outputs are present. The Y input is 0 at the same time the X input is 0, so G_3 and G_6 have their outputs forced to 1. Now two inputs to both G_8 and G_7 are 1, so we have to determine only the last input to completely define the SUM and CARRY OUT outputs of the full adder for the first combination of inputs. In fact, the third input to G_8 (the output of G_1) has already been defined as 1, due to either the X or Y input. Therefore G_8 has each of its three inputs defined as 1, and its output is 0. Thus when $X = 0$, $Y = 0$, and $Z = 0$, the CARRY OUT output of the NAND gate full adder is 0.

The third input to G_7 is supplied by G_5. Actually, two of the three inputs to G_5 are 1s because they come from G_2 and G_3, respectively. So the third input to G_5 determines its output. Working back toward the input, it can be seen that Z is the third input to G_5, and this input was originally defined as 0. The output of G_5 is forced to 1, providing the third 1 input to G_7 and causing its output to become 0. Just as with the CARRY OUT outputs, when $X = 0$, $Y = 0$, and $Z = 0$, the SUM output is 0.* These results correspond to the similar entry in the truth table for the full adder.

---

*If difficulty is encountered keeping the 1s and 0s in mind, writing them down may be of assistance. Placing a sheet of clear plastic over the diagram being evaluated and marking with a crayon or grease pencil provides a means of recalling previously determined values. When the problem is finished, the markings may be wiped from the plastic, and the next problem attacked in the same manner.

With $X = 1$, $Y = 0$, and $Z = 0$ (row 1), different results can be expected. G_1, G_3, and G_6 have an output of 1, due to $Y = 0$. Additionally, G_2 and G_5 have an output of 1, due to $Z = 0$. All of the inputs for G_8 have been derived (G_1, G_2, and G_3), and they are all 1s. The G_8 output is forced to 0 to present no CARRY operation when binary 1 is added to binary 0. Only the G_4 output to feed G_7 has not been ascertained at this time. Investigation shows that G_4 requires X, G_2 output, and G_1 output. All of these signals are 1, forcing the output of G_4 to 0. Now G_7 has one 0 input, and its output becomes 1. Thus when binary 1 is added to binary 0, the SUM is binary 1.

A little additional effort allows some very useful *general statements* to be made. They are listed here and may be used to complete the remaining steps of the analysis.

1. When $X = 0$, G_1, G_2, and G_4 outputs are 1.
2. When $Y = 0$, G_1, G_3, and G_6 outputs are 1.
3. When $Z = 0$, G_2, G_3, and G_5 outputs are 1.

Consider row 2, where $X = 0$, $Y = 1$, and $Z = 0$. The general statements just listed allow establishment of the G_1, G_2, G_3, G_4, and G_5 outputs at the 1 level. Since G_1, G_2, and G_3 supply the inputs to G_8, the CARRY OUT output is 0. Two of the three inputs to G_7 are already described (G_4 and G_5), and all that is necessary is to find the G_6 output to determine the G_7 output. The G_1 and G_3 inputs to G_6 are both 1, leaving the final decision to the value of Y. Since $Y = 1$, all of the inputs to G_6 are 1, resulting in a 0 output. G_7 now has two 1s and one 0 as inputs, and the output is forced to the 1 condition. Thus when $X = 0$, $Y = 1$, and $Z = 0$, the SUM output of the full adder is 1 and the CARRY OUT output is 0. The remaining five rows in the table are left to the reader as an exercise.

Waveform analysis of the full adder may also be used to demonstrate that the logic diagram of Fig. 7-1a actually performs the required operations. The eight different input-signal combinations required for investigation are developed from the logic-signal source used in earlier chapters. Recall that the logic-signal source is made up of *interconnected flip-flops* (FFs), arranged so that the outputs of the individual FFs furnish input signals for logic circuits to be tested. The relationship of the FF-output signals is such that with two FFs all four possible combinations required are generated, with three FFs all eight possible combinations required are generated, and so on. The waveform outputs (X, Y, and Z) of a 3-stage logic-signal source are shown at the top of Fig. 7-5. Recognition of the characteristics of the logic gates used in the full adder allows investigation of the output of each gate under varying input conditions. Waveform analysis of the full adder of Fig. 7-1b is shown in Fig. 7-5. Note that the H and L markings are not provided in Fig.

A Typical Analysis

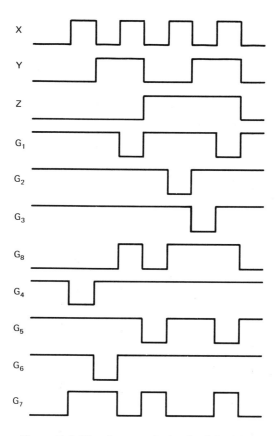

Figure 7-5 Waveform analysis of a full-adder

7-5. (The reader should have sufficient experience with waveforms to be able to recognize the HIGHs from the LOWs, if all furnished problems and examples have been investigated.) The reader should examine each input condition to ensure that the outputs are shown correctly.

7-3 TROUBLESHOOTING COMBINATIONAL LOGIC

The Systems Concept

Few digital systems are strictly combinational in nature. In subsequent chapters, the sequential aspects of digital logic will be considered, and combinational and sequential logic will be combined to form complete systems. The *complete system* (see Fig. 7-6) occupies the highest level of the hierarchy. A number of *subsystems* make up the next level. Each one of these subsystems is usually identifiable by

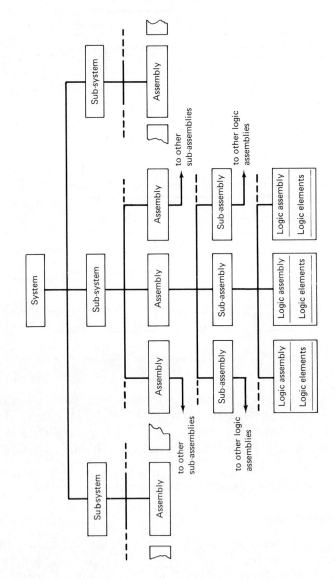

Figure 7-6 The systems concept

its specific functions, and often, it is physically packaged separately. These subsystems are, in turn, made up of *assemblies* that contain a number of *subassemblies,* which are usually removable and capable of being tested as a single unit. The subassemblies consist of *logic assemblies* (such as multifunction integrated-circuit packages), *logic elements* (such as gates), and perhaps even *discrete components* (such as transistors, diodes, resistors, and capacitors).

A modern electronic *telemetering system,* although not completely digital in nature, is a good example of the systems concept. *Telemetering systems acquire data at relatively inaccessible locations, then process, transmit, recover, and display the data at the location where it is to be used.* Figure 7-7 applies the systems concept to a typical telemetering system. Six subsystems are combined to form the complete telemetering system: the *transducer subsystem* (1) senses and converts physical and electrical parameters into suitable electrical quantities for application to the *multiplexer-programmer subsystem* (2), which electrically times and encodes the sensed data from the transducer subsystem into a form that may be transmitted; the *transmitter subsystem* (3) conveys the encoded information to the *receiver subsystem* (4), which acquires and processes the data; within the *demultiplexing-decoding subsystem* (5), the originally encoded data is separated and decoded so that the information originally sensed at the source may be visually presented by the *display subsystem* (6).

Each of the subsystems is identified according to *function,* and, in the case of the telemetering system, is *physically* separated. The transducer, multiplexer-programmer, and transmitter subsystems may be located within a space vehicle, while the receiver, demultiplexing-decoding, and display subsystems are at widely separated locations on earth. Considering only the multiplexer-programmer subsystem shows that it is made of a number of individual assemblies, such as a timer assembly, a multiplexer assembly, an analog-to-digital converter, and a memory. The multiplexer assembly is composed of two digital multiplexer subassemblies and one analog multiplexer subassembly. Finally, an analog multiplexer subassembly consists of numerous integrated-circuit gate packages interconnected to perform the multiplexing of a number of analog signals.

The logic technician can become involved in troubleshooting at any level from the complete system to the logic-assembly level. Working at the lowest level of complexity (the logic-assembly level) requires basic knowledge of individual logic functions combined in relatively simple relationships. As the level of complexity increases toward the complete system, the relationships between logic functions become more abstract, and the technician must become more and more skillful.

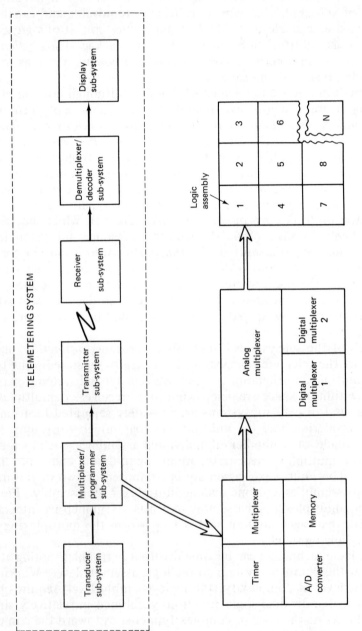

Figure 7-7 A telemetering system

The methods discussed in this chapter are aimed at the isolation of malfunctions at the *logic-element level of a subassembly*. It is at this level that combinational logic is most likely to be encountered. As sequential logic is covered in subsequent chapters, additional methods will be developed so that the overall concepts of system troubleshooting will evolve.

Troubleshooting Philosophy and Concepts

The concepts of troubleshooting combinational logic are little different than the concepts of troubleshooting any type of electronic equipment. Certainly, different troubleshooting equipment is employed, and slightly different viewpoints on interpretation of indications exist, but if you are currently skilled in the techniques of troubleshooting, you will find little new in this section other than a means of *organizing the task*.

Organization of the task is not only the first but probably the most important of all the concepts associated with logic-circuit troubleshooting. If you desire to go from point A to point B, you must first know the location of point A! Then you must determine where point B is and how to get there. Effectively, the same situation exists in troubleshooting logic circuits—you must know from where you are starting, and you must also know where you are going and how to get there. You must organize your task so that you may efficiently determine what information you have available, what additional information is needed, how to get that additional information, and how to analyze the logic circuit to determine the point(s) of malfunction. The method of organization selected must include all of these points, and it must provide a logical plan to accomplish the troubleshooting.

Preliminary to any troubleshooting activity must be the identification of the fact that a malfunction *really* exists. In many cases, an apparent malfunction turns out to be "pilot error": that is, improper use or lack of knowledge of what the equipment should really do. Pertinent questions should be asked: what did it do? what didn't it do? under what circumstances did the apparent malfunction occur? did it smoke? did it blow fuses? what does the operator think is wrong? These types of questions will not only help pinpoint the malfunction but will also reveal potential cases of improper use, perhaps saving many hours of troubleshooting time.

If an *instruction manual* is available, consult it. Become familiar with the theory and operation of the device under examination. Follow the manual when troubleshooting if such procedures are provided. On much modern digital equipment, a self-test provision is available. Perform the self-test to determine if an actual malfunction exists. If no self-test is provided, use the equipment or device to perform an operation with known results. For example, if a digital voltmeter is being checked, measurement of a known voltage will furnish information that can be used to verify proper or improper operation.

At the subsystem, assembly, circuit-card, or subassembly and logic-function levels of effort other approaches must be used. Of course, the same preliminary questions are asked. However it may not be practical to verify the malfunction in quite such an easy manner. Unless special test sets are available to perform detailed checks, we have to assume that the malfunction does exist and proceed with troubleshooting procedures. In cases where a large volume of subsystems or logic cards are encountered, *special test set-ups* are commonly built so that a specific unit may be tested quickly and efficiently.

As equipment becomes more and more modularized (contained in packages or on cards), *automatic test equipment* is being used. Complete systems and subsystems are connected to the test equipment and digital logic circuits sequence tests, providing the correct stimuli for inputs and measuring/recording the output responses. Limits are provided, and when output responses fall outside of the limits, the defective subassembly is identified. If the subassembly is repairable, it is also subjected to automatic testing in another area with another test set performing detailed checks that isolate the malfunction on the subassembly to a replaceable/repairable item.

But there is no need to despair! The automatic test equipment is not about to replace the troubleshooter! Automatic operations are applicable only where *large-volume testing* must be performed on a relatively small variety of subassemblies or assemblies. This situation occurs in some digital computer and data processing applications. Even if automatic testing expands tremendously, the technician is still needed to troubleshoot and repair the test equipment!

The complexion of troubleshooting is changing with the changing technology. Where the electronic technician previously isolated troubles to the discrete-component level (e.g., resistors, capacitors, and tubes), now the trouble may be isolated to complete subsystems. The trend seems to be *packaging complete functions on single, replaceable units.* Failure of the unit results in replacement with a new one—rather than repair. Equipment reliability must be high with this concept, and this appears to be the goal of digital equipment and module manufacturers. From a troubleshooting viewpoint, this means that the technician must begin to think "system level" rather than discrete-component or simple logic-function level. Analysis at this level requires the ability to understand the operation of subsystems and smaller assemblies and to logically evaluate the inputs and outputs of the larger system in terms of what is happening in the smaller assemblies. Skills learned and abilities acquired while evaluating simple logic functions result in more adequate troubleshooting of the subassemblies and complete systems that are commonplace today.

Once the malfunction is verified and it becomes apparent that actual troubleshooting is required, it is necessary to determine *what information is needed* to perform the troubleshooting tasks. Actually, the information required is somewhat dependent on the troubleshooting methods to be applied. For example: if an oscilloscope is used, waveforms that will be encountered will surely be needed. If some type of logic-level indicator such as a voltmeter, logic probe, logic analyzer, or indicator lamp device is used, the logic levels must be available. In both cases, the descriptive Boolean equations may prove quite helpful, and in many cases, it becomes advisable to develop the equations even if they are not immediately available. And, of course, the logic diagram of the device being tested is needed.

Information available for troubleshooting can vary from practically nothing to considerably more than is ever required in day-to-day operation. Instruction manuals may contain everything from English-language descriptions of equipment operation to complete logic diagrams with waveforms and Boolean equations. Troubleshooting guides may be provided to lead to the precise point of the malfunction. The technician should investigate all available information to determine if sufficient data is available to allow troubleshooting of the malfunctioning equipment.

The Flowchart Concept

One way of organizing a troubleshooting task into a logical plan is to borrow a method from the digital computer field. The digital computer, despite all the publicity to the contrary, is not capable of any appreciable problem-solving effort by itself. Every task that it undertakes must be carefully thought out by someone who understands the task. Each step in the task must be identified in terms of actions that the computer is capable of performing and then presented to the computer in the proper sequence for storage and later use. The person who understands the task and can "talk" to the computer is called a *programmer*. When the programmer is called upon to tell the computer to perform some task, he is faced with a situation similar to that facing the technician who is troubleshooting a logic circuit. Both the programmer and the technician must first organize the task and then develop a logical plan to follow. Of course a number of organizational methods exist, but the programmer usually develops a *flowchart* to aid in logically thinking out the solution to the problem. Remembering that the computer must be told step by step what to do to solve a problem, the programmer develops a list of instructions for solving the problem. In mathematical terms, this list of instructions is commonly called an *algorithm*. To make the list of instructions relatively understandable by other programmers and to help organize and check the algorithm, the programmer makes a diagram to represent the algorithm. This diagram is called a "flowchart."

Flowcharts are applicable in areas other than computer programming. In fact, any time a list of instructions has been developed for a task, it can be put into flowchart form. Even a simple recipe from a cookbook (which could be considered as an algorithm) can be put in flowchart form for greater understanding. Organization of troubleshooting procedures for combinational logic is presented by the use of flowcharts in this text. Following the initial presentation of the special symbols and their meanings, a number of typical troubleshooting situations are shown. It will soon become apparent that much effort is saved by the use of flowcharts, and greater assurance that the complete procedure has been accomplished is provided.

In flowcharts, *symbols define the general type of operation that is to take place at each step in the procedure.* Just as with logic diagrams, the *shape* of the symbol is meaningful. Entrance into and exit from a particular flowchart is shown by the symbol in Fig. 7-8a. This symbol is seen only at the beginning or at the end of a procedural flowchart and signifies by the words written inside which of the meanings is to be assigned to it. Usually, the word START signifies the beginning, and the word STOP or EXIT signifies the end (see Figs. 7-9 and 7-10). The rectangular shape of Fig. 7-8b is used to imply that "something is to be done" in that specific step. Typical legends to be seen inside the "do something" shape might be "determine logic level" or "check system power."

Decisions are signified by the diamond-shaped symbol of Fig. 7-8c. As in troubleshooting of conventional electronic circuits, many decisions must be made. Every time a measurement is performed, the technician must determine whether that measurement is within tolerance. Of course, in binary-based equipment such as is implemented with digital logic, measurement decisions are based on simple criteria, e.g., whether the measured parameter is HIGH or LOW or whether the HIGH signal is present for the required period of time. Typical statements seen inside the decision block could read "power supply voltage in tolerance?" or "output gate level LOW?"

The last symbol to be used in troubleshooting flowcharts in this text is the circle in Fig. 7-8d. No action is performed as a result of this symbol, since it serves merely as a connector from one place on the flowchart to another. Due to the physical constraints of paper size, it is often impractical to draw a flowchart in the easily visualized straight line flow of information. When the end of the paper is reached and

Troubleshooting Combinational Logic

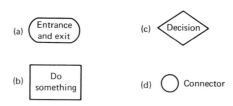

Figure 7-8 Flowchart symbols

additional flowcharting is required, the connector symbol is drawn and an identifying letter is placed inside. Another circle is drawn elsewhere on the paper or on another page to signify the restarting of the flowchart at that particular point. Figure 7-9 demonstrates this technique. If the circle is shown without an identifying symbol, it is merely used to connect or combine two or more branches of the flowchart.

Finally, flowchart symbols are connected by straight lines with arrows at the receiving end of the line to signify direction of information flow. Using combinations of the flowchart symbols and straight lines allows the technician to develop an easily visualized, organized means of troubleshooting combinational logic circuits.

The general concepts behind the construction and use of flowcharts are shown by investigating a flowchart that is relatively common to all troubleshooting situations. Figure 7-9 represents a procedure often followed at the beginning of a troubleshooting problem. Each symbol is numbered for reference purposes so that the following explanation may be easily traced. As we noted, most flowcharts begin with the START symbol (Step 1). The shape of Step 2 indicates "doing something," i.e., visually inspecting the logic device that is malfunctioning. Problems commonly detected by visual inspection are burned or broken discrete components, improperly soldered connections, and loose or broken interconnections. Modern circuit boards are precision devices, and very slight discrepancies in the position of components or interconnections can result in malfunctions. The use of a magnifier helps locate problems that would otherwise go undetected visually.

Decisions concerning the physical condition of the logic device are made in Step 3. Note the shape of the symbol, denoting decision-making. If physical damage was detected during the visual inspection, the NO path is followed to Step 4. Another decision is made at this point. If the maintenance concept for the system of logic being used allows, repair or replacement of damaged components is accomplished according to Step 6. The logic device is replaced (Step 5) if the maintenance concept does not allow repair and/or replacement of damaged or defective components. Many manufacturers feel that replacement of

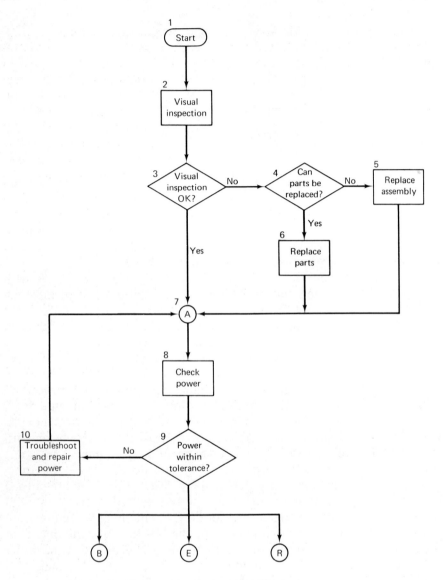

Figure 7-9 "Preliminary" flowchart

the complete logic device enhances system reliability, so repair and/or replacement of damaged or defective components is accomplished at a facility where special precision repair equipment is maintained. The digital logic technician encounters both maintenance concepts (plus many variations) during his troubleshooting assignments.

Step 7 merely serves as a connector, where the results of Steps 3, 5, and 6 come together. If the visual inspection of Step 2 was satisfactory, as indicated by the YES path from Step 3, then the entrance to the connector would be from decision Step 3. If it was not satisfactory, then the entrance to Step 7 would be from either Steps 5 or 6, depending on the maintenance concept. Regardless of the entrance path into the connector, the exit is to Step 8, another "do something" step. All power provided to the logic device under investigation is measured at this time. Some logic devices use as many as three separate voltages, but modern integrated-circuit logic devices generally use only one voltage supply. In any case, all applied voltages and ripple percentages should be measured. The decision about whether the applied voltages and ripples are within tolerance is made in Step 9. One or more power sources out of tolerance can send the technician to Step 10, which indicates that the power supply should undergo troubleshooting and repair. Upon completion of repair, the flowchart returns to connector 7 so that the normal path through the flowchart can be resumed. If the power had been out of tolerance the first time through Step 9, now the decision would be "not out of tolerance," and the YES exit from Step 9 would be to connector B, E, or R.

The combination of Steps 7, 8, 9, and 10 form what is known in flowchart terminology as a "loop." *Loops allow modification of a sequence based on the results of a decision.* In this case, the decision that a defective power supply existed caused the action of "troubleshoot and repair" instead of proceeding on with a known defect. The original action and decision that detected the malfunction are repeated to verify the adequacy of the repair action.

This flowchart demonstrates the use of all the basic symbols and provides an introduction to flowchart techniques. Actually, the flowchart has been generalized so that it may serve as a guide and preliminary chart to all troubleshooting situations. The only prerequisite is that the logic-circuit description be in a form that is compatible with the logic-troubleshooting charts to follow. The flowchart just discussed is not specifically a logic-troubleshooting chart. It is applicable to any electronic troubleshooting situation when it is desired to establish that the malfunction is within the equipment being checked. This chart will account for repair of physically damaged or defective components and improperly operating power supplies.

Common Troubleshooting Situations

In all of the situations we shall investigate, certain general assumptions are made. The logic diagram is assumed to be available. In addition, the logic circuit is assumed to be inoperative, i.e., not performing its required operations as designed. Furthermore, it is also assumed that the logic circuit *had been* operative and *had been* performing its required operations properly.

Test points are usually provided on logic assemblies at circuit locations that allow isolation and replacement of defective components without actual soldering operations. For example, if the logic assembly is so constructed that components are removable and replaceable without soldering, sufficient test points are provided to isolate difficulties to that degree. When a plug-in logic board contains a number of different functions on the same board but none of the components on the board can be replaced, at least the output of each logic circuit (for example, an AND gate or an OR gate) is made available for test purposes. Isolation of the defective logic circuit by use of the test points identifies the defective board, which is removed and replaced. The defective board is then repaired at a facility with the necessary equipment. Similar approaches are used with complex integrated circuit boards. Only sufficient test points are made available to identify the defective integrated circuit, which is replaced at a separate repair facility. The complete integrated-circuit board is usually replaced at the equipment installation and operation location to reduce the requirement for specialized repair equipment at each location.

The choice of which flowchart to use next depends on what type of troubleshooting aids are to be used: one chart assumes the use of built-in indicators; another assumes the use of external logic-level indicators; and the third considers the use of an oscilloscope to troubleshoot malfunctions.

Built-in Indicators

Depending on the type of digital equipment and its degree of complexity, *built-in indicators* may be available to assist troubleshooting operations. At the system level, many types of digital equipment provide panel indicators to allow monitoring of system status. Such indicators are usually gross indications of system malfunction and isolate problems only to major subassemblies. From that point, it becomes necessary to resort either to automatic test equipment to isolate to more detail or to individual logic analysis to determine the amount of repair necessary.

Some manufacturers of digital equipment provide built-in indicators on each subassembly. When this troubleshooting aid is available, it usually allows isolation of malfunctions to a replaceable component. The general trend, however, is to provide only sufficient indicators to isolate malfunctions to replaceable major subassemblies. Instruction manuals commonly contain charts relating combinations of status indicators to malfunctioning subassemblies, so troubleshooting tasks are relatively simple. Skillful analysis of logic circuitry will usually reveal numerous additional combinations of indicators that can extend the malfunction isolation capabilities of the built-in indicators. Figure 7-10 is the flowchart that could be used to troubleshoot digital equipment using only built-in indicators.

Entrance to the flowchart in Fig. 7-10 is from connector B, which previously appeared on the "Preliminary" Flowchart in Fig. 7-9. The first step encountered is a decision (101): can the inputs and outputs be verified? Logically, this is a reasonable question. If the built-in indicators are not properly located to allow verification of the inputs and outputs, there is no need to continue the procedure. The path to follow is from the NO side of the decision block to Step 102 (use another method). Either the "External Logic-Level Indicators" or the "Waveforms" procedure must be accomplished to troubleshoot the malfunctioning equipment if indicators are not available to verify the inputs and outputs. Exit from this procedure is via Step 103.

If the logic circuit inputs and outputs can be verified with internal indicators, the YES path is followed. Connector C, which is the next symbol in the YES path, allows interconnection of a number of additional paths within the flowchart. The logic circuit inputs are checked in Step 105, and the decision as to the correctness of the inputs is made in Step 106. If the correct inputs are not being provided, it is necessary to troubleshoot and repair the circuits providing the inputs (Step 107) before proceeding. The NO path from Step 106 through Step 107 terminates at connector C so that the circuit inputs may be checked again after accomplishing Step 107. When the correct inputs are obtained, the YES path from Step 106 is followed to Step 108 (check outputs). If the outputs are correct after performing Step 109, the YES path is followed to an exit from the procedure at Step 110. Verification of correct outputs with correct inputs is an indication of proper circuit operation.

A number of actions have been performed since the start of the troubleshooting procedures, and any of these actions could have cleared the malfunction. Repair/replacement of defective components, assurance that power inputs are within tolerance, and/or verification of correct inputs could have rectified the original difficulty. Or—and this is not uncommon—handling of the logic assembly during troubleshooting and maintenance has temporarily "fixed" a problem. Extreme care must be exercised to prevent such an occurrence. When replacing components, only those requiring replacement should be disturbed. All other components should be carefully avoided to prevent physically

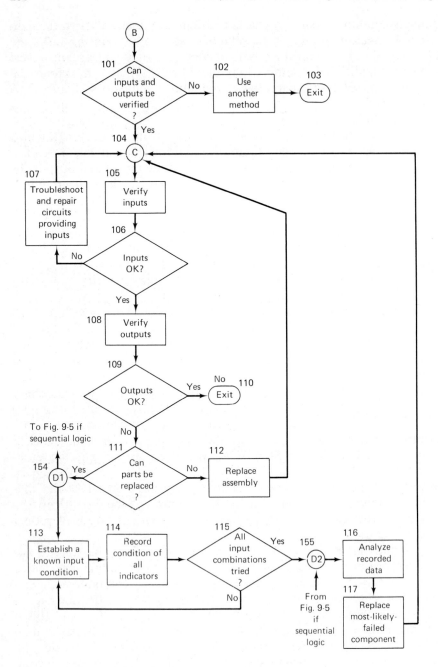

Figure 7-10 "Built-in Indicators" flowchart

displacing them and perhaps causing an intermittent condition to temporarily disappear. Power should be applied to the circuit under test in a manner that closely approximates power application during actual use. Transients that develop due to sudden removal and application of circuit power often cause an intermittent condition to vanish. In addition, shorting of power inputs to each other or to common power should be avoided when making measurements. Additional damage to the circuit under test can result along with a possible "masking" of the real malfunction, due to the power transients. If no actual maintenance actions have been performed on the circuit under test and the exit from the procedure has been taken at Step 110, it is likely that the circuit under test will appear for troubleshooting and repair at some later time. The actual problem may have just been concealed by *any* of the situations described earlier.

Selection of the NO route from Step 109 (outputs *not* OK) results in another decision at Step 111. It is necessary to decide whether components, logic elements, or logic assemblies may be replaced on the assembly being tested. If the answer is NO, there is no need to troubleshoot further. The assembly should be replaced (Step 112) and the flowchart should be reentered at connector C for recheck. This time through, the flowchart should result in YES answers at all decisions (through Step 109) and exit at Step 110 with a properly operating assembly.

If components, logic elements, or logic assemblies can be replaced, the YES path from Step 111 is taken to connector D1, which allows a branch to Step 113 if combinational logic is being repaired or to Fig. 10-5 if sequential logic is being repaired. Reentry to Fig. 7-10 (if necessary) is via connector D2 (155). At Step 113, a known input condition is established. The condition of all built-in indicators is recorded at Step 114, and a question (have all input combinations been tried?) is asked at Step 115. If all input combinations for the number of input variables have been tried, the flowchart continues on to Step 116. If they have not, the flowchart returns to Step 113, and a new combination of inputs is tried. The indicator conditions are recorded, and the decision is made again as to whether all input combinations have been tried. The loop continues until all combinations have been tried and all indicators recorded. At Step 116 the recorded data is analyzed by using *at least* the logic diagram and all other available information. The most likely failed component, logic element, or logic assembly is replaced (Step 117), and the flowchart is reentered at connector C for a complete recheck. If the only difficulty in the logic circuit under test has been caused by the replaced component, logic element, or logic assembly, the YES path is followed through Steps 106 and 109, and the flowchart is completed by exit with a repaired assembly at Step 110.

By using two flowcharts, a complete troubleshooting procedure has been developed—one that allows identification and repair of a malfunctioning logic assembly. The Preliminary Flowchart aided in identification and repair of physically damaged parts and established circuit power within tolerance. Application of the procedures of the Built-in Indicators Flowchart guided the technician through input and output checks and established the conditions for analysis and identification of the most likely failed component, logic element, or logic assembly.

While there is no guarantee that flowcharts will cover every possible case of troubleshooting logic assemblies with built-in indicators, most of the malfunctioning assemblies will submit to these procedures. Modifications may be necessary to cover special cases, but the concepts shown in these and subsequent flowcharts should prepare the reader to construct most flowcharts with a minimum of effort.

External Logic Level Indicators

The scope of logic circuit troubleshooting may be extended to greater detail by using external logic-level indicators. Such devices consist of one or more indicator(s) in a hand-held test instrument that is applied to interconnections in logic circuits to determine the logic level at the point of application. Four general types of troubleshooting probes are used. The typical logic probe is a small hand-held instrument somewhat larger than a ball-point pen. It contains an indicator lamp plus some electronic circuits. When applied to a point in a digital circuit, the lamp indicates the logic level at that point. For example, if the lamp is on brightly, a HIGH is indicated. If the lamp is not on, a LOW is assumed. Advanced versions of logic probes can also indicate single pulses, trains of pulses, and even open circuits.

A companion to the logic probe is the logic pulser. When a "logic signal source" is not available to supply all combinations of inputs, the logic pulser may be used to stimulate the circuit. The logic pulser is a hand-held instrument about the same size as the logic probe. Simple logic pulsers supply either a single pulse or a continuous stream of pulses. More advanced pulsers can also supply an exact number of pulses at a selected rate.

About 75% of digital logic circuit failures can be detected with the logic pulser and the logic probe. Some failures, however, are concerned with short circuits of either interconnecting lines or of components themselves. Short circuits do not allow changes in level, and require other troubleshooting methods. A current tracer may be used to solve short circuit type failures. For example, when many gates are connected to a common point, a short circuited input on just one of the gates keeps the common point "locked" at some level. The current tracer probe allows the defective gate to be identified by tracing the connecting line having excessive current flow. Current tracers physically resemble logic pulsers and logic probes.

The use of conventional multipurpose voltmeters, digital voltmeters, and high-impedance voltmeters should not be overlooked as simple logic level indicators. If the actual magnitude of the voltage level at a given point is of interest, such indicators are very useful. When it is important to observe logic levels at more than one point at a time, logic clips are used. They fit completely over an integrated circuit and contact each of the IC pins. Logic-level indicators then show what level is present on all of the IC inputs and outputs. The effect of changing inputs can also be readily observed. Figure 7-11 shows the various probes and clips in use.

The flowchart for troubleshooting logic circuits using external logic-level indicators has many points in common with the flowchart for built-in indicators. Since the external logic-level indicator may be

Figure 7-11 External logic level indicators (courtesy Hewlett-Packard)

applied directly to the input and output connections of the logic circuit assemblies, we will assume that there is no difficulty in verifying the input and output signals. Therefore the decision block that appeared at the beginning of the Built-In Indicators Flowchart will not be required. In other words, the External Logic-Level Indicators Flowchart (Fig. 7-12) may be directly entered at the "verify inputs" block and developed from that point.

Steps 201 through 208, as a group, perform the same functions as Steps 105 through 112. Inputs are checked and restored if necessary; outputs are verified; and a decision is made concerning replacement or repair of the assembly being tested. The entry to the External Logic-Level Indicators Flowchart is via connector E on the Preliminary Flowchart. If components, logic elements, or logic assemblies can be replaced, the YES path from Step 207 is directly to Step 209. Here the question "are test points available at each logic element?" is asked. If test points are not available at each logic element, it is then necessary to determine if internal connections are accessible (Step 210). Lacking built-in test points, it is necessary to have access to the *wires* or *printed circuit conductors* connecting the various components, logic elements, or logic assemblies. If neither test points nor circuit connection accessibility is provided, then there is no further need to troubleshoot. Malfunctions cannot be isolated to replaceable units. The complete assembly must be replaced (Step 211) and a retest initiated at connector E to verify that the new assembly is operating satisfactorily.

Accessibility of circuit connections leads to Step 212 from the YES path of Step 210. This path is used to select test points to be used in the troubleshooting portion of the flowchart. To select test points, it is first necessary to determine the logic levels (HIGHs and LOWs) at each logic element. If the logic circuit has been analyzed (Step 212), then the logic levels have already been determined, and the YES path from Step 212 leads through connector H to Step 216 (select test points to be used) through connector 215. The NO path from Step 212 leads through Step 213 (analyze circuit) and Step 214 (develop logic levels at each circuit element) to Step 216 via connector H. Either entry to Step 216 via connector H has required logic-circuit analysis. Any of the circuit-analysis methods previously discussed are adequate for application at Step 216. The *truth table form* is probably the most informative, since the information is presented in a way that can be easily correlated with indicators depicting HIGHs and LOWs. Following selection of test points, the flow returns via connector E to the major path through the chart.

The second time through Step 209 test points will be available at each logic element, and the YES path is followed through connector F

Troubleshooting Combinational Logic

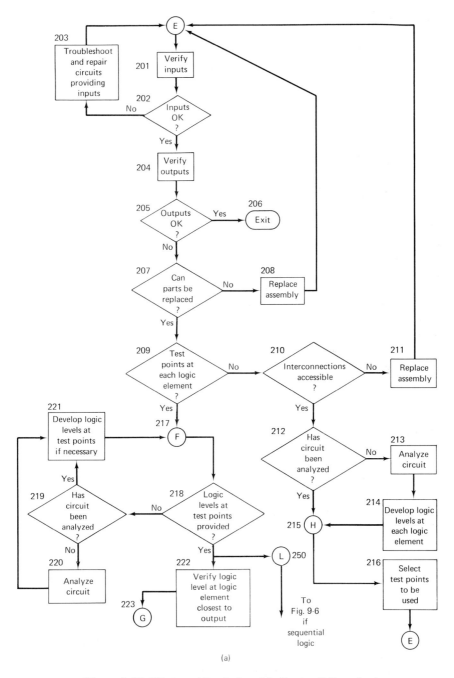

Figure 7-12 "External Logic Level Indicators" flowchart

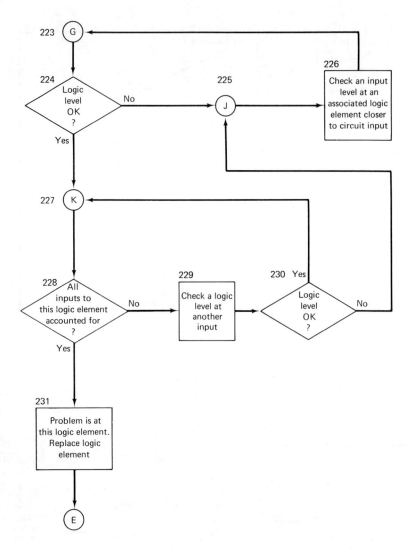

Figure 7-12 "External Logic Level Indicators" flowchart (cont'd)

(Step 217) to the question at Step 218 (are logic levels at each test point provided?). Perhaps test points were provided by the equipment manufacturer, but the information concerning logic levels at the test point either was not furnished or was not available.

The NO path from Step 218 is then followed to Step 219, where the question "has the circuit been analyzed?" is asked. A procedure similar to that discussed in the previous paragraph is followed in Steps

Troubleshooting Combinational Logic

220 and 221, and control of the flowchart is returned to the major path through connector F.

Actual troubleshooting begins with an exit from Step 218 via the YES path. If the logic being evaluated is sequential, an exit via Step 250 (connector L) to Fig. 9-6 is made. Combinational logic troubleshooting continues to Step 222.

A short review of what we know to this point will materially aid us in following the remainder of the flowchart. To get to Step 222, it has been necessary to verify the following conditions:

1. No visible damage.
2. Circuit power is within tolerance.
3. Inputs have been verified.
4. Outputs are not correct.
5. Repairs can be performed when malfunctioning components are located.
6. Test points are available at each logic element.
7. Logic levels are available at each logic element.

Thus the malfunction has been pinpointed to the logic assembly under test, and sufficient information is available to further isolate the problem to a replaceable component, logic element, or logic assembly.

Checking the logic level at an *input to the logic element closest to the output* (Step 222) provides the first bit of information that will lead to detection of the faulty portion of the logic assembly. The decision as to the validity of the logic level just checked is made in Step 224, which has been entered from Step 222 via connector G (Step 223). Assuming a correct logic level, Step 224 is exited via the YES path through connector K (Step 227) to Step 228, where it is determined whether all inputs to the logic element being examined have been accounted for. A single-input logic element such as an inverter amplifier has only one input, and under these circumstances, all inputs would have been accounted for. If the output of a device is incorrect and the input is correct, the device must be faulty. Therefore Step 228 can be exited via the YES route to Step 231, which contains direction to replace the defective logic element. After replacement of the malfunctioning component, the flowchart returns via connector E to completely recheck assembly operation. If the only defective component was the one replaced, the output check is now passed through the flowchart, and an exit from the procedure occurs at Step 206.

If the logic element closest to the output is a multiple-input device such as a gate and if the first input checked is correct, then the NO path from Step 228 is taken. At Step 229, the logic level at another input to the logic element under test is checked. A proper logic level at the Step 229 test causes exit from Step 230 on the YES path to connector K.

The output of connector K feeds the decision block 228, and the resultant loop is continued until all inputs to the logic element have been verified as correct. Once again, if all inputs are correct and the output is incorrect, the difficulty must be with the logic element being tested. Replacement is made at Step 231, and retest is accomplished by returning to the main flow via connector E.

Any time that a logic-level check is not satisfactory, a loop is entered that causes *measurements to be made at an associated logic element closer to the circuit input.* Step 226 provides the directions for this measurement and is entered as a result of either Step 224 or 230 through connector J (Step 225). The loop entered through connector J to Step 226 allows the troubleshooting to advance toward the actual basic inputs to the complete unit under test. Multiple entry is provided to this loop, and the flow is quite complex. A complete explanation is "wordy"—it is better left to a troubleshooting example that will appear later in this chapter. However following the procedure provided will result in detection and isolation of components that have caused the overall unit under test to develop improper outputs. In all cases, the detection and replacement of the faulty components at Step 231 cause recheck via connector E, and an exit at Step 206 with a properly functioning logic assembly.

OSCILLOSCOPE

In strictly combinational logic applications, the *oscilloscope* provides little advantage when compared with other external logic-level indicators. Instead of an illuminated indicator or the position of a pointer on a meter, the position of a trace on the oscilloscope provides the indication of logic level at a particular point in the circuit. The oscilloscope is most useful when one needs to determine the action of a logic circuit at the time that circuit inputs are changing. *Most external logic-level indicators are relatively slow acting, and transient conditions are difficult to detect without the speed of the oscilloscope.*

Special purpose oscilloscopes called *logic state* or *timing analyzers* have been designed for digital troubleshooting. Logic analyzers allow viewing of more than one circuit point at a time (up to 16 or more in some models). Information at the circuit points may be displayed as waveforms or as 1s and 0s actually "written" on the cathode ray tube. Logic analyzers are most often used in complex combinational/sequential circuits to identify a general circuit failure. Logic probes can then pinpoint the actual failed component. More details on logic analyzer applications are given in Chapter 9.

The flowchart for troubleshooting with the oscilloscope and logic state or timing analyzer (Fig. 7-13) is almost identical to the flowchart

Troubleshooting Combinational Logic

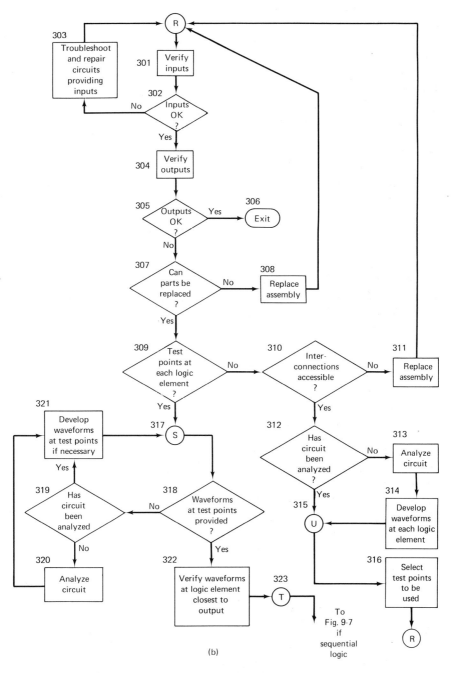

Figure 7-13 (a) "Oscilloscope" flowchart

218 Analysis and Troubleshooting of Combinational Logic

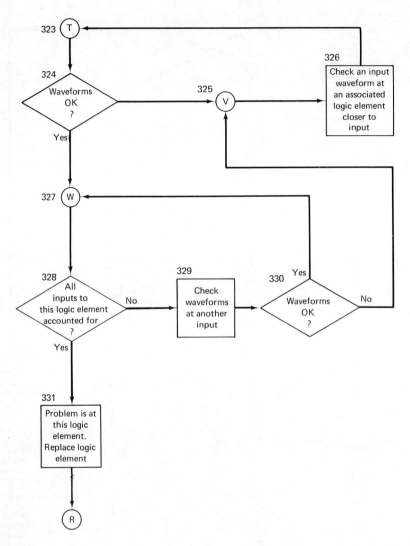

Figure 7-13 (b) "Oscilloscope" flowchart

for troubleshooting with the other external logic-level indicators. Generally, the word *waveforms* replaces the term "logic level," and the overall flow remains the same. The use of the oscilloscope and logic state or timing analyzer will be readily apparent when analysis and troubleshooting of sequential circuits is covered in subsequent chapters. Exit to Fig. 9-7 is provided if sequential logic is being repaired.

7-4 A TYPICAL TROUBLESHOOTING PROBLEM

With the completion of the flowcharts for troubleshooting combinational logic, the guidance is now available for working with actual logic-circuit problems. The sample troubleshooting exercise provided in this section allows the reader to use flowcharts and develop an appreciation for this method of organization. One example alone, however, is not enough to firmly implant the complete concepts. The reader should apply these and subsequently developed flowcharts at every opportunity to fully realize their tremendous potential.

Preliminary Information

When extreme reliability is required in logic applications, as many as three redundant systems may be used. Of course, this requires three times as much hardware, but it does assure accurate and reliable operation. Usually the system is arranged so that when *any* two out of three paths are operating properly, it is considered sufficient indication that accurate information is being provided. A logic arrangement called a *majority circuit* is used to control acceptance of the incoming information, based on the two-out-of-three rule. Thus if three paths called A, B, and C are furnished, either A and B, B and C, or A and C (or all three) should provide an output that indicates the incoming information is acceptable. For maintenance purposes, however, it is necessary to know if *any* of the three paths is *not* functioning. This information is not available from the majority circuit, so a *dissent circuit* is required to furnish the information that one of the three paths has failed. The output of the dissent circuit is an indication that one or more of the inputs is *not* operating correctly.

A hypothetical application of these circuits could be found in modern jet aircraft, which use autopilots for much of their routine flying operations. Three different navigational devices can be used to guide the aircraft along a predetermined path. Experience has shown that any two of the three provide an adequate safety margin as far as maintaining accurate navigation is concerned. Therefore when any two of the three navigational devices are operating satisfactorily, the output of the majority circuit enables the autopilot to accept information. However, if one of the devices has failed, the crew must be made aware of the fact so that alternate procedures may be instituted should a second device fail. The dissent circuit provides the warning that one of the three devices has failed, so that maintenance personnel can use predetermined procedures to isolate and replace the defective navigational devices.

The three inputs to the majority/dissent circuit are logic levels. For example, if a device represented by *A* is operating correctly, *A* is HIGH. If the device is not operating correctly, *A* is LOW. Determination of correct operation is made within each navigational device, and a fault warning light is illuminated on the device when it is operating incorrectly. Logic levels *A, B,* and *C* are merely indications of correct or incorrect operation. The majority output (*M*) is HIGH when any two of three (or all three) inputs are HIGH, and an indicator lamp called the *majority lamp* is illuminated. When one or more (but not all) inputs are LOW, the output of the dissent circuit (*D*) is LOW, and the *D Warning Lamp* is illuminated. Thus either a LOW *M* or a LOW *D* output is indicative of difficulty, and maintenance action is required. Normal indications are the *D Warning Lamp* off, and the *M Majority Lamp* on. This indicates that all three inputs are present (*D Warning Lamp* off), and that the autopilot is connected to the navigational units.

For the purposes of this example, it is assumed that the complete majority/dissent circuit is contained on a removable printed-circuit card. Each of the logic elements is individually removable without soldering, and repair to the plug-in component level is authorized by the maintenance concept. No built-in indications are available on the circuit card, and no intermediate test points are provided. However it is possible to gain access to the interconnection wiring on the circuit board. It is also assumed that a test fixture is available. As we mentioned previously, such an arrangement is common when a large number of circuit cards need to be tested. The test fixture is usually composed of a socket which accepts the circuit card, power supplies to simulate system power, a means to synthesize input signals, and indicators to monitor inputs and outputs. The only other troubleshooting aids available are a logic diagram (Fig. 7-14) and an external logic-level indicator.

The majority/dissent circuit card has been returned to the troubleshooting area for repair with the following information furnished.

> The *M Lamp* went off, and the autopilot automatically disconnected from the navigational units, requiring manual control of the aircraft. After regaining control, it was noted that the *D Warning Lamp* was off and that both *A* and *B* navigational units had fault warning lamps illuminated. Replacing the majority/dissent circuit card caused the *D Warning Lamp* to illuminate with both *A* and *B* navigational units malfunctioning. Replacement of unit *A* caused reillumination of the *M Lamp*, indicating that two of three navigational units were

A Typical Troubleshooting Problem

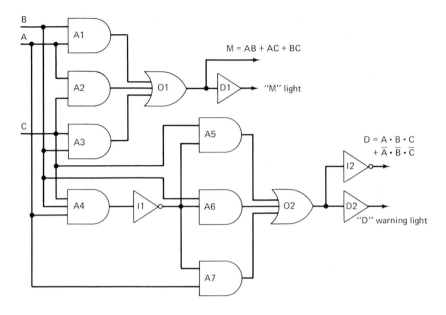

Figure 7-14 Majority/Dissent circuit logic diagram

operable. The *D Warning Lamp* stayed illuminated. Replacement of unit *B* extinguished the *D Warning Lamp,* and the *M Lamp* stayed illuminated as expected.

The Troubleshooting Task

A decision must now be made concerning the actual troubleshooting procedure that is to be used. In all cases, of course, the preliminary flowchart should be followed to repair physical damage and to assure that the power in the text fixture is properly adjusted and connected. The next flowchart to follow is dependent on the troubleshooting aids available. In the case of the majority/dissent circuit card, only the logic diagram and an external logic-level indicator have been furnished. This leads to the 200 series flowchart (Fig. 7-12), which allows the technician to locate and identify the malfunctioning logic element. The complete procedure is now followed from beginning to end to demonstrate the use of flowcharts.

Starting with the Preliminary Flowchart (Fig. 7-9), a visual inspection is performed. With no obvious physical damage apparent, the circuit card is placed in the test fixture, and the applied voltage(s) are checked. When all voltages are properly adjusted, the Preliminary Flowchart is exited, and the External Logic-Level Detector Flowchart is entered. The inputs are checked and placed in the configuration of the original malfunction, i.e., both A and B LOW. Upon checking the outputs, it is determined that they are incorrect for the input conditions, and Step 205 is exited via the NO path. Parts can be replaced according to the maintenance concept, so Step 207 is exited via the YES route to Step 209.

The earlier description of the circuit card indicated that no built-in test points were provided but that interconnections were accessible. A path through Blocks 209 and 210 to Block 212 is established, and the first real troubleshooting/analysis decision is made. Of course, if the logic diagram has already been analyzed, we need only to select the test points to be used (block 216) and proceed back to Step 209 via connector E and intermediate steps. If the circuit has not been analyzed, the route through Steps 213 and 214 must be followed. Note that the logic diagram does not show the minimum logic implementation of the majority/dissent functions, but the circuit arrangement adapts well to the demonstration of troubleshooting techniques. Any of the common analytical methods discussed in this chapter may be used in *circuit analysis*. The most worthwhile analysis when troubleshooting with an external logic level detector is one in which levels at each logic element result. A truth table provides this information, and the expected logic level can be correlated directly with that indicated on the test device. The truth table for both outputs of the majority/dissent circuit is shown in Table 7-2.

Analysis identifies logic levels at various points throughout the circuit so Block 209 test points are available this time through. The YES route to Block 218 is thus taken. If the logic levels at the test points are not known, it is necessary either to analyze the circuit and develop the logic levels or to develop the logic levels from the already analyzed circuit. Blocks 219 through 221 provide the required procedural steps. When test-point logic levels are known, Block 218 exits to Block 222, and actual troubleshooting begins. Although it is not shown on the flowchart, in a logic circuit with more than one output, it is necessary to decide which output to start with. Random selection could be used, but in many cases, sufficient information is available to make one choice better than others. The clue in this case is the malfunction description that came with the circuit card. Apparently the portion of the logic circuit feeding the D *Warning Lamp* is the culprit, since the D *Warning Lamp* did not illuminate even though both A and B navigational units were malfunctioning. The description of the dissent circuit states that D is HIGH, and the D *Warning Lamp* is off when all three inputs agree. With both A and B LOW and C HIGH, obviously the three inputs do not agree. Therefore the D *Warning Lamp* should be on, and the D out-

TABLE 7-2 Majority/Dissent circuit truth table

Row	A	B	C	A_1	A_2	A_3	$O1$	D_1	A_4	$I1$	A_5	A_6	A_7	$O2$	$I2$	D_2
0	0	0	0	0	0	0	0	0	0	1	0	0	0	0	1	0
1	0	0	1	0	0	0	0	0	0	1	1	0	0	1	0	1
2	0	1	0	0	0	0	0	0	0	1	1	1	0	1	0	1
3	0	1	1	0	0	1	1	1	0	1	0	1	0	1	0	1
4	1	0	0	0	0	0	0	0	0	1	1	0	1	1	0	1
5	1	0	1	0	1	0	1	1	0	1	0	1	1	1	0	1
6	1	1	0	1	0	0	1	1	0	1	0	1	1	0	0	1
7	1	1	1	1	1	1	1	1	1	0	0	0	0	0	1	0

put should be LOW. It is with this information that Block 222 is entered.

The input to the logic element closest to the output (INVERTER 2) is obtained from OR 2. If the *D* output should be LOW, then the input to INVERTER 2 should be HIGH. With the input combination that causes the malfunction, the information in Table 7-2 (specifically row 1) is used to determine the required logic level at each test point. The logic level indicator shows the level to be LOW, which is incorrect. This is confirmed by the nonillumination of the *D Warning Lamp*. The NO path is taken from Block 224 to Block 226, and an input to an associated logic element closer to the input (OR 2) is checked. Arbitrarily selecting the output of AND 5, it is found to be LOW. According to row 1 on the truth table, the output of AND 5 should be HIGH with the given input combination. This causes the selection of the NO path again from Block 224 and the checking of an input level at an associated logic element closer to the input (AND 5). AND 5 has two inputs, one of which comes from input *C*. It is known that *C* is HIGH under the given input conditions, so the only way that the AND 5 output can be LOW is for the other input (from INVERTER 1) to be LOW or for AND 5 to have failed with its output LOW.

The output of INVERTER 1 is checked and is found to be LOW. Table 7-2 indicates that the given input combination should result in a HIGH output, so the input to INVERTER 1 is investigated. The input to INVERTER 1 is from AND 4, and the logic-level indicator reveals a HIGH level. Note that the loop consisting of Blocks 223, 224, 225, and 226 is being constantly repeated, striving to locate a logic level that is valid. The HIGH level at the output of AND 4 is incorrect according to row 1 of Table 7-2, and the loop is again repeated. The complete dissent circuit from output to input has been traversed, and when an input to AND 4 is checked, it will be found to be correct, as indicated in the truth table. Block 224 is exited through connector K via the YES path to Block 228.

But all inputs have not been accounted for, and the NO exit from Block 228 requires checking of a logic level at another input. Block 230 makes the decision on the validity of the logic level (which is correct according to the truth table), and an exit along the YES path leads to connector K and Block 228 for a repeat of the loop through Blocks 228, 229, and 230. As soon as all inputs to the logic element (AND 5) have been checked, Block 228 exits via the YES route to Block 231. Since all of the inputs to AND 4 have been verified, and the output is still incorrect, AND 4 is defective, having failed in the HIGH output condition. AND 4 is replaced by direction of Block 231, and the flowchart returns via connector E to recheck the complete majority/dissent circuit card.

A Typical Troubleshooting Problem

PROBLEM APPLICATIONS

7-1. What information should be obtained prior to starting an investigation of a combinational logic circuit?

7-2. Where does the analysis of a combinational logic circuit begin?

7-3. What must be done when a level of logic is reached that requires other inputs?

7-4. What is the most effective method of logic analysis?

7-5. What is meant by the term "systems concept"?

7-6. Why is it desirable to understand the systems concept?

7-7. Discuss the philosophy and concepts of combinational logic troubleshooting.

7-8. What is a flowchart?

7-9. Why are flowcharts used?

7-10. What flowchart symbols are used in each of the following cases?
(a) Begin at this point.
(b) Begin at this point as defined on the previous page.
(c) Is the logic level HIGH?
(d) Check the power supply voltage.
(e) Set the input switches to all 0s.
(f) Go back to the input-level check.

7-11. What information must be obtained before beginning troubleshooting?

7-12. Develop a flowchart for troubleshooting a complete logic system, including a status panel consisting of indicator lamps that are capable of displaying the inputs and outputs of each assembly.

7-13. What is the most effective method of troubleshooting?

7-14. Assume that a full adder is constructed from two half-adders and an OR gate as shown below. The half-adders are implemented in the manner of Fig. 4-24. Analyze the operation of the full adder using
(a) truth tables
(b) waveforms
(c) logic levels (H and L)

Document the analysis thoroughly.

7-15. Assuming that the inputs are correct, the outputs of the first half-adder are correct, and the C_0 output remains HIGH at all times, troubleshoot the logic circuit in Problem 14 using
(a) the external logic-level indicator method.
(b) the oscilloscope method.
Document the analysis thoroughly.

7-16. What symptoms would appear if in Fig. 7-1
   (a) G_1 fails in the HIGH state?
   (b) G_1 fails in the LOW state?
   (c) G_2 fails in the HIGH state?
   (d) G_3 fails in the LOW state?
   (e) G_5 fails in the LOW state?

8

SEQUENTIAL LOGIC

The preceding chapters have been primarily devoted to developing an understanding of combinational logic circuits. Combinational logic circuits have outputs that are solely dependent on the input conditions at the time that outputs are being observed. A particular set of input conditions always establishes the same output, regardless of the past history of the circuit. But changing or removing the inputs also changes the outputs, and no record is provided to show the state of the previous output before the input change. In other words, *the combinational logic circuit has a very poor memory!*

In Chapter 3 we pointed out that two general types of logic circuits exist in digital equipment: combinational and sequential. And, *in contrast to combinational circuits, sequential circuits do possess memory.* Furthermore, *the outputs of sequential circuits* depend not only on *the present input condition* but also on *past input history.* It is this memory capability that allows the sequential circuit to perform its function.

8-1 AN EXAMPLE OF SEQUENTIAL LOGIC USE

A simple example of memory capability requirements may be found in the digital control of machinery in an industrial application such as the drilling and shaping of precision metal parts for high-speed aircraft. The metal-working machines must be told precisely where to position their cutting tools in order to perform each operation. Positioning commands are often as simple as "move left" or "move right." In fact, to take advantage of digital operation, only *simple commands,* capable of being represented by binary notation, may be used. Information is commonly fed to the digitally controlled machine by means of a punched paper tape with hole location and number of holes arranged so that numerous unique combinations exist. (Chapter 13 discusses the use of such digital codes.)

But, since only *one* command may be issued at a time, some sort of *memory capability* is definitely required. If the machine is told to "move the tool right," this information must be stored until it is told "how *far* to move." A combinational logic device cannot remember which direction the tool is to move when it is told how far to move. Issuance of the "how far to move" command removes the "move the tool right" command, and the machine could move the tool in the wrong direction. In just such a simple situation as this, it is apparent that a requirement exists for retention of information in digital systems.

Although similar, the memory requirement discussed in this chapter should not be confused with the meaning of the term "memory" as applied to digital computers. In describing the memory function in sequential logic, a more general description is applied. *Memory* in sequential logic serves the purpose of retaining information about the *immediate past* so that the basic sequential-logic element may perform based on that information and the current incoming information. Having furnished information on the immediate past, the sequential-logic element then stores the results of the interaction of that information and the current incoming information as "immediate past" information and awaits the next sequence of incoming information. This suggests a *time dependency* for sequential logic. It is the memory and time dependency properties of sequential-logic circuits that differentiate them from combinational logic circuits. Thus *in a sequential logic circuit, the output(s) depend not only on the input(s) immediately present but also on the past history of both.*

8-2 THE GENERAL SEQUENTIAL ELEMENT

The "Latch" Function

Combinational-logic circuits may be *interconnected* to provide a sequential-logic function. The term *latch* is used to describe a logic function that is similar to a mechanical operation. A mechanical latch, such as that used on a fence gate, requires two separate actions, usually applied at two separate physical locations. One action occurs upon closing the gate, when the latch is physically raised to allow the latch pin to enter and then falls into position to lock the latch pin in position.

Further attempts to close the gate have no effect, since the latch pin is locked in position and cannot move. The only way to unlatch the gate is to raise the lock and allow the latch pin to be released. The unlatch operation occurs at a different physical location than the latch operation, i.e., raising the latch lock to unlatch was a different operation than raising the latch lock to allow entry of the latch pin.

A logic latch operates in a similar manner—*an input action must be applied to a specific input to cause the device to assume one of two logic conditions.* Further application of the latch input signal has no effect. The latch remains in the latched condition. Release of the latch must be accomplished by application of an input signal to another input location.

A very primitive latch (Fig. 8-1) may be constructed using only two inverting amplifiers. Single-input NOR gates are used to demonstrate the principle of operation. (Actually, the circuit shown is not a latch at all but is merely a circuit that exhibits some of the properties of a latch.) The single-input NOR gate acts as an inverter. When the input of the top inverter is HIGH, the output is LOW. The LOW output of the top inverter fed to the input of the bottom inverter causes its output to be HIGH. Thus the circuit is locked into a specific configuration. Which inverter initially possesses the HIGH output on power application depends on individual circuit difference, but one or the other is HIGH and initiates the locked-up condition. As we can see, no provision is made to control this circuit. It is merely representative of the general concept of the latching principle.

A more practical latching circuit is shown in Fig. 8-2. Cross-coupled NOR gates are still used, but another input is shown on each gate. The initial feedback configuration is retained for both NOR gates, and the added inputs are used for control purposes.

For the purposes of discussion, it is assumed that the circuit of Fig. 8-2 has the LATCHED output LOW and the UNLATCHED output HIGH, following initial power application. Furthermore, it is assumed that both the UNLATCH and LATCH inputs are at a LOW level. With the given inputs and outputs, the feedback input to NOR 1 is HIGH, due to its connection directly to the UNLATCHED output. And, since any HIGH input to a NOR gate results in a LOW output, LATCHED is LOW. If the UNLATCH input transitions from LOW to HIGH (either permanently or momentarily), no change occurs at the outputs. (Any or all HIGH inputs to a NOR gate result in a LOW output.) Examination of NOR 2, however, shows another situation. The feedback input to NOR 2 from the LATCHED output is a LOW level. As in the primitive latch, the LOW input forces the output HIGH if no other inputs are present. In Fig. 8-2, however, the state of the LATCH input must be considered. As initially defined, the LATCH input is LOW, and when

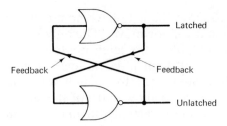

Figure 8-1 A primitive latch

combined with the LOW input from the LATCHED output in the NOR gate, furnishes a HIGH level at the UNLATCHED output. If the LATCH input now transitions from LOW to HIGH (either permanently or momentarily), NOR 2 has one LOW and one HIGH input. Any HIGH input to a NOR gate results in a LOW output, and the UNLATCHED output transitions to a LOW level. The change from HIGH to LOW at the UNLATCHED output is fed back to the input of NOR 1, and removal of the HIGH input causes the output of NOR 1 to transition to the LOW level. The outputs are now reversed from the original state—the LATCHED output is HIGH and the UNLATCHED output is LOW.

Return of the LATCH input to a LOW level does not affect the state of the circuit, since the feedback from the LATCHED output effectively locks the circuit into the LATCHED state. The only means of changing the state of the circuit is to apply a HIGH-going input to the UNLATCH input. NOR 1 operates in exactly the same manner as NOR 2, and the circuit returns to the UNLATCHED state. Thus, just as in the mechanical example, actions must occur at *different positions* to change from a latched to an unlatched condition and vice-versa.

Nomenclature and Symbology

A generalized approach to the nomenclature and symbology of *flip-flops* (FFs) is almost mandatory in discussing the different types and many means of operational implementation.* Perhaps the most basic requirement is to be able to discuss FFs without becoming confused about the relationship between inputs, outputs, and the "state of the circuit."

The basic shape of the FF symbol is retained, but the input and output designations are modified to fit the specific type of FF being discussed. One of the simplest configurations, the *R-S* FF, is used as an

---

*Slight modifications must be made to MIL-STD-806B symbology to accommodate the many different circuit configurations.

The General Sequential Element

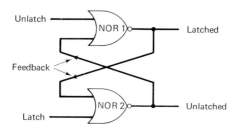

Figure 8-2 A practical latch

example. The symbols shown in Fig. 8-3 can be recognized as *R-S* FFs by the input labels. When this symbol is used, it indicates that the FF represented behaves in accordance with a specific set of rules peculiar only to the *R-S* FF.

Defining the rules of operation for the FF in a nonambiguous manner is somewhat of a problem. Just classifying the state of the circuit is not easy, due to the many different terms used for this purpose. SET and CLEAR (RESET) are excellent descriptive terms, but they tend to imply a particular type of FF. 1 and 0 work well, but what about the inevitable confusion generated with binary 1 and binary 0? TRUE and FALSE are good but are more commonly applied to Formal Symbolic Logic. Q and $\bar{Q}$ are usually used to define FF outputs, not states of FFs. What, therefore, is the least ambiguous means to define the rules of operation for a FF?

Quoting from MIL-STD-806B:

> The flip-flop is a device which stores a single bit of information. It has three possible inputs, set (S), clear (reset)(C), and toggle (trigger)(T), and two possible outputs, 1 and 0. When not used, the trigger input may be omitted. The two outputs are normally of opposite polarity. A "1" is stored in the flip-flop when the "1" output level is active, and the "0" output level is inactive. A "0" is stored in the flip-flop when the above condition is reversed. The flip-flop assumes the "1" state when an active signal appears at the "S" input regardless of the original state. It assumes the "0" state when an active signal appears at the "C" input regardless of the original state. . . .

One possible approach is to use the SET and CLEAR (RESET) notation to describe FF states, despite the possible confusion that could be generated when discussing other types of FFs. Although the

(b)

Figure 8-3 Flip-flop symbols

The General Sequential Element

initial rules described in Ch. 2 indicated that HIGH-going signals activated the FF and caused the Q output to be HIGH when the FF was SET and the $\bar{Q}$ output to be HIGH when the FF was CLEAR (RESET), MIL-STD-806B offers some alternatives. The use of the terms *active* and *activating*, along with the basic symbology of the attached circle indicating the LOW level as the activating or active level (depending on whether it is attached to the input or the output of the symbol) allows almost universal representation of the FF function. Figure 8-3 shows the most common combinations of FF symbols normally encountered.

One further important point must be discussed. *Despite* the resulting confusion, the SET state of a FF is almost universally considered as the "1" state and the CLEAR (RESET) state as the "0" state. Of course, such convention may generate further confusion. As far as the FF is concerned, it is storing a 1 when the Q output is active, whether that active condition is HIGH or LOW! It is storing a 0 when the $\bar{Q}$ level is active and the Q level is inactive.

Tabular Representation of Sequential-Logic Circuits

Sequential-logic circuits may be represented in tabular form in a manner similar to combinational circuits. However conventional truth tables cannot be directly used due to the memory aspects of sequential circuits, the dependence on previous circuit conditions, and the feedback requirements. A variation of the conventional truth table may be employed to describe sequential logic circuits. Before showing the most commonly used form of the sequential-logic truth table, however, we should collect all of the known information about the circuit being examined. This has been done in Table 8-1.

Rows 1 through 4 examine the response of the latch to various combinations of input signals when the assumed condition following power application is the latched state. Latch operations described earlier are reproduced in rows 1, 2, and 3. Row 4 poses a problem. Note that when both LATCH and UNLATCH are HIGH, the transition from the latched state is indeterminate. The actual resulting condition may not be the same each time. When both inputs are HIGH, both outputs will be LOW. The final state is determined by which of the inputs is removed (or returned to the LOW level) first and by circuit constants. Thus additional logic circuits must be provided to guard against both inputs going HIGH at the same time. Without such safeguards, a latch with both inputs going HIGH at the same time might be looked upon as a defective light switch—every time it was turned on, the light *might* or *might not* come on! In digital logic, *maybe* is not allowable! Rows 5 through 8 duplicate the operations of rows 1 through 4, except the initial condition is assumed to be unlatched.

TABLE 8-1 Latch Operation

Assumed conditions		Signal Inputs		Resulting conditions	
(LATCHED)					
LATCHED OUTPUT	UNLATCHED OUTPUT	LATCH	UNLATCH	LATCHED OUTPUT	UNLATCHED OUTPUT
1 H	L	L	L	H	L
2 H	L	L	H	L	H
3 H	L	H	L	H	L
4 H	L	H	H	Indeterminate	
(UNLATCHED)					
5 L	H	L	L	L	H
6 L	H	L	H	L	H
7 L	H	H	L	H	L
8 L	H	H	H	Indeterminate	

Table 8-1 seems to duplicate functions and information. Combining functions alone reduces the size of the table, but another point must be taken into consideration. The latch's state following an input-signal change is dependent upon its state prior to application of the input signal. A conventional truth table does not provide this type of information, so modifications must be accomplished. The information from Table 8-1 is rearranged into the form of Table 8-2 and is renamed a *detailed truth table*. HIGH is replaced by 1, and LOW is replaced by 0. The first column is renamed $latched_n$, and the last column is renamed $latched_{n+1}$.* Only the LATCHED output is used, since it is known that the UNLATCHED output is always opposite. Evaluation of Table 8-2 reveals that the rules previously defined for operation of the latch are valid.

Duplication still exists in the detailed truth table, and further simplifications may be made. In the *limited truth table*, Table 8-3, the redundant entries are not shown. Also, since it is known that $latched_n$ represents the LATCHED output prior to the change in inputs defined in the table, no need exists for the $latched_n$ column. Table 8-3 is the most commonly encountered form of table used to describe sequential-circuit operation.

As you may have observed, the tables used to describe the latch are not really "truth tables" in the strict sense of the word. The mem-

---

*The subscripts n and $n + 1$ are used to indicate a time differential. If $latched_n$ is considered to be the logic condition at some time n, then $latched_{n+1}$ is the logic condition at some later time, usually following a change in input conditions.

The General Sequential Element

TABLE 8-2 Detailed truth table, Latch

Unlatch Input	Latch Input	Latched$_n$ Output	Latched$_{n+1}$ Output
0	0	0	0
0	0	1	1
0	1	0	1
0	1	1	1
1	0	0	0
1	0	1	0
1	1	0	*
1	1	1	*

* Indeterminate

TABLE 8-3 Limited truth table, Latch

Unlatch Input	Latch Input	Latched$_{n+1}$ Output
0	0	Latched$_n$
0	1	1
1	0	0
1	1	*

* Indeterminate

ory and time dependency of sequential-logic elements place other requirements on tabular representations. Other names commonly encountered include *characteristics table, state table, modified truth table,* and *flow table* (usually reserved for specific types of tabular displays). The term *state table* is used in this text to describe the tabular representation of sequential-logic element characteristics. Justification for this choice appears during initial discussion of the *R-S* FF.

Algebraic Representation of Sequential Logic Circuits

Boolean equations may also be used to represent sequential-logic elements. Determination of the *Boolean expression* for sequential-logic elements (from state tables) follows closely the extraction of combinational-logic expressions from truth tables. The state table for the latch circuit (Table 8-4) is used as a starting point, and necessary modifications are made to allow easy development of the Boolean expression for operation of the latch.

Replacement of the input names on the left side of the state table with letter designations is the first step. The unlatch column is labeled *U*, and the latch column labeled *L*. Since the input columns define the states of the inputs at a reference time *n*, the labels are called U_n and

L_n. At this point, the inputs are in a form that can be easily used in Boolean expressions. The output column is now relabeled in a similar manner. It is common practice to provide sequential-logic elements with a "name" and to use that name as an identifier in Boolean expressions. When such a "name" technique is used, the true or HIGH output is assumed. For example, if the latch circuit is called "A", then when the circuit is in the latched position, it is so identified by calling the output A. When in the unlatched position, it is identified as $\bar{A}$. The output column is then labeled A_{n+1} to denote that it is the state of the output *following* the change in input conditions defined on the same row. Table 8-4 is the relabeled state table.

Examination of the A_{n+1} column discloses that a 1 exists in the second row; therefore A_{n+1} is true (HIGH). The input requirements necessary to obtain this condition are read from the input side of the state table as $U_n = 0$ and $L_n = 1$, or $\bar{U}_n L_n$. Thus one possible Boolean equation describing a means of obtaining the true state for the latch called A is $A_{n+1} = \bar{U}_n L_n$. Another possibility exists. Note that $A_{n+1} = A_n$ if $U_n = 0$ AND $L_n = 0$. If $A_n = 1$ prior to U_n becoming equal to 0 and L_n becoming equal to 0, then A_{n+1} would equal 1 at the time of transition. Thus A_{n+1} counted equal 1 if $U_n = 0$, $L_n = 0$, and $A_n = 1$, or if $A_{n+1} = \bar{U}_n \bar{L}_n A_n$.

Now two means of obtaining the true state for the latch called A are available, that is $A_{n+1} = \bar{U}_n L_n + \bar{U}_n \bar{L}_n A_n$. But one additional fact must be taken into consideration. According to the truth table, at no time can both U_n and L_n be 1 at the same time. As we noted, the results are indeterminate, and such a situation is not allowed in digital logic. Another way of showing this condition is $U_n L_n = 0$. It may be verified from the state table that all of the allowable combinations of U_n and L_n do meet this criterion.

The two equations that fully describe the operation of the latch circuit are

$$A_{n+1} = \bar{U}_n L_n + \bar{U}_n \bar{L}_n A_n \qquad (8-1)$$

$$U_n L_n = 0 \qquad (8-2)$$

Algebraic simplification of Eq. (8-1) would be

$A_{n+1} = \bar{U}_n L_n + \bar{U}_n \bar{L}_n A_n$	Original expression
$= \bar{U}_n L_n + \bar{U}_n \bar{L}_n A_n + 0$	Theorem 7
$= \bar{U}_n L_n + \bar{U}_n \bar{L}_n A_n + U_n L_n$	Substitution Eq.(8-2)
$= U_n L_n + \bar{U}_n L_n + \bar{U}_n \bar{L}_n A_n$	Commutative Law
$= L_n(U_n + \bar{U}_n) + \bar{U}_n \bar{L}_n A_n$	Factoring
$= L_n + \bar{U}_n \bar{L}_n A_n$	Theorems 12 and 13

The General Sequential Element

TABLE 8-4 State table, Latch circuit

U_n	L_n	A_{n+1}
0	0	A_n
0	1	1
1	0	0
1	1	*

* Indeterminate

Waveform Representation of Sequential Logic Circuits

Waveform diagrams are also an accepted method of displaying sequential-logic element operation. In fact, the waveform *diagram* was the method encountered in Ch. 2, where a sequential-logic element, called the FF, was used to demonstrate the binary counting process. The waveform diagrams of Fig. 8-4 should be studied carefully and compared with other methods of sequential-logic element presentation to assure that correlation between methods is established.

PHYSICAL CONFIGURATION OF SEQUENTIAL ELEMENTS

FFs may be obtained in any of the physical configurations previously mentioned during discussion of combinational-logic element packaging. The *Dual In-line Package* (DIP) is in common use. DIP integrated circuits (ICs) generally appear mounted on epoxy-glass circuit boards. Each IC may be either permanently mounted, or inserted in a socket which allows easy removal and replacement. Interconnections on the circuit boards are of the "printed" or "etched-conductor" type and lead to multicontact plugs which are often an integral part of the circuit board.

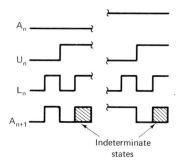

Figure 8-4 Sequential circuit waveform diagram

Circuit boards may be from 1/32 to 1/4 in. thick and vary from less than 2 × 3 in. to larger than 4 × 5 in. More than 100 external connections are found on some circuit boards with upwards of 50 ICs mounted. The actual physical configuration of circuit boards is the task of the design engineers. In fact, many manufacturers provide the mechanical design, layout, and interconnection of their logic devices directly from a user's logic diagram.

In subsequent chapters we shall see sequential logic elements (and combinational elements) combined in IC form so that complete functional logic devices are contained in a single package. Numerous FFs, already connected as counters, are commonplace. At even greater levels of component density, complete systems containing both FFs and gates give the logic user remarkable capability in a single IC.

8-3 THE *R-S* FLIP-FLOP

Just as the word *gate* is universally used to describe combinational-logic elements, so is the word *flip-flop* (FF) universally used to describe sequential-logic elements. Different kinds of gates are identified by additional words describing the function or kind of operation performed (e.g., AND gate, OR gate). FFs also employ this type of designation. Terms like *R-S* and *J-K* are used to *define how a sequential-logic element responds to input signals,* just as AND gate defines how a specific combinational-logic element responds to input signals. New words and definitions must be learned, but in so doing, sequential-logic operations become clearer and easier to use.

Characteristics

The *R-S* FF (Fig. 8-5a) usually consists of electronic circuitry with two input and two output connections. One input connection is labeled *R* (RESET) and, when activated, causes the *R-S* FF to assume the RESET (CLEAR) condition. Activation of the second input, labeled *S* (SET), causes the FF to assume the SET condition. As we indicated previously, the outputs are labeled Q and $\overline{Q}$ and are complements of each other, i.e., when one output is HIGH the other is LOW and vice versa. By generally-agreed-upon definitions, the *R-S* FF behaves according to Table 8-5.

Implementing the *R-S* Flip-flop with Combinational-Logic Elements

NOR gate implementation of the latch function has already been discussed and the logic diagram shown in Fig. 8-2. By relabeling the inputs and outputs, this same logic diagram may be used to describe the

The R-S Flip-flop

(a) Symbol   (b) State table

Figure 8-5 *R-S* flip-flop, symbol and state table

R-S FF. Of course, many other combinational-logic element arrangements perform the same function. The only requirement is that the circuit perform in accordance with the general rules and the table of operating characteristics.

Consider now the *R-S* FF operation with NAND gates. It seems reasonable that such an implementation is practical, since NOR gates have already been used with good success. Figure 8-6 shows a cross-coupled NAND gate version of the *R-S* FF and the MIL-STD-806B symbol. Investigation of the symbol reveals the following information.

1. Activation of the S input with a LOW-going signal causes the Q output to become HIGH.
2. Activation of the R input with a LOW-going signal causes the $\bar{Q}$ output to become HIGH.

The first step in analysis of the NAND gate version of the *R-S* FF is the establishment of an arbitrary starting condition. Choosing the situation where $\bar{Q}$ is HIGH (RESET state) is a conventional approach. If $\bar{Q}$ is HIGH, then Q must be LOW. For Q to be LOW, both inputs to G_1 must be HIGH, since a NAND gate requires all inputs to be HIGH for the output to be LOW. Since one input is derived from the $\bar{Q}$ output, which is HIGH, the second input ($\bar{S}$) must be HIGH at the same

Table 8-5 General Operating Characteristics, $R - S$ Flip-Flop

Input		If the Q output			
R	S	was	it becomes	was	it becomes
Inactive	Inactive	Inactive → Inactive		Active → Active	
Inactive	Active	Inactive → Active		Active → Active	
Active	Inactive	Active → Inactive		Active → Inactive	
Active	Active	*		*	

* Indeterminate

time. G_2 then receives a LOW input from the Q output of G_1, and since any LOW input to a NAND gate results in a HIGH output, the circuit is latched into the RESET state. As long as we remain in the RESET state, the $\bar{R}$ input level is not important. It may be either HIGH or LOW, although with the NAND gate version of the R-S FF, it is usually maintained in the HIGH level condition when the FF has completed its transsition from one state to the other. One of the differences between the NOR gate and the NAND gate version of the R-S FF should now be apparent. The static conditions for the NAND gate version require the inputs to be at HIGH levels, while the NOR gate version requires them to be at LOW levels.

With the FF in the RESET state, any change in the $\bar{R}$ input has no effect. The FF is latched, due to the LOW input to G_2 furnished by G_1. The only way to unlatch the circuit is to cause the Q output of G_1 to go HIGH. Changing the $\bar{S}$ input from HIGH to LOW causes this change to occur, since both inputs to G_1 are no longer HIGH. The HIGH Q output combined with the quiescent HIGH $\bar{R}$ input at G_2 causes the $\bar{Q}$ output to go LOW. G_1 now has a permanent LOW input from the G_2 output, and the circuit remains in this (the SET) state. Note that it is the *LOW-going change* that causes the FF to change state, not the HIGH-going change that is usually considered to be the activating signal. The MIL-STD-806B symbol of Fig. 8-6 makes this fact immediately apparent, due to the "inverting" circles at both the S and R inputs.

The NAND gate R-S FF thus performs the same functions as the NOR gate R-S FF—only the activating input signals are different. (This is seen by perusal of the symbol used to represent the FF.) Numerous other arrangements of combinational-logic elements may be used to implement the R-S FF, but they all perform the same basic operation. The symbol used describes the input and output conditions necessary to perform the required functions.

R-S Flip-flop Applications

Although it is the basic FF, the pure R-S FF sees little application in modern digital equipment. It is generally used in situations where only a memory function is required or where sequential operations can be implemented that do not necessitate synchronization with other parts of the sequential circuit. Most other applications of the R-S FF require external gating to modify its input/output (I/O) characteristics.

The R-S FF could be used in the machine-control example mentioned in Sec. 8-1. One state of the FF is assigned the "move-left" meaning and the opposite state to the "move-right" meaning. The choice of a state for a specific meaning ("move-left" or "move-right") is a matter for the circuit designer to decide.

The Clocked R-S Flip-flop

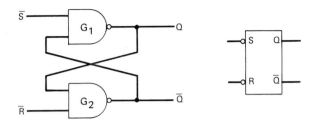

Figure 8-6 NAND gate implementation of the R-S flip-flop

Another common application for the simple latch, or R-S FF, is in the area of mechanical switch replacement. Actually, the mechanical switch is not replaced, but the function is improved by use of the FF. Due to physical construction, most mechanical switches do not provide a "clean" make or break. Spring tensions and mechanical linkages tend to cause a number of intermittent makes and breaks at every activation. When digital equipment is switched, it responds to each of the makes and breaks and often provides unwanted and erroneous operations.

Figure 8-7 shows a typical logic diagram of a device that provides a *bounceless* output for a single-pole, double-throw switch. In the position shown (A), the Q output is HIGH and the $\bar{Q}$ output is LOW. Any erratic contact with A as the switch moves toward B can have no effect, since the NAND gates cannot change state until at least one input to G_2 becomes LOW. Upon first contact of the switch arm with B, one input to G_2 is LOW, and $\bar{Q}$ becomes HIGH. G_1 now has two HIGH inputs, and the Q output is LOW. Any bounce that occurs at contact B following initial contact has no effect, since the circuit must be in the opposite state for any change to occur. The operation of the R-S FF as a contact-bounce eliminator is exactly as indicated by the state table for the

Figure 8-7 Contact bounce eliminator

NAND gate implemented function. Once in the SET state, further application of S inputs has no effect. Similarly, once in the RESET state, further application of R inputs has no effect.

8-4 THE CLOCKED R-S FLIP-FLOP

Although the R-S FF, or latch circuit, is the basis for almost all other FFs, it does not appear in great numbers in modern digital devices. The basic R-S function is commonly required, but implementing it directly with the R-S FF presents a fundamental problem. Any spurious input signals that appear on either the R or the S inputs cause the FF to change state at a time when such a change is not desired.

To alleviate this problem, the FF is often conditioned (enabled) first with a level and then allowed to make its required transition *after* it receives another input from a different source. The addition of two more NAND gates to the R-S FF of Fig. 8-6 allows the R-S FF to be controlled by *both* the SET/RESET inputs *and* the added input (commonly called the *clock input*). The clock-input signal (called a *clock pulse*) is normally of short duration compared with the time between FF state changes, so the SET and RESET inputs may be changed almost any time between clock pulses (the FF does not change state until the arrival of the clock pulse). Operation of the FF is now isolated from spurious changes of the SET and RESET inputs, except for the extremely short time that the clock pulse is present.

Clock pulses are provided by a clock generator. In *clocked systems,* clock pulses are distributed throughout the system and are used to synchronize equipment operation.

The clocked R-S FF responds exactly as the standard R-S FF, except for the function of the clock inputs. The clock pulse is assumed to be present when required, and the transition from one state to the other happens upon its occurrence.

Figure 8-8 shows the clocked R-S FF symbol, the state table, waveforms, and a typical implementation using combinational-logic elements. It should be noted that numerous implementations of the clocked R-S FF are available and that the method shown in Fig. 8-8 is merely representative.

The clocked R-S FF finds application most often in *simple* memory circuits, just as the standard R-S FF. Reduction in response to spurious input signals causes the clocked R-S FF to be more acceptable where reliability of stored information is a requirement. In addition, the clocked R-S FF also serves as the basic circuit from which the T, D, and J-K (some versions) FFs are developed.

The Clocked R-S Flip-flop

R	S	CP	Q_{n+1}
0	0	0	Q_n
0	0	1	Q_n
0	1	0	Q_n
0	1	1	1
1	0	0	Q_n
1	0	1	0
1	1	0	Q_n
1	1	1	*

$=$

R	S	Q_{n+1}
0	0	Q_n
0	1	1
1	0	0
1	1	*

If the clock pulse is assumed to be present

*Indeterminate

Figure 8-8 The clocked R-S flip-flop

8-5 THE MASTER-SLAVE PRINCIPLE

Adoption of the clocked R-S concept alleviates the problem of spurious input signals triggering the FF into operation at unwanted times. However, if the R-S FF (either basic or clocked) is used in other than simple storage applications, other difficulties arise. In many counting (Ch. 10) and shift-register (Ch. 11) applications, the Q and $\bar{Q}$ outputs of one FF are connected to the S and R inputs of another FF. The change in state of the first FF tends to cause an immediate and sometimes undesired change in state of the next FF, depending on clock-pulse width and internal delays. External gating may be used to prevent such undesired changes, but common practice is to employ the *master-slave principle*.

The basic concepts of the master-slave principle are seen in Fig. 8-9. Information on the S and R inputs is stored in the master FF at the HIGH-going transition of the clock input and transferred to the slave FF at the LOW-going transition of the clock. The S and R inputs are isolated from the Q and $\bar{Q}$ outputs, since the slave control logic is disabled when the master control logic is enabled and vice versa.

Symbology and state tables for the master-slave (sometimes called *dual-rank*) R-S FF are identical to the conventional clocked R-S FF. It is necessary, however, to realize that the actual change in the outputs does not occur until the "trailing edge" of the clock pulse. The waveform diagram in Fig. 8-9 should be carefully studied to verify operation of the master-slave FF.

8-6 THE T FLIP-FLOP

Although the R-S FF is perhaps the basis for all sequential-logic memory circuits, many other forms of FFs exist. Most of these circuits use the R-S FF as the basic component. For example, a T (*toggle*) FF, which changes state *every* time an activating signal is applied, is often used in counting circuits. T FF variations are numerous, although the basic logic circuit is quite simple. It is constructed from the simple R-S binary element with the addition of gates (or additional inputs to existing gates) to route the incoming signal (trigger) to the appropriate side of the FF. Figure 8-10 shows the logical symbol used for the T FF, its state table, construction using combinational-logic elements and typical waveforms.

Referring to the logic diagram discloses that NOR 1 and NOR 2 comprise the basic binary element, while AND 1 and AND 2 "steer" the input to the proper NOR gate to operate the T FF. Assuming an initially RESET condition, we see from the waveforms drawing that AND 2 has one input HIGH from NOR 2 and the other input LOW from the normally LOW trigger source. If the T input momentarily transitions from LOW to HIGH under these conditions, both inputs to AND 2 are HIGH, and the resulting output supplies a HIGH to NOR 2 input. Any HIGH input to a NOR gate results in a LOW output, and $\bar{Q}$ goes LOW. NOR 1 no longer has any HIGH inputs, and all LOW inputs to a NOR gate provide a HIGH output. Q goes HIGH, locking the binary into the new state (SET) and furnishing a HIGH input to AND 1. AND 2 is now locked out (disabled) with a LOW input from $\bar{Q}$, while AND 1 is enabled with its HIGH input from Q. The next momentary LOW to HIGH transition of the T input is routed through AND 1 to NOR 1, and the binary returns to the RESET state. Thus the T FF changes state for every HIGH-going transition that is received on the input line.

Figure 8-9 The master-slave principle

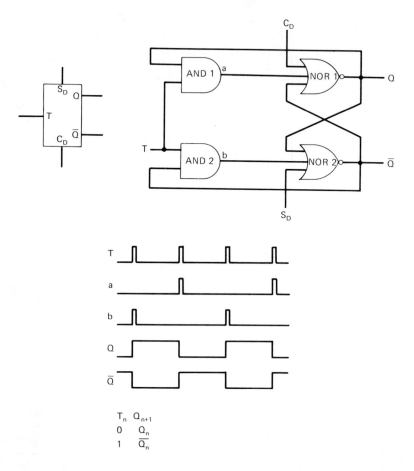

Figure 8-10 The T flip-flop

The T FF finds application in simple asynchronous (non-clocked) counting and dividing circuits, such as the logic signal source used to generate truth table combinations in earlier chapters. Many applications require the FF in the logic circuit to be in a specific state prior to the start of the circuit function. Direct SET (S_D) and/or direct CLEAR (C_D) are made available in many FFs. These inputs are used to establish the initial state of the circuit. When S_D and C_D are provided, the state table describing that portion of the FF's operation is identical to the R-S FF (Fig. 8-5). The symbol indicates the availability of the direct SET and direct CLEAR as shown in Fig. 8-10.

A retractable-point ballpoint pen acts in a manner similar to a T FF. If the operating button of the pen is pushed while the point is re-

The D Flip-flop

tracted, the point is lowered to the operating position. If already in the operating position when the operating button is pushed, the point retracts. The "output" changes each time the button is operated.

8-7 The D FLIP-FLOP

In all of the FFs discussed (except the T FF), a major problem exists. A specific combination of inputs has resulted in an indeterminate output condition, which must be guarded against with external circuitry. One way of ensuring that no indeterminate state occurs in the circuit operation is to provide only *one* conditioning input, which can be either HIGH or LOW. This input is called the D (*DATA*) input. Whatever logic level is present at the D input prior to and during the clock pulse appears at the Q output when the clock pulse occurs. Since the Q output does not assume the D input level until *after* the arrival of a clock pulse, this configuration is often called a *DELAY* FF.

Figure 8-11 shows the symbol, state table, logic diagram, and waveforms of the D FF.

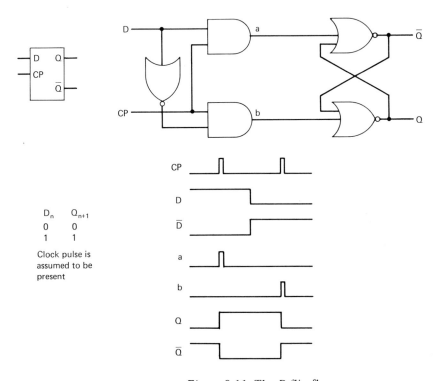

Figure 8-11 The D flip-flop

The *D* FF is used most often to delay a change in operation so that the change can occur in synchronization with the clock pulse. For example, the *D* FF may be used to store output information from a shift register or counter until readout has occurred. This allows the register or counter to begin its next cycle of operation while readout from the previous cycle is taking place.

8-8 THE *J-K* FLIP-FLOP

The *J-K* FF is perhaps the closest approach available to a truly *universal* FF. It was developed primarily to overcome the ambiguity of the *R-S* operation and is available in both *asynchronous* (nonclocked) and *synchronous* (clocked) versions. Since the synchronous version is commonly used, it is the subject of this discussion. The same state table applies for both versions (the clock pulse is assumed to be available in the synchronous mode).

The characteristics of the *R-S* and *T* FFs are combined in the *J-K* FF. New designations for inputs are chosen so that the *J-K* characteristics will not be confused with the *R-S* characteristics, but the *J* input may be considered equivalent to the *S* input and the *K* input the same as the *R* input. However the *J-K* FF may have both inputs activated at the same time—in contrast to the *R-S* circuit. Activation of the *J* and *K* inputs simultaneously causes the FF to change state upon activation by the clock pulse, just as though the *T* input of a *T* binary had been activated. The indeterminate state of *R-S* operation is not present in the *J-K* FF.

J-K symbology, state table, logic diagrams, and waveforms appear in Fig. 8-12. Note the availability of the direct SET and direct CLEAR operations on the symbol.

Due to the flexibility of the *J-K* FF, it may be found in almost all applications of sequential logic. In fact, many designers use only *J-K* FFs to take advantage of quantity purchase prices and to reduce spares requirements. If a *T* FF is needed, the *J* and *K* inputs are merely connected together as may be seen in many of the counter circuits shown in Chapter 10. The *D* FF can be obtained by applying the input signal to the *J* input and connecting an inverter between the *J* and *K* inputs. Thus the versatility of the *J-K* FF allows the user to have both *T* and *D* FFs along with *J-K* FFs by merely reconnecting inputs.

J-K, and many other types of flip-flops, may also be classified as to the manner in which they are activated. Some flip-flops are *level-triggered,* which means the state of the clock or triggering input determines when information can be transferred from input to output. Thus, when the triggering input is active, whatever information is present at

The *J-K* Flip-flop

J	K	Q_{n+1}
0	0	Q_n
0	1	0
1	0	1
1	1	$\overline{Q_n}$

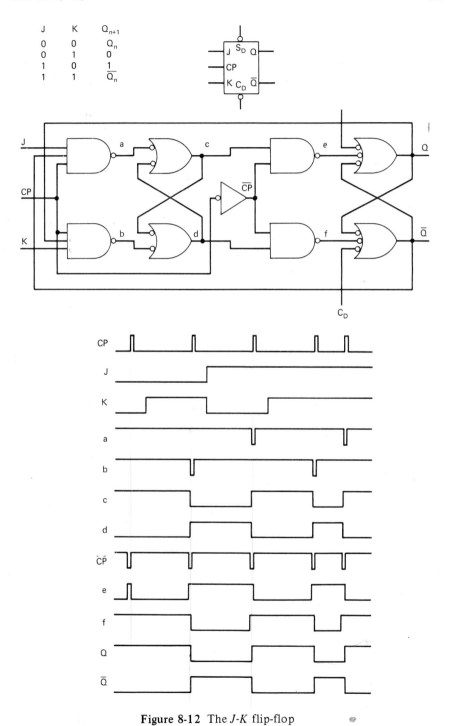

Figure 8-12 The *J-K* flip-flop

the input appears at the output. If input information changes when the triggering input is active, then the output information also changes. This can result in timing problems. Level-triggered flip-flops should be avoided if any possibility exists for input changes during the time the flip-flop is triggered.

Edge-triggered flip-flops transfer information from input to output *only* at the beginning (or end) of the clock or triggering input. It is the *change* in clock or trigger input that accomplishes the information transfer, not the level. Edge-triggered flip-flops require 0-to-1 (or 1-to-0) transitions that occur typically within 150 nanoseconds. Slower transitions generally do not allow the flip-flop to operate. Those devices that operate with 0-to-1 transitions are called *positive-edge-triggered flip-flops*; those using 1-to-0 transitions are called *negative-edge-triggered flip-flops*. The timing problems encountered with level-triggered flip-flops do not exist with edge-triggered flip-flops.

8-9 SEQUENTIAL LOGIC ANALYSIS

Modern sequential circuits, such as flip-flops, tend to be constructed as complete integrated circuits, and offer little opportunity for analysis. However, a sequential circuit will occasionally be found that is composed of interconnected gates. This section provides some insight into how such a sequential circuit may be analyzed.

Establishing Initial Conditions

Now for a typical sequential circuit. We know that the logic diagram of Fig. 8-13a represents a sequential network. Troubleshooting has isolated a malfunction to the module that contains this circuit. However we do not know what the module is supposed to do or how it functions. So we must first explore the circuit and establish its normal operation, before we can identify its abnormal operation.

If the Boolean expression is not easily ascertained through the logic diagram, alternate methods may be used. In the case of the sequential network of Fig. 8-13a, a number of feedback connections exist, thus complicating the derivation of the Boolean expression. The method shown herein is helpful under these conditions.

One approach is to make some assumptions about the inputs and outputs. Hence we may assume that the input is initially at a LOW level and that the output of G_2 is at a HIGH level. The inputs and outputs of all gates in the circuit are developed using these assumptions.

With the input initially LOW, the output of G_3 is HIGH, and when combined with the HIGH output of G_2 in G_1, it results in a LOW out-

Sequential Logic Analysis

(a) Logic diagram

(b) Waveforms

Figure 8-13 A sequential circuit

put from G_1. G_5 receives inputs from G_1 (LOW) and G_3 (HIGH), and a HIGH G_5 output is noted. One input to G_6, then, is a HIGH from G_5. The other input to G_6 is from G_4, where the LOW input level has generated a HIGH output. Therefore G_6 has both HIGH inputs and also furnishes a LOW input to G_4, thus assuring that its output remains HIGH. G_2 has a LOW input from G_1 and a HIGH input from G_4, so the initially assumed HIGH G_2 output is verified. Although many circuits provide outputs that are complementary, this is also *one* of the characteristics of the Q and $\bar{Q}$ outputs of a FF. Thus at least a small piece of evidence exists to justify identification of the circuit of Fig. 8-13a as a sequential network.

Using Waveforms

The preceding information may be displayed on a waveform diagram; it is seen in the time period t_0 to t_1 in Fig. 8-13b. Reaction of both outputs to changes in inputs is shown in the waveform diagram. Only one external input is provided, and an investigation of the internal inputs reveals that the output of G_1 is indirectly applied to both G_3 and G_4. One additional internal input should be considered so that the complete flowtable may be developed. The choice lies between G_4, G_5, and G_6. G_6 is selected because it derives its output from *both* G_4 and G_5. The reader should investigate each of the input transitions and verify that the waveform diagram adequately describes the circuit operations.

From the information furnished, it is apparent that the outputs are complementary and that they change state with every activating signal input. The circuit thus performs according to the definition of a T FF.

Boolean Expressions

Derivation of the Boolean expressions is not exactly "intuitively obvious," but the reader should now be able to perform such derivations. Each gate output is listed, and the selected internal inputs (G_1 and G_6) are derived as follows.

$$G_1 = \overline{G_2 G_3} = \overline{G_2} + \overline{G_3}$$
$$G_3 = \overline{TG_4 G_5} = \overline{T} + \overline{G_4} + \overline{G_5}$$
$$G_5 = \overline{G_1 G_3} = \overline{G_1} + \overline{G_3}$$
$$G_1 = \overline{G_2} + \overline{G_3}$$
$$= \overline{\overline{G_1 G_4}} + \overline{TG_4 G_5}$$
$$= G_1 G_4 + \overline{TG_4 G_5}$$
$$= G_1 G_4 + T(G_4 G_5)$$
$$= G_1(\overline{T} + \overline{G_6}) + T\overline{G_6}$$
$$\therefore G_1 = \overline{T}G_1 + T\overline{G_6} + G_1 \overline{G_6}$$
$$\text{(Eq. 9-5)}$$

$$G_2 = \overline{G_1 G_4} = \overline{G_1} + \overline{G_4}$$
$$G_4 = \overline{TG_6} = \overline{T} + \overline{G_6}$$
$$G_6 = \overline{G_4 G_5} = \overline{G_4} + \overline{G_5}$$
$$G_6 = \overline{G_4} + \overline{G_5}$$
$$= TG_6 + G_1 G_3$$
$$= TG_6 + G_1(\overline{TG_4 G_5})$$
$$= TG_6 + G_1[\overline{T} + (\overline{G_4} + \overline{G_5})]$$
$$= TG_6 + G_1(\overline{T} + G_6)$$
$$\therefore G_6 = TG_6 + \overline{T}G_1 + G_1 G_6$$
$$\text{(Eq. 9-6)}$$

8-10 SUMMARY

Each of the major classifications of FFs have been discussed in this chapter. Different methods of presenting the operating characteristics were used so that the reader could choose the most understandable approach. However one method of presentation is common to almost

Table 8-6 Master State Table

Inputs							R-S	Clocked R-S	T	D	J-K	Clocked J-K
S	R	T	D	J	K	CP	Q_{n+1}	Q_{n+1}	Q_{n+1}	Q_{n+1}	Q_{n+1}	Q_{n+1}
0	0						Q_n					
0	1						0					
1	0						1					
1	1						*					
0	0					0		Q_n				
0	0					1		Q_n				
0	1					0		Q_n				
0	1					1		0				
1	0					0		Q_n				
1	0					1		1				
1	1					0		Q_n				
1	1					1		*				
		0							Q_n			
		1							$\overline{Q}_n$			
			0							0		
			1							1		
				0	0						Q_n	
				0	1						0	
				1	0						1	
				1	1						$\overline{Q}_n$	
				0	0	0						Q_n
				0	0	1						Q_n
				0	1	0						Q_n
				0	1	1						0
				1	0	0						Q_n
				1	0	1						1
				1	1	0						Q_n
				1	1	1						$\overline{Q}_n$

* Indeterminate

all specification sheets, textbooks, etc. The *state table* precisely defines the operating characteristics of any sequential logic device. Table 8-6 is a master state table that combines all of the information from the individual state tables. It summarizes the information in this chapter and should be carefully studied so that the identifying characteristics of each type of FF are understood.

One additional fact is worth mentioning: *most FFs have provision for direct SET and RESET without the need for clock pulses or internal gating.* This allows preconditioning of FFs when used in counter and shift-register applications so that all stages start in the required logic state.

PROBLEM APPLICATIONS

8-1. Define sequential logic as opposed to combinational logic.
8-2. Why is sequential logic required?
8-3. What is a logic latch?
8-4. Explain the response of a latch circuit to LATCH and UNLATCH inputs.
8-5. Is it possible to determine the operating characteristics of a FF from its MIL-STD-806B symbol? How?
8-6. What is a state table? How is it used?
8-7. List two additional methods of representing sequential functions and explain how they are used.
8-8. Show the characteristics of the R-S FF and explain how these characteristics may be applied to a practical sequential circuit.
8-9. Explain why the R-S FF locks into one or the other state when both the R and S inputs are activated.
8-10. Define asynchronous sequential circuit. List the advantages and disadvantages.
8-11. Define synchronous sequential circuit. List the advantages and disadvantages.
8-12. Is the clocked R-S FF synchronous or asynchronous? Why?
8-13. Discuss the operation of the clocked R-S FF, explaining the operation of each of the gates in Fig. 8-8.
8-14. In the circuit of Fig. 8-8, what is the effect of changing the R and/or S input levels when the clock pulse is not present?
8-15. Why is the master-slave R-S FF preferable to the standard R-S FF?
8-16. Discuss the T FF and its applications.
8-17. What are the minimum number of inputs required for operation of the synchronous D FF? The asynchronous D FF? Explain.
8-18. Does changing the input level of the D FF during the activating time of the clock pulse affect the output after the transition? Explain.
8-19. What is a J-K FF? What are its advantages and disadvantages?
8-20. Why are there no indeterminate states in the operation of the J-K FF? Explain by reference to Fig. 8-12.

9

TROUBLESHOOTING COMBINATIONAL/SEQUENTIAL LOGIC

Having developed combinational logic in Chs. 3 through 7 and sequential logic in Ch. 8, our next step is an examination of circuits containing *both* combinational and sequential logic elements. In practical applications, most functions performed by digital logic combine both types of circuits. The analysis and troubleshooting of *combined circuits* closely parallels the techniques discussed in Ch. 7. Due to the increased complexity, however, some modifications are necessary.

9-1 A PROPOSED PROCEDURE

Chapter 7 (Sec. 7-1) describes the preparatory steps required to investigate a combinational logic circuit. These preparatory steps apply to practically all situations, whether they are combinational, sequential, or a combination of both. So before we begin our actual analysis, we must obtain *the overall logic-function description, define the inputs* and *outputs, establish timing* (if any), and *define the rules of logic* (if nonstandard symbols are used).

At this point, a deviation from the procedure in Ch. 7 is required. The sequential and combinational functions should be separated *first*. This permits each type of logic circuit to be discussed separately and reduces the confusion of working with circuits that operate in different manners. It may be difficult, however, to decide what portions of the circuit are sequential and what portions are combinational. No hard and fixed rules are used, but some clues appear in most logic diagrams.

If the symbol for the flip-flop (FF) is noted, that portion of the circuit is definitely sequential. Logic circuits that have outputs connected *back* to inputs within the same general circuit *could* easily be sequential and should be perused closely. In many cases, the sequential functions will be grouped separately from the combinational functions —an obvious advantage. Most of all, *experience* with a generous serving of common sense must be liberally applied to adequately separate the two types of functions. The reader should investigate as many logic circuits as possible to gain this experience.

Actual analysis of combinational logic functions presents little difficulty if the material in Chs. 3 through 7 has been carefully studied and well understood. Sequential logic circuits may be a bit more challenging. Within Ch. 8, sequential logic was approached as interconnections of combinational logic circuits. Little was mentioned, however, concerning the interconnection of sequential logic circuits in performing time-dependent operations. Such discussions properly belong in the chapters concerned with counters and shift registers. Simple time-dependent sequential circuits are discussed in this chapter, however, to point out their application in everyday logic circuitry.

After the combinational and sequential circuits have been investigated separately, the functions may be recombined to develop an overall description of the circuit. All of the common descriptive methods may be applied to combined circuits.

It should be recognized that combinational and sequential circuits combined in the manner shown in this chapter are most often found in small, individually designed devices. When used in commercial applications, where hundreds or thousands of units are built and sold, a special *Large Scale Integration* (LSI) circuit chip would probably be designed and used. *All* functions would be included on one integrated circuit chip, and troubleshooting would be limited to removal and replacement of that chip. More versatile implementations of this type of function are performed using microprocessors. (See Chapters 15 and 16.)

9-2 A TYPICAL MIXED FUNCTION

The Logic Diagram

Practically all logic-circuit analysis requires a *logic diagram*. Usually, the analysis is performed because of a circuit malfunction, and from a troubleshooting standpoint, the problem must be isolated to a specific portion of the overall system. The logic diagram for this part of the system is then reviewed, and if other descriptions are not available, investigation begins at this point. Figure 9-1 is a logic diagram that must be analyzed. It represents a control unit for positioning a worktable

Figure 9-1 Control unit logic diagram

carrying metal parts to be machined into shape for use in aircraft. A description of the function of the control unit follows.

The control unit responds to two limit switches (A and B) which are used to detect the position of the worktable. In addition, the unit reacts to an internally generated function (C_n), which is a function not only of the internal function's immediate past history (C_{n-1}) but also of the present state of the limit switches. Operationally, the control unit provides an output I_n (an *indexing pulse*) whenever any one (and *only* one) of the three inputs (the two limit switches and the internally generated input) are logic 1; also an output is produced when all three inputs are logic 1. The state of the internally generated function C_n is determined by the requirement that it be logic 1 when any two of the inputs provide a logic 1 and also when all three inputs are logic 1. The control unit also provides an indexing direction capability. For example, the table may be moved four steps to the left, then reversed and moved six steps to the right. Each index pulse must therefore be counted, and a direction-change signal generated after every four and every six index pulses.

The logic description of the circuit was just provided in English-language form. But, as shown previously, such descriptions tend to be ambiguous and open to individual interpretation. The assignment of letter symbols to each of the variables, common practice in previous chapters, is continued here.

$$\begin{aligned}A &= \text{One limit switch}\\ B &= \text{The other limit switch}\\ C_n &= \text{The internally generated function}\\ I_n &= \text{The index pulse (output)}\\ I_L &= \text{Index-left direction}\\ I_R &= \text{Index-right direction}\\ C_{n-1} &= \text{The immediate past state of } C_n\end{aligned}$$

The equations for I_n and C_n become

$$I_n = \overline{AB}C_{n-1} + \overline{A}B\overline{C}_{n-1} + A\overline{B}\overline{C}_{n-1} + ABC_{n-1} \qquad (9\text{-}1)$$

and
$$C_n = \overline{A}BC_{n-1} + A\overline{B}C_{n-1} + AB\overline{C}_{n-1} + ABC_{n-1} \qquad (9\text{-}2)$$

To minimize these equations, either algebraic manipulation or logic maps may be used, yielding.

$$I_n = ABC_{n-1} + \overline{C}_n(\overline{\overline{AB}C_{n-1}}) \qquad (9\text{-}3)$$

and
$$C_n = AB + BC_{n-1} + AC_{n-1} \qquad (9\text{-}4)$$

The inputs and outputs have tentatively been described, but further clarification may be in order. A and B represent switches that may be either on *or* off. For the purpose of this discussion, when a switch is *on* (actuated), it furnishes a true, 1, or HIGH input; when it is *off* (not actuated), it furnishes a false, 0, or LOW input. Further, since *pulsed outputs* are desired, the portion of the circuit associated with the output is actuated by a continuously operating string of clock pulses. Outputs I_L and I_R are pulses occurring at the appropriate time.

Special rules of logic are not necessary, since standard logic symbology is used. Also, the timing functions become apparent as the discussion progresses.

Identification and Breakout of Combinational and Sequential Functions

Determination of a starting point for separation of the combinational and sequential logic circuits may be the most difficult part of mixed-functions investigation. In many cases, the development of a *block diagram* from the English-language description proves helpful. At this point, a complete logic-diagram approach is not necessary and may even complicate matters.

From the initial description, note that the three inputs (A, B, and C_{n-1}) are combined to produce I_n and C_n. The terms used to describe the manner of combination are *combinational terms,* implying that combinational logic is used to perform these functions. Figure 9-2 is the block diagram developed from the description of the control unit.

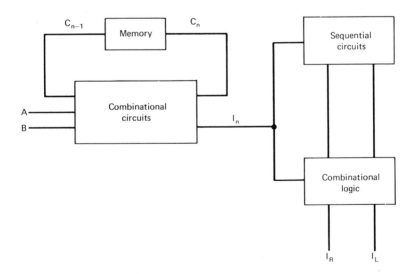

Figure 9-2 Control unit block diagram

To determine the immediate past state of C_n, a memory function is required. C_n is the input, and C_{n-1} is the output. By earlier definition, the basic memory function is sequential in nature, and a separate block is shown for this function. Counting of index pulses I_n is performed with FFs (sequential circuits) in a manner similar to the logic-signal source counter discussed in earlier chapters. Detection of the number of counts is furnished by combinational decoding circuits in much the same manner as the decoder in Ch. 3. Reference to the block diagram should ease the difficulty of identifying the combinational portions of the logic diagram.

The *symbology* used in the logic diagram aids identification of the combinational and sequential portions of the circuit. The FF symbols shown in Part III of Fig. 9-1 are indicative of sequential circuits, and by combining the description and block-diagram knowledge, the major portion of the sequential circuits may be easily isolated. The function of the ONE-SHOT blocks is clarified during the discussion of Part III of Fig. 9-1. Although Part V is actually combinational logic, its close functional tie-in to Part III requires investigation at the same time as Part III. However, due to the implementation of the memory function it *may* not be easily identified.

In any logic circuit, a good indicator of possible sequential circuitry is the *cross-coupled gate arrangement* shown in Ch. 8. Such a connection may be seen with G_{15} and G_{16}. In addition, a review of the block diagram reveals that an output from the combinational circuits is fed through the memory circuits and then back to the input of the combinational circuits. Close perusal of the logic diagram, tracing back from the signal fed to the sequential circuits, reveals that an output from G_8 feeds gates G_{12}, G_{13}, and G_{14}, which are connected to the cross-coupled gates G_{15} and G_{16}. The cross-coupled gates then feed back into gates G_4, G_5, and G_6, which are used to combine inputs A and B. Thus the circuitry between G_{12} and the output of the cross-coupled gates G_{15} and G_{16} may temporarily be considered to be the memory circuit (see Part II, Fig. 9-1). If needed, this decision may be modified as more information is gathered during the exploration of the combinational circuits.

Parts I, IV, and VI remain to be identified. No apparent cross-coupling exists in Part I, and the gates seem to be combining inputs. Part I will thus be tentatively classified as combinational. Part IV contains only two gates, and the logic diagram shows that they are used to *combine* the Q and $\overline{Q}$ outputs of FF B with the delayed output of G_{11}. The function of the DELAY block is discussed during the analysis of Parts III, IV, and V. Part IV is considered combinational.

Part VI contains FFs and is therefore classified as sequential. Its function is to store the direction of worktable travel and to light indicator lamps that show that direction.

Investigation of Combinational Circuits (Part I)

The combinational circuits that combine A, B, and C_{n-1} are shown in Part I of Fig. 9-1. As might be expected, Eqs. (9-1) and (9-2) are not implemented directly from the truth table. More efficient methods may be found, and the minimized forms of Eqs. (9-3) and (9-4) are used. The reader is urged to use Boolean algebra and/or logic maps to verify the validity of these equations. Each gate's output is shown below so that it may be verified that the C_n and I_n equations have been properly implemented.

Gate	Output
1.	$\overline{A}$
2.	$\overline{B}$
3.	$\cdot \overline{\overline{A}\,\overline{B}} = A + B$
4.	$\overline{\overline{B}\,\overline{C}_{n-1}} = B + C_{n-1}$
5.	$\overline{\overline{A}\,\overline{C}_{n-1}} = A + C_{n-1}$
6.	$\overline{\overline{A}\,\overline{B}\,\overline{C}_{n-1}} = A + B + C_{n-1}$
7.	$\overline{ABC_{n-1}}$
8.	$\overline{(A+B)(B+C_{n-1})(A+C_{n-1})}* = \overline{C}_n$
9.	$\overline{\overline{C}_n(A + B + C_{n-1})}$
10.	$\overline{\overline{C}_n(A+B+C_{n-1})(\overline{ABC_{n-1}})} = \overline{C}_n(A+B+C_{n-1}) + (ABC_{n-1}) = I_n$
11.	$\overline{I_n C_p}$

The clocked I_n output of Part I is also used in its negated form in the remainder of the logic circuit; that is, the FFs in Part III and DELAY in Part IV are activated by a LOW-going input (as indicated by the symbols) rather than a HIGH-going input.

I_n is "clocked" so that the operation of the sequential portions of the logic circuit may be synchronized. Also, unwanted changes in the combinational portion of the circuit will not cause changes in the sequential circuits, except during the clock time. Immunity to noise and external influences is thus provided.

In summary, the combinational logic circuit shown in Part I of Fig. 9-1 combines inputs A and B with an internally derived input C_{n-1}. It provides an output C_n for Part II according to Eq. (9-4). In addition, an output I_n is derived according to Eq. (9-3) and combined with an external clock signal to furnish LOW-going pulses to be used by the sequential circuits of Part III.

---

*Note that this is the *negation* of the AND-OR form of Eq. (9-3). Thus it represents $\overline{C}_n$. If C_n is required, it may be obtained by the use of an inverter. However it is used in memory in its negated form.

Investigation of Sequential Circuits (Part II)

Evaluation of Part II, which has tentatively been identified as a sequential circuit, utilizes the methods in Ch. 8. Two inputs (C_n and C_p) and two outputs (C_{n-1} and $\overline{C}_{n-1}$) are evident. Boolean equations and the waveforms for the logic circuitry of Part II are shown in Fig. 9-3. The cross-coupled gate configuration usually results in the outputs being complements of each other.

The information in Fig. 9-3 reveals that the logic circuitry of Part II functions according to the standard definition of the D FF; that is, the logic state at the input (C_n) is reflected at the output (C_{n-1}) following the arrival of the next clock pulse. The output is thus delayed by *one* clock period, and the output reflects what the input was during the *previous* clock period.

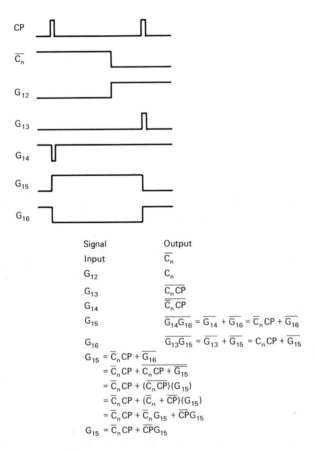

Figure 9-3 Analysis of Part II, Figure 9-1

A Typical Mixed Function   263

Analysis of Sequential Circuits (Parts III, IV, V, and VI)

With the information given, perhaps the most appropriate method for examining the logic circuitry of Part III is the use of *waveforms*. In fact, since Parts IV and V are so closely aligned operationally with Part III, a composite analysis can be performed. In sequential circuits, it is common to establish a "starting point" from which to start analysis. G_{17} and G_{18} perform this function. A *preset input* is provided, which occurs upon initial power turn-on or when it is necessary to restart the control unit. Application of this LOW-going pulse to G_{17} and G_{18} results in HIGH-going pulses at the direct SET inputs of all four FFs, and the FFs all begin operation in the SET condition. The function of the ONE SHOT is discussed shortly. Figure 9-4 shows the preset waveforms and all other waveforms pertinent to the operation of Parts III, IV, and V of the logic circuit of Fig. 9-1.

Investigation of the waveforms and interconnections of the FFs shows that the circuit closely resembles the configuration and operation of the logic-signal source circuit discussed in Ch. 2. The resemblance stops, however, with the arrival of the ninth input pulse. At this time, a rather startling thing happens. FFs *B* and *C* are not allowed to assume their natural states for the input sequence; they are forced to the SET state. Since the requirements of the control unit dictate four discrete states for the control of worktable movement to the left and six discrete states for control of movement to the right, only ten states are required. Without the forcing operation, the FF connections shown in Part III would result in sixteen different states, and improper table movement would result.

The FFs of Part III function like a *simple binary counter*, except for the feedback connections. Binary counting was adequately discussed in Ch. 2, so only the forcing function is discussed at this time. After normal operation with the first eight input pulses to be counted, the ninth pulse establishes a sequence of events that ultimately results in the outputs of FFs *B* and *C* being forced to the SET state, despite the fact that with the arrival of the 9th pulse, their normal states would be the CLEAR state. The inset shown in Fig. 9-4 shows this sequence on a vastly expanded time scale. FF *D* cannot change state simultaneously with arrival of the ninth input pulse, due to delays inherent in the gates making up the FF. After the short delay, however, the Q_D output goes LOW and causes FF *C* to start changing state. Another delay occurs in this FF, but soon its Q output also goes LOW. FF *B* follows suit shortly thereafter, triggering *A* into action. $\bar{Q}_A$ goes LOW for the first time in the sequence, causing the ONE SHOT device to operate. *The* ONE SHOT *is merely a special kind of FF that develops a single output pulse every time it is actuated.*

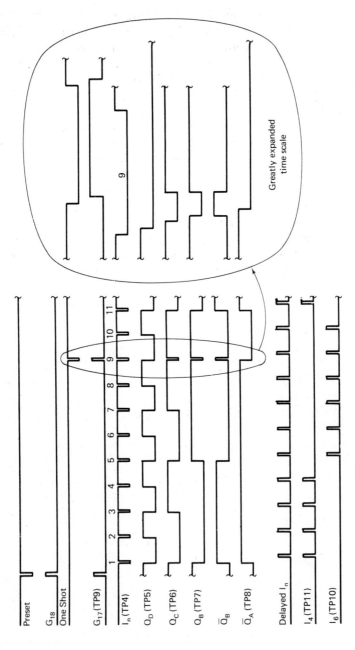

Figure 9-4 Waveforms for Part III, Figure 9-1

TABLE 9-1 Count progression of counter in Part III

Input pulse	Counter	Remarks
Initial condition	1111	Established by PRESET or counter recycle
1	0000	
2	0001	
3	0010	
4	0011	
5	0100	
6	0101	
7	0110	
8	0111	
9	1110	Forced count to eliminate unnecessary states
10	1111	Initial condition—counter starts over

The output of the ONE SHOT is applied to the input of G_{17}, and the inverted pulse from G_{17} sets FFs *B* and *C*. Thus the result of the ninth input pulse is to place FFs *B* and *C* in the SET condition rather than the CLEAR condition. Counting continues in the usual manner, until the equivalent of 15_{10} is reached. The next pulse to be counted clears all FFs by propagating a LOW-going signal to each binary, and the count sequence starts over again at 0000.

Actually, the count sequence in Part III is a bit unconventional. Although it is not absolutely necessary, many binary counters begin the counting sequence with the equivalent of 0 and progress with binary equivalents of decimal numbers in the conventional sequence. Such a method is easy to visualize, and conversion to decimal numbers is simplified. Other sequences result in certain advantages, which will be discussed in Ch. 10. The counting sequence used in Part III behaves in a conventional manner part of the time, and the remainder of the time it is forced to a *different sequence* to fit the requirements of the circuit. Table 9-1 lists each state of the counter with comments (when required).

G_{19} and G_{20} (Part IV) are AND gates with inputs from $\bar{Q}$ and Q outputs of FF *B* and a common input from the DELAY block. Due to the inherent delays encountered in the FFs of Part III, the gates of Part IV become enabled from FF *B* at some finite time following arrival of the pulses to be counted. Since we desire to develop output pulses to move the worktable, application of the pulses to be counted to the gates could result in no or limited output, due to the delays. Therefore an element called a DELAY is inserted, and the pulses to be counted are delayed an amount of time sufficient to allow the FFs to stabilize. It may be seen from the symbol (nonstandard functions are symbolized by a rectangle with an aspect ratio of 2:1 or greater according to MIL-STD-806B) that the LOW-going input to the DELAY element results in a HIGH-going output pulse of the same length at some time later than the occurrence of the original pulse. Under these circumstances, it is assured that input pulses to G_{19} and G_{20} arrive *after* the gates become enabled from FF B.

Figure 9-4 shows the development of the four pulses required for left movement of the table and the six pulses for right movement of the table. The complete operation of Parts III, IV and V should now be reviewed with the complete waveform diagram, and the reader should verify that the sequential requirements of the system have been met.

Part VI provides visual display of the operation of the control unit. FF E stores the fact that the first "move left" command has occurred, while FF F serves a similar function for the "move right" command. The I_L pulse output not only sets FF E but also clears FF F each time it occurs. I_R pulse output sets F and clears E. When the LEFT indicator is illuminated, it means that *at least one* "move left" command has occurred and that no "move right" commands have yet been generated. The first "move right" command clears E, turns off the LEFT light, sets F, and turns on the RIGHT light. Thus the indicator lights show the direction that the worktable should be moving.

Recombination of Circuits

Now that the operation of the combinational and sequential portions of the logic circuit are developed, the complete system operation should be investigated to ensure that the analysis matches the system requirements.

In the case of the *control unit*, it is advantageous to start from the output and work back toward the inputs to integrate the combinational and sequential logic circuits. The overall output requirement was to provide control signals that would move the worktable four steps to the left and then reverse it to move six steps to the right. Both the number of steps and directional movement have been developed by Parts III, IV, and V. Four outputs pulses are seen at the output of G_{19}, while six output pulses are made available at the output of G_{20}. It is now merely necessary to connect these outputs to the electromechanical mechanism that moves the worktable.

Next we must verify *the criteria for development of the pulses* that are counted by Parts III, IV, and V. Comparison of the equation for I_n, developed from the English-language statement, with the equation for I_n that results from the logic implementation should be sufficient verification. C_n must be developed in a similar manner prior to development of I_n, and C_{n-1} is necessary for both C_n and I_n. Previously provided detail is sufficient to verify that C_{n-1} actually represents immediately past values of C_n. C_n provided by the logic circuitry is compared

A Typical Mixed Function

TABLE 9-2 Comparison of C_n Requirements with Logic Implementation of C_n

A	B	C_{n-1}	C_n	From logic diagram $\overline{C_n}$	C_n
0	0	0	0	1	0
0	0	1	0	1	0
0	1	0	0	1	0
0	1	1	1	0	1
1	0	0	0	1	0
1	0	1	1	0	1
1	1	0	1	0	1
1	1	1	1	0	1

—— Identical ——

$C_n = \overline{A}BC_{n-1} + A\overline{B}C_{n-1} + AB\overline{C_{n-1}} + ABC_{n-1}$ (From English language statement)

$\overline{C_n} = \overline{(A+B)(B+C_{n-1})(A+C_{n-1})}$ (From Logic diagram)

$= \overline{(A+B)} + \overline{(B+C_{n-1})} + \overline{(A+C_{n-1})}$

$\overline{C_n} = \overline{A} \cdot \overline{B} + \overline{B} \cdot \overline{C_{n-1}} + \overline{A} \cdot \overline{C_{n-1}}$

with C_n described by the English-language description in the truth table of Table 9-2. A similar proof for I_n is left to the reader.

After proving that the logic implementation of I_n matches the initial system requirements, the clocked I_n pulse fed to Part III is used by the sequential portion of the logic circuit and produces the required system outputs. The complete control unit has now been explored, and sufficient information is available to troubleshoot any malfunction that may appear in the logic circuitry. Subsequent sections in this chapter will use this data as the basis for troubleshooting.

9-3 TROUBLESHOOTING MIXED FUNCTIONS

Introduction

Troubleshooting techniques for mixed functions closely approximate the combinational logic techniques. Some minor variations are necessary to compensate for the *time dependency* of the sequential logic circuits. The technique of separating combinational and sequential circuits during analysis allows the use of *flowcharts* (as modified in this chapter) in troubleshooting the combinational circuits. Within this chapter, sequential logic troubleshooting techniques are developed and then integrated with combinational techniques, resulting in an overall digital-logic troubleshooting guide.

The preparatory steps for troubleshooting mixed functions are also similar to those steps used in combinational logic. Chapter 7 supplies the basic information concerning these preparatory steps, which must be implemented prior to any attempt at troubleshooting mixed functions; that is, complete analysis is assumed along with collection of all available technical data concerning the circuit.

Using Built-in Indicators

When troubleshooting logic circuits that contain both combinational and sequential logic, it is quite important to have a complete grasp of the meaning and function of each of the *indicators*. If such information is not furnished with the equipment, analysis should be performed to define each indicator's function.

The flowchart in Fig. 7-10 is followed when troubleshooting logic circuits containing both combinational and sequential logic, but one important fact must be pointed out. When sequential elements are used, *their outputs are a function of not only the present input conditions but also past input conditions*. Complete analysis must be performed so that the required combination of inputs and the number of times these combinations must be provided are obtained for the purpose of verifying complete circuit operation.

One of the problems encountered in troubleshooting logic circuits that contain sequential elements is the *speed* at which such elements and circuits operate. Circuit conditions change rapidly—so rapidly that indicator lights may not even have time to illuminate. If the circuit may be operated at a slow speed for test purposes (which is common), then the effect of a single input change may be viewed before going on to the next input change. Certain critical functions, however, may require constant monitoring. In these cases, the indicators are supplied by latching circuits (usually FFs), which maintain the indicator in the ON condition until reset. An example may be seen in the LEFT and RIGHT indicators of Fig. 9-1.

The basic difference in the use of Fig. 7-10 during troubleshooting of the sequential part of a circuit is in the *type of information* that is to be recorded and used to determine the malfunctioning logic element. With combinational elements, the inputs were varied until all possible combinations of the input variables had been tried. With sequential circuits, the input that causes the sequential circuit to change state must be applied a sufficient number of times to cause operation of the circuit throughout its design regime. In other words, if the device is a counter, enough input applications must be furnished to allow the counter to operate completely through its design range and to perform all of the assigned operations.

A Typical Mixed Function

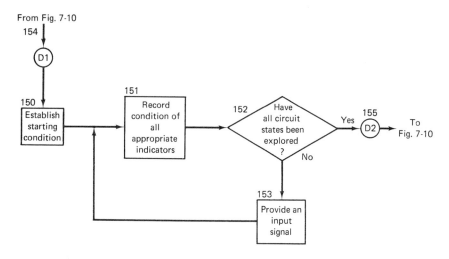

Figure 9-5 Addition to flow chart for troubleshooting logic circuits (Figure 7-10)—sequential circuits

Due to the fact that the inputs to sequential circuits usually are supplied by combinational circuits (as in the example used in this chapter), it is preferable to troubleshoot *each type of circuit* separately. If the information available from the built-in indicators does not make the nature of the malfunction immediately obvious, then every attempt should be made to isolate the combinational and sequential circuits for separate troubleshooting. Figure 7-10 should be used first on the combinational circuits to ensure that they are capable of furnishing the correct inputs to the sequential circuits. The sequential circuits should then be approached using the modified flowchart (Fig. 9-5).

A starting condition, such as the PRESET condition of the circuit in Fig. 9-1, is established in Block 150. The condition of the built-in indicators is recorded in Block 151 to serve as a reference point. If all possible states of the circuit have not yet been examined (Block 152), an input signal is applied to cause the circuit to assume another state (Block 153). As soon as all possible states of the circuit have been examined and the data recorded (Block 151), operations return to Fig. 7-10. These concepts will be applied in Sec. 9-4, where an example of malfunction location in the diagram of Fig. 9-1 is shown.

Using External Indicators

When insufficient information is furnished by built-in indicators, the use of *external indicators,* discussed in Sec. 7-3, is a useful troubleshooting technique. The use of external indicators allows a more detailed investigation of the logic circuits. Whereas built-in indicators allow isolation of malfunctions only to a *general* area of circuitry, external indicators allow investigation of *individual* circuit elements. Combinational and sequential sections of the circuit should be separated so that the time dependent operations may be more efficiently investigated. Figure 9-6 shows the modifications required to the flowchart in Fig. 7-12 so that sequential circuits may be examined.

The only new concept used in Fig. 9-6 is that of *opening feedback loops* to investigate circuit operation (Block 267). This allows the sequential circuit to operate independent of the feedback, thus reducing the complexity of the troubleshooting. The remainder of the blocks should be self-explanatory.

Using Waveforms

The use of *waveforms* as a troubleshooting aid finds considerable application in mixed function circuits—especially in the sequential portions. Oscilloscope presentations correlate directly with the waveform diagrams and allow quick identification of malfunctions. The *cause* of the malfunction may not be immediately apparent, but the reaction of the sequential circuit to inputs and the changing of circuit elements rapidly detects the malfunctioning component. The flowchart in Fig. 7-13 (as modified in Fig. 9-7) is applicable to sequential circuits.

Note that troubleshooting sequential circuits using Fig. 9-7 reduces the number of steps required to locate malfunctioning components. The oscilloscope's ability to show information on a time scale makes it readily applicable to analysis and troubleshooting of sequential circuits.

Perhaps one of the most effective ways to troubleshoot logic circuits is with logic analyzers. These are special-purpose oscilloscopes designed for logic purposes. In practical effect, logic analyzers are multi-channel oscilloscopes that allow the investigation of more than one input at a time. Thus it is possible to connect to a number of different locations in a logic circuit and see the relationship of each location's operations as the circuit performs its functions.

With a typical single-channel oscilloscope, it is necessary to move the input connection to each location, trying to match up each group of waveforms. With a logic analyzer, these relationships are seen immediately. The flowchart of Fig. 9-7 is still used, but becomes much more effective with a number of different locations available for analysis on the analyzer.

Modern logic analyzers have several features that make troubleshooting very easy. Each location being viewed can have its sequence of changes stored in an internal memory. When a preselected event

A Typical Mixed Function

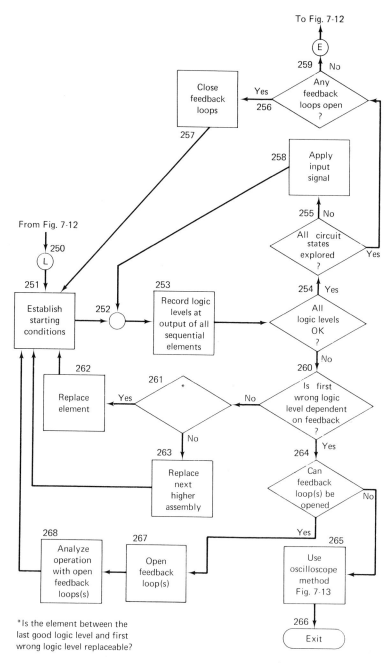

Figure 9-6 Flowchart for troubleshooting sequential circuits using external indicators—use with Figure 7-12

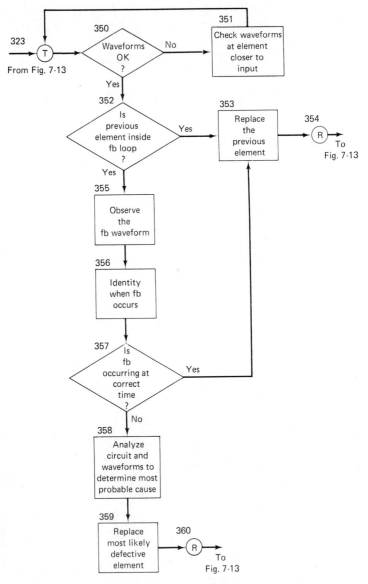

Figure 9-7 Flowchart for troubleshooting sequential circuits using waveforms—use with Figure 7-13

occurs, the information in memory is displayed. In fact, the information may be selected in time relationship to the event, so that what has happened before and after the event can be seen. Such time-related information is very valuable while troubleshooting sequential logic circuits.

Troubleshooting Mixed Functions 273

Figure 9-8 Logic analyzer (courtesy Hewlett-Packard)

The information actually displayed by the logic analyzer can be in a number of different forms. Many technicians and engineers are familiar with waveforms and use that form of display. When it is more appropriate, the highs and lows of the waveform display can be replaced by 1s and 0s on the analyzer. Some analyzers even display information in hexadecimal or octal form. When extremely complex digital systems are encountered, more detailed displays may be employed. Fig. 9-8 shows a logic analyzer in use.

When the logic analyzer's primary function is to investigate circuit waveforms, it is called a *logic timing analyzer*. If the changes in circuit states (1s and 0s) as the device operates are important, a *logic state analyzer* is used. Many logic analyzers are capable of operation in either mode.

Another form of complex logic circuit troubleshooting is accomplished with a *signature analyzer*. Signature analysis requires knowing what logic pattern of 1s and 0s exists at every available test point in a device. The common way to implement signature analysis is to build in a self-test capability for each device. The known self-test then causes known results at each test point. The results are converted into a 4-digit hexadecimal number and printed as part of the device's schematic diagram or as a troubleshooting flowchart. Thus it is only necessary to connect the signature analyzer to various test points and start the self test. The display on the signature analyzer then shows the "signature" of that test point. When an incorrect "signature" is noted, the technician follows the schematic or flow chart to the source of the problem, signature by signature. Satisfactory use of signature analysis requires major effort by the manufacturer during design of a digital device. Although signature analysis can be used on digital devices without internal self-test, it is too time-consuming usually. Signature analysis shows great promise for maintaining future digital devices and computers because it reduces the amount of training required for technicians.

9-4 A TYPICAL TROUBLESHOOTING PROBLEM

Preliminary Information

The circuit used in this troubleshooting example is the control unit of Fig. 9-1. To simplify the troubleshooting effort, all the information obtained during the circuit analysis in Sec. 9-2 is available, including the initial description of the unit's requirements, the algebraic expressions, tables, the logic diagram, and the waveforms.

Physically, the control unit consists of a printed circuit card with the logic elements plugged into sockets mounted on the card. Test points are provided on one edge of the card so that they are available when the card is plugged into its socket or into a test fixture. The LEFT and RIGHT lights are also mounted on the edge of the card and are visible either during actual operation of the control unit or during a test. Test points are indicated on the logic diagram in Fig. 9-1.

An equipment malfunction was noted when the worktable ceased to move, although the LEFT light was illuminated. The RIGHT light remained off. Since the maintenance concept for the machine was to remove the apparently defective card and replace it with a known good card, this was done. The machine then worked satisfactorily, and the apparently defective card was returned for repair. Testing and troubleshooting of the control unit card is to be accomplished in a test fixture. The test fixture provides power, individual control of each input signal, an oscilloscope, and a logic probe to detect and display logic levels. These are the given conditions, and it remains to verify the malfunction, locate the cause of difficulty, and repair the unit for future use.

Verification of the malfunction is accomplished by determining if the indicated conditions can be duplicated. The logic diagram shows that it is necessary to obtain at least one I_4 pulse to turn on the LEFT light. An I_4 pulse may be obtained by presetting the sequential counter (Part III) to 1111 at power turn-on, receiving an I_n pulse and, subsequently, a delayed I_n pulse. The presetting operation is *automatic*. An I_n pulse is obtained whenever the combinational logic is activated by proper input signals and the clock pulse occurs. It is thus necessary to establish *one* of the input conditions that yields an I_n signal; then allow

A Typical Troubleshooting Problem

one clock pulse to occur. The clocked I_n signal clears FF D, which in turn clears C, B, and A. $\bar{Q}_B$ is HIGH under these conditions, and G_{19} is enabled. Arrival of the delayed I_n pulse activates G_{19}; FF E is set by the resulting pulse; and the LEFT light illuminates. Thus it is necessary merely to set up the required A and B inputs that will generate the I_n signal and then allow one clock pulse to occur. The LEFT light should then illuminate. As long as the A and B inputs are not changed, additional clock pulses may be generated to determine the operation of the sequential counter of Part III. If the initial malfunction is present, the LEFT light will remain illuminated after five or more clock pulses are provided.

The Troubleshooting Task

Following verification of the malfunction, troubleshooting is initiated. It is preferable to use *built-in indicators* to isolate the malfunction. Previous circuit examination should provide sufficient information to decide how to approach the problem. The waveform diagram showed that four I_4 pulses are followed by six I_6 pulses in response to the proper combination of inputs A and B and the presence of clock pulses. Thus the LEFT light should be illuminated for the first four clock pulses and the RIGHT light for the next six. Ten clock pulses then should illuminate the LEFT light 4/10 or 2/5 of the time and the RIGHT light 6/10 or 3/5 of the time. The LEFT light should illuminate first. If the LEFT light does not extinguish and the RIGHT light illuminate upon arrival of the fifth clock pulse (slightly delayed), it probably means that G_{20} has not furnished an I_6 pulse. This could occur because of either a failure in G_{20} or a failure in the counter. Of course, the delayed I_n pulse could be missing, but this is highly unlikely because the LEFT light has already been turned on.

Using built-in indicators and a knowledge of the operation of the control unit has isolated the malfunction to the general vicinity of the counter of Part III and the decoder of Part IV. If we wish to attempt repair without further identification of the malfunctioning elements, G_{20} and the FFs in the counting chain may be replaced (one by one) until the malfunction vanishes. In the interest of rapid repair, such a method is sometimes used when difficulties may be isolated to a relatively few replaceable elements.

A more reasonable approach is to use either *external logic level indicators* or *the oscilloscope.* Use of the external-logic level indicator flowchart (Fig. 7-12) is shown first. The malfunction has already been isolated to a general area, so it is not necessary to apply the flowchart to the complete control-unit circuitry. Parts III and IV may be considered as a separate circuit and examined in detail by using the flowchart.

According to Fig. 7-12, after test-point logic levels have been determined, the test point (TP) nearest the output is checked. In this case, TP_{10} at the output of G_{20} is the first test point to be examined. Use of the built-in indicators and previous investigations have shown that at the fifth clock pulse and through the tenth clock pulse, an output pulse should appear. The procedure to be used is generally the same as the flowchart shows, except that test points are not changed until sufficient input pulses have been generated to allow the counter to go completely through its cycle. For the malfunction under discussion, TP_{10} would remain at a LOW logic level throughout the complete counter cycle time.

Assuming that the delayed I_n pulse is present because of the functioning of the LEFT light, the Q_B output (TP_7) is the next test point closest to the input. The waveform diagram shows that Q_B should be HIGH following preset, go LOW for four input pulses, and then go HIGH for the next six input pulses. With the external logic level indicator attached to TP_7, the control unit is preset by turning power on, and ten clock pulses are generated. The output at TP_7 is investigated following each clock pulse, and it is noted that Q_B presets properly, goes LOW at the first I_n pulse, and remains LOW for the remaining input pulses. This accounts for the lack of I_6 pulses, since G_{20} is continuously disabled and the delayed I_n pulse is not allowed through the gate.

TP_6 is now observed for ten clock pulses to determine if FF C is supplying triggers to FF B. It is noted that FF C also presets properly, but its output (Q_C) goes LOW and remains LOW following the first clock pulse. Thus FF B could not change state because FF C was not furnishing input triggers.

Investigation of TP_5 reveals that FF D responds according to the waveform diagram. FF C is provided with the proper input triggers, yet it does not change state following the initial preset and response to the first clock pulse. When placed into the CLEAR state by the application of a trigger from D, it locks into the CLEAR state. Replacement of FF C and a complete retest of the counter verifies that the malfunction has been cleared and that the counter and decoder circuits are functioning properly.

Before the control-unit card is returned to use again, it should be subjected to a complete test from inputs to outputs. Due to the numerous combinations of A and B that can result in an I_n output, and due to the requirement that the counter be actuated ten times to fully prove its operation, it is advisable to check only the combinational logic at this time. The sequential logic has been proven, so all that remains necessary is to assure that TP_2 shows the correct I_n signal for the proper combinations of A and B. TP_1 and TP_3 are available to check intermediate signals, C_n and C_{n-1}, respectively.

Once the malfunction has been isolated to the sequential portion

A Typical Troubleshooting Problem 277

of the control unit, the oscilloscope method of troubleshooting might be more expeditious. The clock input could be allowed to operate at its normal speed, and the oscilloscope could be connected to the necessary test points until the improper waveform is noted. Since the oscilloscope pattern and waveform pattern are very similar, the malfunctioning component would be obvious.

An even better troubleshooting method is by use of the logic analyzer. Since the built-in indicators have allowed initial problem isolation to parts III and IV of the circuit, the analyzer can be connected to test points 4 through 11. Then the circuit is operated through its normal sequence. The display on the analyzer can be compared with the normal waveforms to identify the malfunctioning component immediately. Use of a logic analyzer in such a simple task is unusual, but the analyzer's capabilities are quite apparent.

To generalize, the logic analyzer and oscilloscope methods of troubleshooting are usually superior to other methods when sequential circuits are under examination. They may also be used to troubleshoot combinational logic if input signals are changing rapidly and following a repetitive pattern. Otherwise, the external logic level indicator is usually more satisfactory. As mentioned earlier in this chapter, signature analysis may also be used if the logic device has been so-designed.

9-5 SUMMARY

Troubleshooting logic circuits that contain both combinational and sequential elements is adequately summarized in the appropriate flowcharts. However no such visualization of analysis procedures has been shown. Figure 9-9 summarizes the investigative methods for mixed logic circuits.

Since the purpose of analysis is to determine the circuit's operating characteristics, these are shown as the end result of the procedure. As we mentioned earlier, the logic diagram is broken into combinational and sequential parts, and examined separately. The "tools of analysis" for the combinational logic portions (truth tables, Boolean equations, waveforms, and maps) have been used throughout this text. Similarly, the sequential logic tools (state tables, Boolean equations, waveforms, and flowtables) have been employed where applicable.

The similarities between the methods of analysis should be carefully noted. For example, truth tables and state tables perform very similar functions. The only difference is that the *state table depicts circuit states at different times,* while *the truth table shows circuit states with different input combinations at a specific time.* The sequential analysis procedures are thus *time oriented,* whereas the combinational analysis procedures are concerned with *combinations of inputs.*

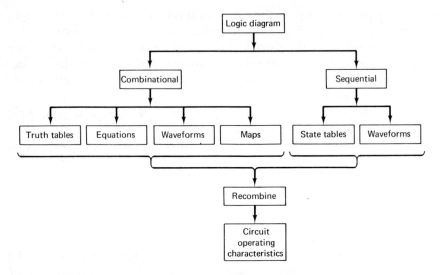

Figure 9-9 Logic circuit analysis methods summary

PROBLEM APPLICATIONS

9-1. Why are combinational and sequential functions separated prior to starting analysis of logic circuits?

9-2. List common identifying characteristics of combinational logic circuits.

9-3. List common identifying characteristics of sequential logic circuits.

9-4. List the common analysis methods used with combinational logic and explain the advantages and disadvantages of each.

9-5. List the common analysis methods used with sequential logic and explain the advantages and disadvantages of each.

9-6. Why is the interface between combinational and sequential circuits sometimes clocked?

9-7. Why must the troubleshooting flowcharts of Ch. 7 be modified for mixed-function logic?

9-8. Discuss the changes necessary to troubleshoot sequential and mixed logic using the built-in indicators flowchart.

9-9. Discuss the changes necessary to troubleshoot sequential and mixed logic using the external indicators flowchart.

9-10. Explain why logic/state analyzers might be superior to logic probes for logic circuit analysis.

9-11. Perform the algebraic and/or map manipulations required to obtain Eq. (9-4) from Eq. (9-2).

9-12. Perform the algebraic and/or map manipulations required to obtain Eq. (9-3) from Eq. (9-1).

9-13. Using algebraic and/or map manipulations, verify that the logic implementation of I_n is the same as Eq. (9-1).

10

COUNTERS AND THEIR APPLICATIONS

A *counter is a device that is composed of properly connected sequential* (sometimes *combinational*) *elements which perform the function of counting.* Counters are binary in nature, but they may be modified to count in other sequences (e.g., base 7 or base 10). Almost all logic systems require counters. Digital computers use counters to control the sequence and execution of their program steps. Industrial controls use counters to determine the number of steps in machine operation, such as the control unit discussed in Ch. 9. Measuring instruments for the scientific disciplines employ counters to count frequency, voltage, number of events, and so forth. Some typical counter applications will be explored in this chapter, following a discussion of both binary and special-purpose counters. A sample analysis of a typical counter will also be performed—in this case showing the techniques used to investigate multiple flip-flop (FF) circuits.

10-1 THE ASYNCHRONOUS BINARY COUNTER

Basic Concepts

Counters may be divided into two general types: *synchronous* and *asynchronous.* The *asynchronous counter* (sometimes called the *series* or *ripple counter*) is probably the simplest device in the counter family. Since it was discussed in detail as the "logic-signal source" in Chs. 2 and 3, a review of Sec. 2-6 will lay the groundwork for the more detailed discussions to follow.

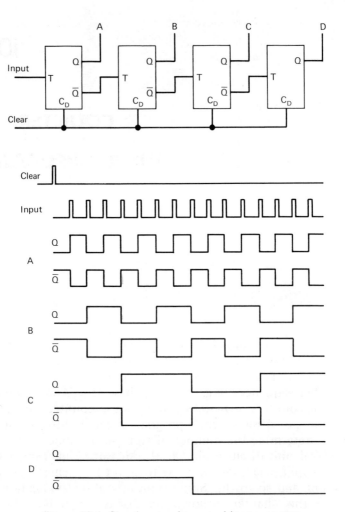

Figure 10-1 Simple asynchronous binary counter

Simplicity is the major advantage of asynchronous counters. As we can see in Fig. 10-1, simple *T* flip-flops (FFs) with the capability of direct CLEAR may be connected to count in the straight binary sequence. The major disadvantage of these counters is the delay encountered as the changes in each FF in the series string propagate. The difficulty is that each FF takes a finite time to change state. This means that a change in state in the first FF of the counter is not coincident with a change of state in later FFs. With enough FFs and a fast enough

The Asynchronous Binary Counter

counting rate, actual counts can be "lost," or unstable operation can result. Delays vary from as much as 400 nanoseconds to as little as 1 nanosecond, depending on the type of logic circuits used. Thus cumulative delays may cause errors when outputs from widely separated stages in the counter are combined in gating circuits for the purpose of decoding specific counts.

Waveforms used to explain counter circuits are idealized in this book for the purpose of clarity. Propagation delays are not shown, but should be taken into consideration as high counting rates are discussed.

The UP Counter

Figure 10-1 is a typical example of an asynchronous, binary UP counter. The name adequately describes the principle of the counter's operation—it starts counting from some predetermined point (often all FFs cleared) in ascending order. The count is accumulated in the binary system; however the ability to convert binary numbers to decimal numbers has already been applied, so no difficulty should exist. Note that the output of each FF represents one binary digit. The first stage, to which the inputs to be counted are applied, represents the number 2^0 or 1. The following stage has a positional value of 2^1 or 2; the next stage has a value of 2^2 or 4; and the final stage represents 2^3 or 8. By adding the values of all stages in the "1" condition, the total count is obtained. Thus n FFs can store any binary number from 0 to $2^n - 1$. The counter in Fig. 10-1 stores the binary equivalents of all decimal numbers from 0 to 15. In the UP counter configuration, these numbers accumulate in the familiar manner shown in Fig. 2-16.

The waveforms should also be closely observed so that the relationship between the input pulses and the outputs of each stage can be seen. The first-stage output, for example, goes through one complete cycle (from LOW to HIGH and back to LOW again) one time for each two input pulses. Effectively, it has an output frequency that is *one-half* the frequency of the incoming signals. The second stage has a frequency that is one-half the frequency of the first stage or one-fourth of the incoming frequency. Due to this characteristic, *counters are sometimes called frequency dividers,* and as frequency dividers, they are extensively used in applications requiring submultiples of input frequencies.

Of course, the FFs used to implement the binary UP counter are not restricted to the *T* type. *D, R-S,* and *J-K* FFs are all easily worked into ripple counters, merely requiring input connections that allow them to perform as a *T* FF. Flip-flops may be found two, four, six, and even eight per IC package. In addition, complete multi-stage binary counters may be obtained in the 14-pin DIP and other multiple-pin packages. Fig. 10-2 shows the logic diagram for such a counter, which

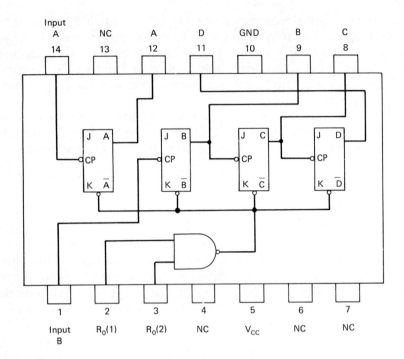

Figure 10-2 Type 7493 four-bit counter

is identified as a 7493* 4-bit binary counter. Note that the FFs require LOW-going inputs for activation and that a direct RESET capability is available to start the counting at 0000.

Representing Counters in Graphic and Tabular Form

Often, it is advantageous to consider counter circuits in terms of the complete counter rather than individual FFs. We have shown that *sixteen separate states* exist in the simple binary UP counter. But how these states are represented in terms of individual FF outputs *may* not be important. The natural binary sequence (see Fig. 10-3a) is both the most common and most convenient because the arbitrarily assigned state numbers coincide with the decimal equivalents of the binary numbers stored.

---

*The numbers used to identify specific logic-circuit packages are purposely abbreviated in this text. A certain amount of standardization does exist in industry, but prefixes and suffixes to the actual numbers vary among manufacturers. The *numbers alone* may generally be used to identify a specific type of circuit, regardless of its prefixes and suffixes.

The Asynchronous Binary Counter 283

Input pulse number	Present state time = n				Next state time = n + 1			
	D	C	B	A	D	C	B	A
0	0	0	0	0	0	0	0	1
1	0	0	0	1	0	0	1	0
2	0	0	1	0	0	0	1	1
3	0	0	1	1	0	1	0	0
4	0	1	0	0	0	1	0	1
5	0	1	0	1	0	1	1	0
6	0	1	1	0	0	1	1	1
7	0	1	1	1	1	0	0	0
8	1	0	0	0	1	0	0	1
9	1	0	0	1	1	0	1	0
10	1	0	1	0	1	0	1	1
11	1	0	1	1	1	1	0	0
12	1	1	0	0	1	1	0	1
13	1	1	0	1	1	1	1	0
14	1	1	1	0	1	1	1	1
15	1	1	1	1	0	0	0	0
0	0	0	0	0		etc.		

Note: Pins 12 and 1 are connected to obtain this count sequence

(a) State table

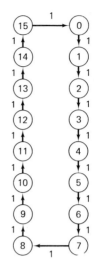

(b) Flow diagram

Figure 10-3 State table and flow diagram

However the complete operation of the binary UP counter of Fig. 10-1 may be shown by merely drawing a *state flow diagram* (Fig. 10-3b), which displays the sequence of counter states. Each circled number represents one of the states, while the lines and arrows define the transitions taking place. The designation attached to the line shows the input signal necessary to cause the transition from one state to the next. We see that a "1" input will cause the counter to change from state number 0 to state number 1. Likewise, the next "1" input will result in a change from state number 1 to state number 2, and so on. The "1" input that appears following establishment of state number 15 causes the counter to recycle back to state number 0, and the counter starts over.

The use of 1 instead of HIGH or LOW to define inputs is not out of place. Since a FF is a storage device, its state may be considered as storing either a 1 or a 0. Convention dictates that when a FF is SET it is storing a 1, and when RESET, it is storing a 0. Whether a 1 is HIGH or LOW depends on the implementation of the logic circuit. *The transition from one level to the other is what causes action in the counter.* Thus, regardless of the counts accumulated in the counter, the state flow diagram shows the next state to be encountered. This method of representation is especially valuable when other than natural binary sequences are encountered.

Algebraic expressions may also be used to represent asynchronous binary counters. However, due to the means of triggering each FF from the preceding FF, interpretation of these equations becomes difficult. Therefore waveforms and state tables will be used here to explain the asynchronous operations.

The DOWN Counter

The binary DOWN counter may be implemented in much the same manner as the UP counter. The DOWN counter, however, *decreases* its stored count every time an input pulse is received. One of the easiest ways to obtain the descending count operation is to use the *complement outputs* of the FFs in the UP counter. For example, since the Q outputs were used in the counter in Fig. 10-1 to obtain the ascending count, we need only complement the Q outputs (use the $\bar{Q}$ outputs) to obtain the descending count. Thus, in the waveforms of Fig. 10-1, all of the $\bar{Q}$ outputs are HIGH, following the initial CLEAR operation, and the count represented at that time (using the $\bar{Q}$ outputs) is 1111. Following the first input pulse, the $\bar{Q}$ output of FF A goes LOW, while the other FFs' $\bar{Q}$ outputs remain HIGH. The number represented is now 1110 or 14_{10}. The first pulse, then, has decreased the stored count by 1. Additional pulses should be examined by the reader.

Further verification of the DOWN counter operation may be made by complementing the outputs shown in the state table in Fig. 10-3. We see that as the output of the UP counter increases, the output of the DOWN counter (its complement) decreases. Another method of obtaining a binary DOWN counter is to use the Q outputs of each FF to activate the following FF. Waveforms, a state table, a state flow diagram, and a simple logic diagram of a DOWN counter are all shown in Fig. 10-4.

Bidirectional Counters

The characteristics of both the UP and DOWN counters may be combined into a single UP-DOWN counter. From a practical viewpoint, all that is necessary is to determine whether *the input to each FF in the series string* comes from the Q or the $\bar{Q}$ output. The logic diagram of such a *bidirectional* device is shown in Fig. 10-5. (The dotted lines should be disregarded for the moment.)

The sequence of operation is similar to the other binary counters described in this section. Following an initial CLEAR signal to establish a starting point, input pulses are applied to the T input of FF A. If the UP COUNT input is HIGH, all of the gates connected to the $\bar{Q}$ outputs of the FFs are enabled, and counting will proceed in the UP direction, just as in the UP counter in Fig. 10-1. Lowering the UP COUNT input and establishing the DOWN COUNT input in the HIGH condition

The Asynchronous Binary Counter

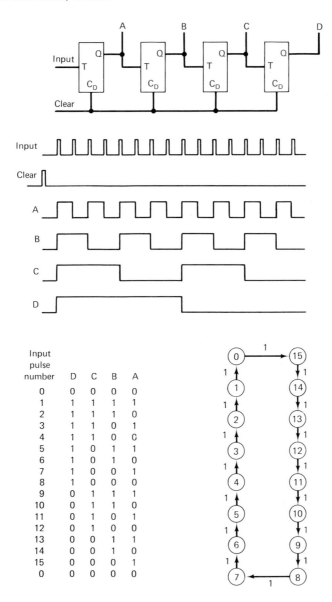

Figure 10-4 An asynchronous binary DOWN-counter

causes the counter to perform in the same manner as the DOWN counter in Fig. 10-4. While both the UP COUNT and the DOWN COUNT inputs may be LOW at the same time (this prevents any FF other than A from changing state), they are *not* allowed to be HIGH simultaneously. The resulting count sequence is somewhat unusual!

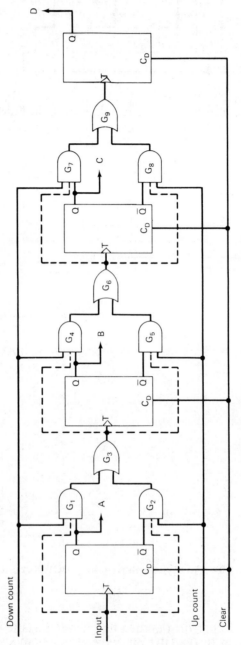

Figure 10-5 Asynchronous binary UP-DOWN counter

Although the restriction on both count-control inputs being HIGH at the same time is observed, other difficulties can arise in the UP-DOWN counter. If the UP COUNT input is HIGH at the time the CLEAR signal is applied, the high-going transition of the $\overline{Q}$ output of FF A will be passed through G_2 and G_3 to the T input of FF B. FF B then attempts to toggle under control of the T input at the *same time* that it is being cleared by the C_D input. In most FFs, the direct CLEAR and direct SET inputs override the T inputs, but difficulties can occur if the T input is HIGH when the C_D input returns to the LOW level. Furthermore, under certain count conditions, lowering the UP COUNT input and raising the DOWN COUNT input without *first* clearing the counter results in an abnormal count sequence.

Both problems can be solved by assuring that all of the AND gates are disabled, *except* during the time an input pulse is being counted. This is easily accomplished by adding the connections shown by the dotted lines in Fig. 10-5. Consider G_1 and G_2. Since the input is at a LOW level except for the short time that the pulses to be counted are present, both G_1 and G_2 are disabled, and they furnish LOW inputs to G_3. G_3 output is then LOW, disabling G_4 and G_5. The LOW output of G_6 disables G_7 and G_8, and no changes on either of the count-control inputs or the FF output may be transferred to following FFs. Clearing operations or changes in count direction may be established without upsetting the count stored in the counter. The only precaution to be exercised is to assure that *such changes do not occur during the time the clock pulse is present.*

10-2 THE SYNCHRONOUS BINARY COUNTER

Basic Concepts

All FFs are *clocked* (triggered) *at the same time* in the synchronous (parallel) counter. Combinational logic elements (gates) are used to control FF state changes so that the stored count progresses only *one* unit each time a clock pulse is received.

The synchronous counter is better suited to *high-speed* operation than is the asynchronous counter. Although the concept of the synchronous counter indicates that all FFs change state at the same time, this is not strictly true. Delays still exist in the gating structure of the circuit (see Fig. 10-6); thus the top counting speed is considerably below the theoretical speed. Synchronous operation requires the use of relatively high-power clock circuits, since the clock must drive all FFs. Heavily loaded clock circuits may also increase propagation delay and reduce counter speed. Despite these factors, synchronous counters still

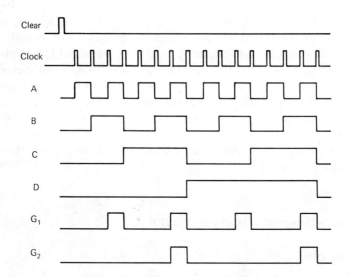

Figure 10-6 The synchronous binary UP-counter

tend to be superior in speed characteristics when compared to asynchronous counters.

But, as is so often true, speed costs money! The additional gates required to obtain synchronous operation, plus the more complex FF needs, place the synchronous counter in a cost category above the asynchronous counter. And, of course, more complex circuitry usually requires greater power supply capability and so the overall system cost increases.

The Synchronous Binary UP Counter

An UP counter showing the principles of synchronous operation is seen in Fig. 10-6. Clocked *J-K* FFs are used, and all FFs are triggered simultaneously. The FFs are interconnected by gating so that the only time a specific FF can change is when all the *least significant* FFs are in the SET state.

Although the waveforms in Fig. 10-6 are almost self-explanatory, the operation of the synchronous binary UP counter is partially discussed to develop an understanding of the use of waveforms with synchronous circuits. Counter operation is based on the already familiar characteristics of the *J-K* FF. Specifically, the fact that the *J-K* FF changes state with every clock pulse if both the *J* and the *K* inputs are HIGH at the same time is used to allow the FFs to perform the required toggling operation.

Following the initial CLEAR operation, all FFs are in the CLEAR state and all *Q* outputs are LOW. The *Q* output of FF *A* disables FF *B*. (When both *J* and *K* are LOW, no change occurs in FF state upon arrival of a clock pulse.) *C* is similarly disabled, due to the LOW output of G_1, and *D* is disabled by G_2. The only FF that can change state with the next clock pulse is *A*, since its *J* and *K* inputs are permanently HIGH. Therefore the first clock pulse changes *A* from the CLEAR to the SET condition, and Q_A goes HIGH. *B* is enabled by Q_A, and the next clock pulse changes not only the state of *B* but also that of *A*. Investigation of the remainder of the waveforms shows that *C* can change state only at the time both *A* and *B* are SET (G_1) and that *D* can change state only at the time *A*, *B*, and *C* are SET (G_2). The effect of each clock pulse should be determined, and the operation of the counter should be investigated at least through count 16.

We see that the waveforms for the synchronous binary UP counter are identical to the waveforms for the asynchronous binary UP counter. Thus the circuits of Fig. 10-1 and 10-6 are functionally equivalent. The state diagram and state table are the same for both asynchronous and synchronous counters. When speed and power parameters are met, either circuit may be used to perform simple binary counting. However, certain limitations exist for each type of circuit; the asynchronous counter may not be able to replace the synchronous counter in all cases.

Synchronous Binary DOWN Counter

Like the asynchronous DOWN counter, the only requirement to convert from an UP to DOWN operation is the use of the *opposite outputs* to feed the following stages—the UP counter of Fig. 10-6 becomes

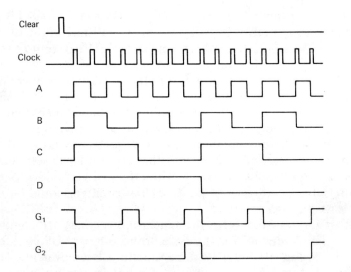

Figure 10-7 The synchronous binary DOWN-counter

a DOWN counter by merely connecting the $\bar{Q}$ outputs to the following gates and/or FFs. Actual outputs continue to be identified with the Q output terminals. Explanatory information in the form of a waveform diagram appears in Fig. 10-7. The $\bar{Q}$ outputs of the UP counter may also be used as DOWN counter outputs in the same manner as in the asynchronous counter.

Synchronous Binary UP-DOWN Counter

Bidirectional counting may be accomplished in the synchronous mode, just as it was in the asynchronous mode. The logic diagram of a synchronous UP-DOWN counter (type 8284) is shown in Fig. 10-8b. More modern versions of this type of counter (discussed later) are available. The 8284 counter, however, has many of the basic features necessary to the understanding of more complex counters.

Because of the complexity of the logic diagram, the "package" format is often used in system diagrams. Fig. 10-8a is the 14-pin DIP form commonly seen in such diagrams. Obviously, it is necessary to have either a complete description of the counter operation (or the logic diagram) available to analyze system diagrams using this form of representation.

The counter in Fig. 10-8 provides the output sequence 0 through 15 in the natural binary code (8-4-2-1), when counting in the UP direction, or 15 through 0 when counting in the DOWN direction. SET and RESET inputs allow a count of either 15 or 0 to be placed in the counter so that a convenient starting point may be established. When these inputs are present, they disable the counting process. LOW-going signals are required to perform the SET and RESET operations.

Input to the counter is via the CARRY IN and the COUNT ENABLE lines. Both must be present (but SET and RESET must *not*) for the counter to operate in either the UP or the DOWN configuration. A LOW level at the UP-$\overline{\text{DOWN}}$ input results in a DOWN count, while a HIGH level results in an UP count. All four Q outputs are available, as is the $\overline{Q}$ output of the final FF. Another output, called CARRY OUT, is provided to indicate that a complete count has occurred.

Although each FF is labeled with a T input, the FFs behave in the same manner as the FFs of Figs. 10-6 and 10-7. For convenience, the UP-DOWN counter FFs may be considered as "clocked T FFs!" A rather unusual designation, but there is no reason why such a name shouldn't be applied. Just like the J-K FF with the J and K inputs connected together, when the T input is HIGH, the FF changes state with every clock pulse; when it is LOW, no change occurs.

The T input of each FF is gate controlled, so all previous stages must be in the 1 state (for UP counting) or in the 0 state (for DOWN counting) *before* the following FF is allowed to change state on the next clock pulse. (Note the increasing complexity in gating structure as count value increases.)

One more point is worth noting on the logic diagram of the 8284 UP-DOWN counter. The CLOCK and the UP-$\overline{\text{DOWN}}$ inputs have inverting amplifiers inserted prior to use of the signal. These amplifiers are

(a) Package diagrams:
 (left) Q package
 (right) AF package

(b) Logic diagram

Figure 10-8 The synchronous binary UP-DOWN counter (Type 8284) (courtesy of Signetics, a subsidiary of Corning Glass Works, Sunnyvale, CA.)

required not only to obtain the correct logic level for circuit operation but also to provide power amplification. The power amplifiers provide sufficient power to operate all of the required circuits, and they also isolate the complete circuit from causing interference with preceding and following circuits. Such isolation is commonly necessary in complex digital systems.

10-3 SPECIAL PURPOSE COUNTERS

Basic Concepts

All of the counters discussed so far have been based directly or indirectly on the binary system or *modulus 2*. *Modulus* is a term used to describe *the maximum number of counts that a counter can achieve*. When using natural binary counting such as that shown in Secs. 10-1 and 10-2, the modulus of a counter may be determined by the equation

$$\text{Modulus} = 2^n \tag{10-1}$$

where n = the number of FFs.

Thus a single stage counter is modulus 2, a 2-stage counter is modulus 4, a 3-stage counter is modulus 8, and so forth. The counters discussed in Secs. 10-1 and 10-2 are modulus 16.

The "real" world is not necessarily base 2 oriented, however. Time of day uses counts of 6, 10, and 12; distance is measured in inches, feet, and yards (or millimeters, centimeters, and meters); and quantity is measured in multiples of 12 (dozen or gross). Therefore a need exists for a counting system that is based on something other than modulus 2 or multiples of modulus 2. So this next section investigates methods of counting in numerous bases, using both asynchronous and synchronous circuits.

Divide by *N* Counters

Although both asynchronous and synchronous counters may be used to obtain counters with moduli other than direct multiples of 2, asynchronous methods are examined first to develop the general concepts; and then synchronous counters are investigated.

The most commonly employed counting element is the *J-K* FF. Actually, complete counters such as the 7493 package (see Fig. 10-2) and its successors are commonly used in place of separate *J-K* FFs to reduce overall logic element requirements. The basic package consists of four *J-K* FFs and a 2-input NAND gate. Internal connections provide a divide-by-2 and a divide-by-8 capability. External connections allow shortening of the basic count cycle and development of various counters with moduli from 2 through 16. The basic logic diagram showing the divide-by-2 (*A* FF) and the divide-by-8 (*B, C,* and *D* FFs) connections is seen in Fig. 10-2. Natural binary counting to base 16 is provided by externally connecting A_{out} (pin 12) to B_{in} (pin 1). At this time, the counter performs exactly as the asynchronous binary UP counter of Fig. 10-1, except for the LOW-going trigger requirement.

The NAND gate is used to RESET or CLEAR all stages of the counter—either for an initial starting count or to return all stages to the reference count following completion of the count cycle. The same speed and readout restrictions mentioned in Sec. 10-2 apply to the asynchronous counters discussed in this section.

To operate the *J-K* FFs in the 7493 or similar counters as *T* FFs, both the *J* and *K* inputs must be enabled. The method of design in the FFs in this counter allows both *J* and *K* inputs to be left unterminated to accomplish enabling.

In all count applications, the frequency of the incoming triggers is divided by 2 in the first FF. The natural binary equivalent of the required modulus is decoded by the reset gate (and any additional gating necessary), and the complete counter is reset to 0000. A divide-by-12 connection that could be used to count inches of measurement or hours of the day is an excellent example of this technique. The logic diagram and waveforms of Fig. 10-9 should be self-explanatory. The count progresses in natural binary sequence from 0_{10} through 11_{10}. When count 12_{10} occurs, the reset gate is enabled and all FFs are returned to the 0 state. The actual count of 12 exists only for as long as it takes to reset all FFs, so the counter can be considered to be counting from 0 to 11 or modulus 12.

A divide-by-6 circuit, constructed in the same manner, is shown in Fig. 10-10. The reader is encouraged to verify the waveforms used to explain counter operation; note especially the use of LOW-going logic signals at both the CLEAR (RESET) and CP inputs in both counters.

Since the divide-by-12 counter is used so often in digital systems, another IC example (type 8288, similar to type 74176) is shown in Fig. 10-11. A divide-by-2 and a divide-by-6 connection is made internally. Divide-by-12 operation is accomplished by externally connecting output *A* to the clock-2 input. Hence type 8288 (74176) and type 7493 circuits are similar. Input/output designations are different for the 8288 and the 74176, however.

The 8288 circuit has what is known as *strobed parallel data entry,* which allows the counter to be preset to any desired output state. Considering a 1 as HIGH and a 0 as LOW, a 1 or a 0 at any data (*D*) input is transferred to the associated output (stored in that FF) when the strobe input goes LOW. The usual common reset is also avail-

Special Purpose Counters

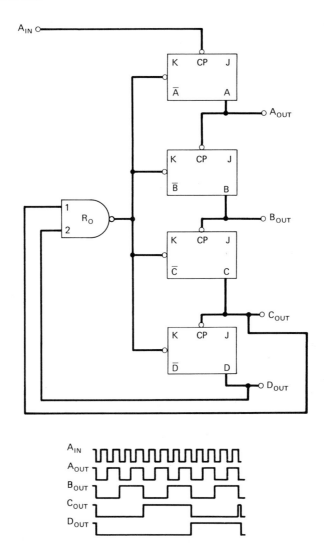

Figure 10-9 Binary divide-by-twelve ripple counter

able to place 0s in all four FFs. Both the strobe and reset inputs are asynchronous with respect to the clock input; that is, they override the clock input.

FF output waveforms for the divide-by-12 portion of the circuit are included in Fig. 10-11c. Strobed parallel data entry is also shown for one of the four FFs (Fig. 10-12) to demonstrate this common technique.

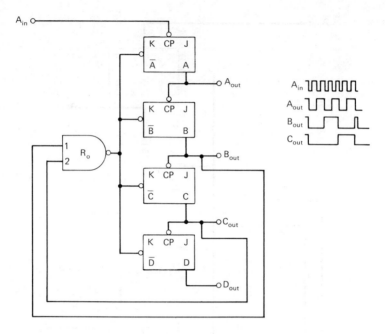

Figure 10-10 Binary divide-by-six ripple counter

Due to the relative timing of the strobe, reset, and data inputs, a truth-table explanation of circuit operation is shown instead of the usual waveforms. One important point must be emphasized in this analysis: *both the S_D and R_D inputs require* LOW-*going activating signals,* as indicated by the FF symbols. Therefore the outputs of G_3 and G_4 that are important are the 0 outputs—not the 1 outputs. With this point in mind, we see that S_D is activated *only* when $Z = 1$, $Y = 1$, and the strobe pulse is LOW. Even if the strobe pulse is LOW and $Z = 1$, Y must also be 1 in order to SET the FF. If $Y = 0$, RESET takes place, since the RESET input overrides any attempt to preset the counter. The reset functions are predictable according to the description of the circuit operation.

Binary Coded Decimal (BCD) Counters

The most common of the divide-by-N counters is the *modulus*-10 *circuit.* Most measurements (physical and electrical) plus numerous mathematical calculations use base 10 numbers, so it is no wonder that many IC packages contain binary coded decimal counters in one form or another. These special packages will be discussed later. Right now, we will present only the simple asynchronous divide-by-10 connection.

Special Purpose Counters

Figure 10-11 Divide-by-twelve counter (Type 8288) (courtesy of Signetics, a subsidiary of Corning Glass Works, Sunnyvale, CA.)

Figure 10-12 Strobed parallel data entry for Type 8288 counter (courtesy of Signetics, a subsidiary of Corning Glass Works, Sunnyvale, CA.)

Just as with the divide-by-12 and divide-by-6 circuits, the count representative of the counter modulus is decoded and used to reset all FFs. Figure 10-13 shows the FF interconnections and the associated waveforms of the commonly used 8-4-2-1 BCD (*Binary Coded Decimal*) counter. Other possible arrangements of FF states are used and have valid applications in special circuits. But the 8-4-2-1 weights provide simple decoding and excellent correlation with the already discussed natural binary counter.

The 8290 type IC package (also similar to the 74176) is a presettable decade (base 10) counter that is partially asynchronous and partially synchronous. It may be connected in the standard BCD counting mode or in the divide-by-5/divide-by-2, biquinary mode (see Ch. 13). The logic rules concerning strobe, reset, count (clock), and data inputs are identical with the rules for the 8288 package discussed earlier.

Counter-circuit Analysis

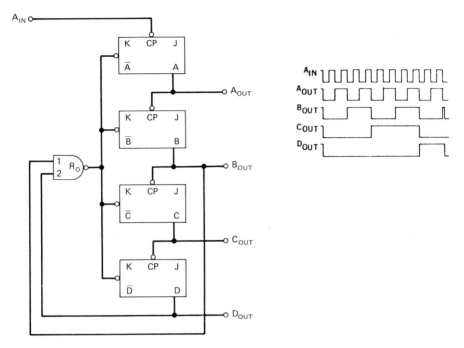

Figure 10-13 Divide-by-ten ripple counter (courtesy Sprague Electric Company)

Figure 10-14 shows the logic diagram for the 8290 counter, along with state tables and waveforms for both standard BCD and biquinary counting. Standard BCD counting is achieved by using Clock 1 as the circuit input and connecting A_{out} to Clock 2. Connection of D_{out} to Clock 1, using Clock 2 as circuit input, and A_{out} as circuit output results in the biquinary mode of operation. Only the FF outputs are shown in the waveform diagrams, since the intermediate waveforms are included as problems at the end of this chapter.

One more counter circuit may be encountered occasionally—a *shift counter*, which uses the principles of *shift registers*. Shift registers are discussed in Ch. 11.

10-4 COUNTER-CIRCUIT ANALYSIS

A Modulus 5 Counter

A counting circuit with unknown characteristics is shown in Fig. 10-15. Sufficient background has been provided in this and previous chapters to determine the count sequence and waveforms, as necessary.

(a) Package diagrams

(b) Logic diagram

(c) Waveforms

Figure 10-14 Type 8290 decade counter (courtesy of Signetics, a subsidiary of Corning Glass Works, Sunnyvale, CA.)

Counter-circuit Analysis

Figure 10-15 Counter circuit to be analyzed

The first step entails a determination of the *characteristics of the logic elements* used in the counter. Assuming MIL-STD-806B symbology, the AND gate follows conventional rules. The *J-K* FF operating characteristics were detailed in Ch. 8 and are repeated in Table 10-1 (as a limited state table). Manufacturers' specification sheets should be consulted for logic elements that have not previously been encountered.

TABLE 10-1 *J-K* **Flip-Flop Limited State Table**

J	K	Q_{n+1}
0	0	Q_n
1	0	1
0	1	0
1	1	$\bar{Q}_n$

Since the clock pulses are made directly available to each of the FFs in the counter, the circuit appears to operate in the synchronous mode. As the equations for each of the FFs are developed, it is assumed that the clock pulse is present. This results in a simpler analysis. Inspection of the logic diagram results in the individual FF equations (*J* and *K* inputs) shown here.

$$J_A = \bar{C}_n \tag{10-2}$$
$$K_A = 1 \tag{10-3}$$
$$J_B = A_n \tag{10-4}$$
$$K_B = A_n \tag{10-5}$$
$$J_C = A_n B_n \tag{10-6}$$
$$K_C = 1 \tag{10-7}$$

A complete state table for the counter may now be constructed, using the individual FF equations and the characteristics table for the

TABLE 10-2 State Table for Counter Circuit of Fig. 10-15

State Number	n Present state			$n+1$ Next state		
	C	B	A	C	B	A
0	0	0	0	0	0	1
1	0	0	1	0	1	0
2	0	1	0	0	1	1
3	0	1	1	1	0	0
4	1	0	0	0	0	0
5,0	0	0	0	etc.		

J-K FF. The present state (time n) of each FF is shown on the left side of Table 10-2, while the state following the next clock pulse (time $n+1$) is shown on the right side. Note that arbitrary decimal-number state designations are used for reference purposes.

All FFs are established in the CLEAR condition prior to the start of the analysis, resulting in entries $C = 0$, $B = 0$, and $A = 0$ for the present state in row 0. From both the logic diagram and the FF equations, we see that if FF *C* is in the CLEAR state, its $\overline{Q}$ output is HIGH and the *J* input of FF *A* is HIGH. The *K* input is also HIGH (permanently connected), so FF *A* changes state with the arrival of every clock pulse as long as *C* is in the CLEAR state. Thus, in row 0, the *A* entry for the $n+1$ portion of the state table is 1.

Prior to the arrival of the first clock pulse, FF *A* was in the 0 state and the *Q* output was LOW, thus disabling FF *B*. When the first clock pulse arrives, no change occurs in FF *B*, and the B_{n+1} entry is 0. Also, since both the *A* and *B* FFs are in the 0 state prior to arrival of the first clock pulse, G_1 has a LOW output, which disables the *J* input of FF *C*. According to the characteristics table of the *J-K* FF with $J = 0$ and $K = 1$, the FF goes to or stays in the CLEAR state, depending on its previous condition. It is thus assured that the state of FF *C* is 0.

The states of all FFs at time $n+1$ are now transferred to the present state side of the state table, and the same process is repeated to obtain the next state, following the arrival of another clock pulse. The analysis continues until repetition of the counting pattern is noted. In the preceding example, this occurs when the fifth clock pulse arrives, recycling the counter to the starting count of 000. Thus the circuit is a modulus 5 counter, operating in the synchronous mode.

Upon completion of the state table, both the state flow diagram (Fig. 10-16a) and complete algebraic descriptions (Fig. 10-16b) may be developed. A waveform diagram may also be developed (Fig. 10-16c) to supplement the other forms of circuit description.

Counter-circuit Analysis

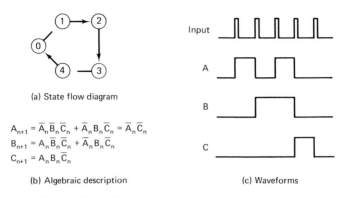

Figure 10-16 Description of counter of Figure 10-15

A Modulus 17 Counter

The increased complexity of modern digital equipment has resulted in the availability of IC packages that perform complete system operations. *Multiple-stage counters* have already been discussed at length in this chapter and will be examined further in this section. The logic diagram of Fig. 10-17 shows a device that uses a type 8290 (74176) presettable decade counter (discussed in Sec. 10-3), a type 8291 (74197) presettable binary counter with natural binary counting sequence and the same presettable inputs as the 8290 unit, and a number of NAND gates. Input/output designations differ in the 8290/8291 and the 7476/74197 ICs, however.

As with any problem, the first step is to obtain as much information as possible concerning the circuit. The characteristics of the 8290 (74176) counter, obtained from manufacturer specification sheets, were used in Sec. 10-3. Type 8291 (74197) counter operation is identical, except for the counting sequence. The NAND gates are conventional, except for the unusual connection between G_1 and G_2. This type of symbology is used to indicate an *expansion input,* which allows the output of one gate (prior to inversion) to be connected to the gating structure of another gate. G_3 and G_4 are used in the conventional latch configuration.

The basic circuit input is called *clock* and consists of a sequence of pulses to be counted. *Out* is an indication that the circuit has performed its intended function. The *strobe* is used to preload the counter, just as in previous applications. $\overline{Clock}$ is an output that may be used in other logic circuits, while *W* is a test point. Some of the individual counter-stage outputs are also available as test points or for use as inputs to other circuits. The only known fact is that the circuit is a counter that

(a) Logic diagram

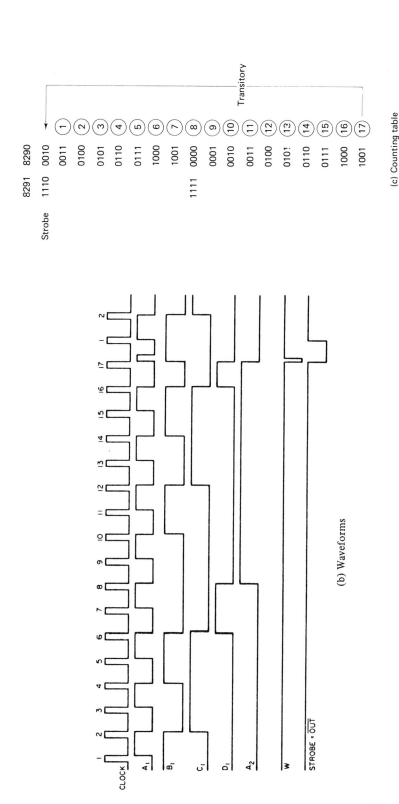

Figure 10-17 Counter circuit using IC packages (courtesy of Signetics, a subsidiary of Corning Glass Works, Sunnyvale, CA)

305

counts to a number (which may be preselected) and then recycles to start counting over again.

A bit of constructive thinking about the circuit connections sometimes helps this type of analysis. For example, the specifications for the 8290 (74176) and 8291 (74197) ICs state that the active condition of the strobe input is LOW-going, so it must be at a HIGH level when the strobe function is not being performed. Since the strobe input is directly connected to the output of G_4, this point must be HIGH most of the time. The latch connection of G_3 and G_4 decrees that G_3 output must be LOW most of the time. The indication of circuit operation then must be a HIGH signal at *out*. Such an indication can occur only if W goes LOW, and this happens only when A_1, D_1, A_2, B_2, C_2, and D_2 are HIGH (along with the normally HIGH strobe input).

Such circumstances exist when the following binary count is in the counter.

$$\begin{array}{ccccccccc} D_2 & C_2 & B_2 & A_2 & D_1 & C_1 & B_1 & A_1 & \\ & & & & & 0 & 0 & & \\ 1 & 1 & 1 & 1 & 1 & & & 1 & \text{(The state of } C_1 \text{ and } B_1 \\ & & & & & 1 & 1 & & \text{is unimportant.)} \end{array}$$

Note once again that the physical drawing of the counter and the actual order of the binary number are reversed, as is common in logic diagrams. A_1 is the least-significant digit, while D_2 is the most-significant digit. Upon occurrence of any of the counts shown above, W goes LOW and the output of G_3 goes HIGH. The strobe output goes LOW, due to the cross-coupled configuration of G_3 and G_4, thus forcing W HIGH and leaving *out* HIGH. Of course, when the NAND-gate FF changes state, the output connected to the strobe terminal goes LOW and strobing action is set up within the 8290 (74176) and 8291 (74197) circuits.

Remembering that a LOW input to the strobe terminal of the ICs transfers the levels on the data lines into the respective FFs leads to another interesting discovery. The data inputs may be either HIGH or LOW, depending on the position of the connected switches. If the data terminal is connected to a HIGH potential, a 1 is stored, and if it is connected to a LOW potential, a 0 is stored. Circuit investigation shows that the data inputs D, C, and B on the 8291 (74197) and B on the 8290 (74176) are connected to the 1 level. Thus, upon occurrence of the strobe pulse, the counter is automatically loaded with the binary number 11100010.

The first clock pulse that arrives following strobing action resets the output latch and advances the counter one count to 11100011. Each subsequent clock pulse advances the counter one more count, until $11111^{00}_{11}1$ is reached, at which time W goes LOW and activates the output latch. The cycle then repeats, as shown in the counting table and waveform diagrams in Fig. 10-17. Note specifically that the 8290 (74176) counter recycles when its maximum count (10) is achieved at clock-pulse number 8, while the 8291 (74197) counter never reaches its full count. It recycles and is reloaded as a result of the decoding action of G_1 and G_2.

Analysis has thus shown that the circuit in Fig. 10-17 is a modulus 17 counter, when the control switches are set as shown. Furthermore, the counter decodes the maximum count capability of the circuit, having reached that count from a preloaded count that was equivalent to the maximum count minus the desired count. This type of operation reduces the time required to restart the counting process, since a conventional reset to 00000000 may require the reset signal (under certain circumstances) to ripple the full length of the counter.

The analysis of logic circuits using prepackaged functions such as those in Fig. 10-17 is usually a rather simple process. However the reader should be on the alert for unusual applications of prepackaged functions. Many design engineers take full advantage of the capabilities of such packages, and the specification sheets may not discuss such applications.

10-5 COUNTER APPLICATIONS

Counter applications are many fold—*any system that requires accumulation of a series of events is a candidate for the digitally implemented counter.* Measuring instruments, digital clocks, and digital computers are but a few of the potential uses for the circuits discussed in this chapter. Block diagrams of a few of these applications are presented in this section.

Measuring Instruments

The basis for most direct-reading digital instruments is the simple *binary counter connected in the BCD configuration.* Of course, the count in the individual decades must be decoded and displayed (see Ch. 14) to be useful, but applications may be discussed without actual knowledge of this process.

Precise measurement of the number of events per unit time (whether random or periodic) *may be made with a device called a* "counter." Other names such as totalizer, frequency timer, scaler, time-interval meter, frequency counter, and digital counter are also encountered. Practically all of these devices operate in a similar manner (Fig. 10-18). A source of precision timing is used to open and close a gate that feeds the events to be counted to the counting portion of the instrument. The

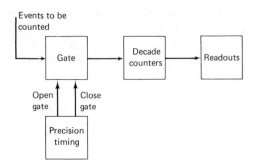

Figure 10-18 Principles of operation for a "Counter"

usual gate open time is 1 second, although most instruments allow the choice of multiples and submultiples of 1 second to improve the instrument's versatility. The events to be counted may be pulses developed from mechanical operations such as switch closures, interruptions of a light beam on a production line, or signals from a radio transmitter.

Digital voltmeters also use counter circuits. The voltage to be measured is converted by some means into a frequency that may be counted. Conventional decade-counting circuits then count and display that frequency in terms of a voltage. Figure 10-19 is a block diagram of a simple digital voltmeter. This voltage-to-time or *ramp* type of voltmeter measures the period of time required for a linear ramp to change from zero to a level that is coincident with the unknown input voltage.

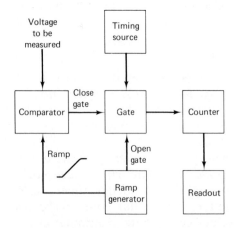

Figure 10-19 Digital voltmeter block diagram

Counter Applications

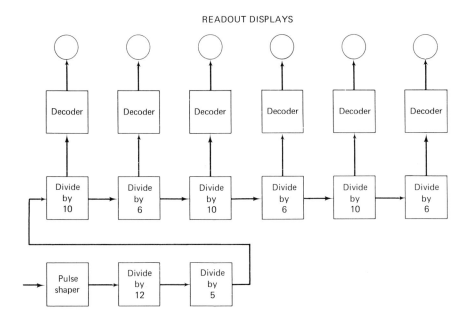

Figure 10-20 Digital clock block diagram

During this period, an internal oscillator is gated on and its output is totalized in a counter. With proper selection of ramp slope and oscillator frequency, the total count during the sample period will numerically indicate the unknown voltage.

Digital Clocks

Most digital systems used to collect data or record events require some means of indicating *the time at which the data was received or the event occurred.* Such indications may be furnished by *digital clocks.* A block diagram of a typical digital clock is shown in Fig. 10-20.

The 60 Hz input signal is derived directly from the power line and is filtered and shaped into a form that will accurately trigger the FF circuits that follow. A divide-by-12 and a divide-by-5 combination is used to obtain one pulse per second. The divide-by-10 circuit that follows develops one pulse every 10 seconds, while the divide-by-6 circuit further divides the output frequency to obtain one pulse every 60 seconds. A similar arrangement is used to develop the minutes and tens of minutes count. Hours and tens of hours are provided by the divide-by-12 and divide-by-3 circuits. A means to preset the clock to any given time is provided. Readout may be visual, in the form of a special code,

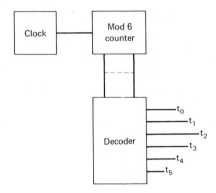

Figure 10-21 Operations sequencer (conceptual)

which is recognized by digital printing devices, or both. The complete digital clock now exists on a single IC.

Digital Computers

Although digital computers are not discussed until Ch. 15, counter applications may be discussed without detailed knowledge of the computer. The key fact is that the digital computer operates in an *organized, sequential manner* in performing its functions. The basic sequence of operation is controlled from a *system clock,* which is merely a source of precision timing. All of the sequencing within the computer develops from the system clock. The information used within the computer (whether it is data to be operated upon or operations to be performed) is divided into *words.* Words are usually composed of a fixed number of binary digits (bits); the computer must keep track of these bits so that it will know when one word is completed and the next one starts. The counter in Fig. 10-17 could do this job quite adequately, for example, if the word contained 17 bits. A change in level at *out* would signify the end of one word and the beginning of another.

The sequential nature of a computer requires that operations be performed in a specific order within given time periods. The counter, together with decoding circuits, provides the necessary information. If, for example, an operation such as addition requires six sequential operations, a modulus 6 counter with decoding circuits is used. A simple block diagram of such a circuit appears in Fig. 10-21.

These examples merely scratch the surface of counter use in digital applications. Since some of these areas (e.g., digital computers) have been pointed out, the reader should experience no difficulty in recognizing many other applications.

Counter Applications

PROBLEM APPLICATIONS

10-1. What is a counter?

10-2. Define asynchronous counter operation.

10-3. Define synchronous counter operation.

10-4. Compare the advantages and disadvantages of asynchronous and synchronous counters.

10-5. What are the identifying characteristics of an asynchronous binary UP counter? DOWN counter?

10-6. List three methods of representing asynchronous counters and explain the use of each.

10-7. Draw waveform diagrams for the UP-DOWN counter of Fig. 10-5 (without the dotted lines) under the following circumstances.
   (a) Initial conditions—all stages CLEARed, UP COUNT HIGH and DOWN COUNT LOW.
   (b) Five input pulses, followed by UP COUNT LOW and DOWN COUNT HIGH.
   (c) Three input pulses, followed by both UP COUNT and DOWN COUNT HIGH.
   (d) Five input pulses, followed by DOWN COUNT LOW, UP COUNT HIGH, AND CLEAR HIGH at the same time.
   (e) Determine the count accumulated at the end of Step (d).

10-8. Repeat Problem 7—with the dotted lines connected.

10-9. Why are the synchronous and asynchronous binary UP counters in Figs. 10-1 and 10-6 equivalent?

10-10. Using the type 7493 IC package in Fig. 10-2, draw logic diagrams and waveforms of modulus 3, 8, 9, and 11 ripple counters. (External gates may be used if necessary.)

10-11. Why is the 8-4-2-1 code the most popular of BCD codes?

10-12. Develop the intermediate waveforms for the type 8290 counter in Fig. 10-14.

10-13. Make a list of procedural steps required to analyze a counter circuit.

10-14. List as many applications of counter circuits as possible and explain each of them.

11

REGISTERS AND

THEIR APPLICATIONS

A *register is a group of logic circuits, usually flip-flops* (FFs), *arranged in a manner that allows the storage and processing of information in binary form.* Single FFs *may* be classified as registers, since they are capable of both storing and processing binary information. However registers usually consist of a number of FFs. The number of memory elements in a register is determined by the requirements of the specific application. For example, a register in a small digital computer might have sixteen stages, since a word (the unit of information handled by many minicomputers) contains 16 bits.

One possible classification of registers is functional: *a storage register retains information,* while a *shift register processes information.* Registers can be further classified in terms of the input–output (I/O) methods used. Information is entered into and taken from most registers either *bit-by-bit* (serial form) or *all bits at once* (parallel form). If entered and removed in serial form, the device is categorized as *serial-serial.* If entered and removed in parallel form, the classification is *parallel-parallel.* (Serial-parallel and parallel-serial designations are also commonly encountered.) Thus registers are classified by both their purpose and their method of operation.

As with other digital-logic elements, registers are difficult to identify *physically,* since they appear in the same types of packages as gates, FFs, counters, and other prepacked logic functions. Once again, the manufacturers' specification sheets must be consulted to obtain the characteristics of particular registers. This chapter will illustrate logic diagrams of registers constructed from both individual memory elements (FFs) and complete integrated circuit (IC) packages.

All digital systems that require the *storage and handling of binary information* use registers. The digital computer, of course, is a prime example. Registers are used to perform arithmetic operations, store commands, and transfer data from one portion of the computer to another. In digital measuring equipment, registers store data to be displayed so that the counters may proceed with the next measurement without losing the previous information. The transmission of digital data from one location to another requires conversion from parallel form to serial form—and back to parallel form. Registers perform these functions. Since the reader will encounter registers in many different forms, the discussions in this chapter should provide sufficient information to recognize and understand the most common implementations of storage, shift, and counting registers.

11-1 THE STORAGE REGISTER

Basic Concepts

Each FF in a register can store *one piece of binary information* (one bit). Since the concept of information storage in FFs was discussed in Ch. 8, it should require little additional coverage here. When storage registers must store more than one bit at a time, additional FFs are used. The interconnections between FFs determine the functional and I/O characteristics of the register.

By the strictest definition, if the FF inputs are connected to the source of data to be stored and the outputs are connected to the circuit that uses the stored information, then a *storage register* exists. Such a definition limits the scope of discussion in this section, but it *does* define a specific, separable class of register applications. Many variations of input and output connections exist and will be discussed in this chapter.

Storage registers are used whenever *temporary storage of information* is needed. A practical example is seen in the block diagram in Fig. 11-1. Many frequency counters have the capability to display the "last" count while accumulating the "next" count. As soon as a counting period is completed, the outputs of the counter stages are immediately transferred to the storage register. Once the transfer is complete (which can be accomplished in less than 50 nanoseconds), the counter can be released to start another counting cycle. The information in the storage register (an exact replica of the information that was in the counter) is decoded into a form acceptable to the display device employed and is displayed for use while the counter is obtaining the latest

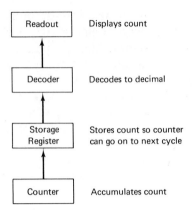

Figure 11-1 A register application

information to be used in updating the display. A similar application exists in digital computers, where data from the computer's memory is transferred temporarily to a storage register before it is distributed to other functional parts of the computer.

Implementation

Any device that can (a) *assume a state representative of its input,* (b) *retain that state after the input is removed,* and (c) *make available an output that may be used to represent its state* can function as a storage register. Not only do FFs possess these characteristics, but certain magnetic, electromagnetic, and semiconductor devices may also be so used. FF implementation of the storage register function is the method discussed in this chapter.

R-S, J-K, and *D* FFs all see application as storage registers. The clocked (gated) versions are employed so that the input data may be transferred to the FF and then leave the FF unable to respond to any further input changes until it is time to store new data. Figure 11-2 shows the general concept of data transfer from a FF in a counter to a *J-K* FF used as a storage register. An *R-S* FF functions in the same manner as a *J-K* FF under the input conditions shown in Fig. 11-2a and may be used in place of a *J-K* FF. Of course other circuit parameters, such as supply voltage and logic levels, must also be compatible.

D FF implementation of the storage function (Fig. 11-2b) is usually simpler than either the *J-K* or *R-S* implementation. Inverted data input is not required due to the characteristics of the *D* FF (see Sec. 8-7). A single connection between the *Q* output of the counter FF and the *D* input of the storage register causes a 1 to be stored in the register when

The Storage Register

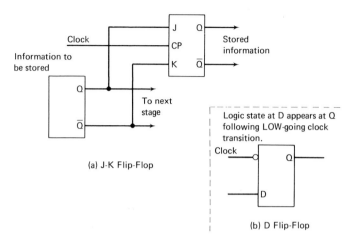

Figure 11-2 Flip-flop implementation of a storage register

the counter FF is in the 1 state and 0 when it is in the 0 state. The clock pulse is assumed to be present.

Whether FFs or other methods are used to furnish the storage register function, the same actions must take place. Whatever data is presented to the input must be stored on command and be made available at the output when needed. If new data is presented at the input, it must not disturb the existing stored data until the next storage cycle takes place. Any medium (electrical, electronic, or mechanical) that performs in this manner is usable in storage register applications.

Storage Registers at Work

FLIP-FLOP STORAGE OF COUNTER OUTPUT

Storage registers, as defined in this section, fit into the *parallel input-parallel output* category. That is, all storage register stages receive their respective inputs at the same time, and each stage supplies a separate output. Such an application using *J-K* FFs as storage elements is shown in Fig. 11-3a. A counter composed of four *T* FFs is operating in the natural binary sequence, as seen in the waveforms of Fig. 11-3b. The Q and $\bar{Q}$ outputs of each counter stage supply the *J* and *K* inputs, respectively, of accompanying storage-register stages. A *read pulse* is supplied to the *CP* inputs of all register stages shortly after the occurrence of every fifth pulse to be counted. The first read pulse samples the output of each counter stage and transfers its logic state to the appropriate register stage. Since Q_A is HIGH at this time, the *J* input of SR_1 is HIGH, and SR_1 goes to the SET state. Its Q output becomes

(a) Logic diagram

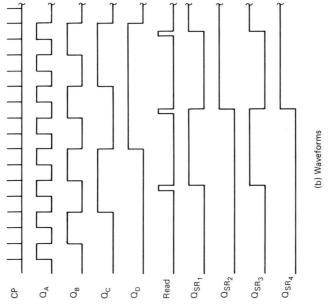

(b) Waveforms

Figure 11-3 *J–K* flip-flop storage register

HIGH, and the state of the A FF is stored in SR_1. Similar actions occur for the other stages, and after the read pulse occurs, the 0101 count of the counter is stored as 0101 in storage register stages SR_4, SR_3, SR_2, and SR_1, respectively. The 1010 count replaces the 0101 count after the second read pulse, and 1111 is stored as a result of the third read pulse.

This example merely demonstrates the operation of storage registers. Without additional FFs, gating, and control, the values placed in the storage register would be representative of *only* the four least-significant digits of the accumulated count. However it *has* been shown that *J-K* FFs can perform storage register functions.

IC Memory Buffer Register

In the design of equipment that requires storage register functions, prepackaged IC registers are common. The 74LS374 (Fig. 11-4) is an array of eight clocked D FFs that may be used in any parallel input-parallel output application. *Buffer register is a term used to describe a register that serves as a temporary store when transferring information between two units that are asynchronous, operating at different speeds, or performing independent tasks.* The package diagram in Fig. 11-4 shows the simplicity of the 74LS374 device. Each of the D FFs has a D, a clock, and an enable input. Only the Q output is provided for each FF. The register is enabled when the enable input is LOW. Input information is loaded into the register on the HIGH-going transition of the clock, making that information available at the output immediately.

A practical (though fictitious) computer application of the 74LS374 buffer register is shown in Fig. 11-5. All of the instructions for the solution of a problem and the data required for the problem are placed in the computer memory prior to performing the steps of the problem. The computer works toward the problem solution in an automatic sequential manner (unless told to change the sequence), selecting one step after the other and performing the operations required. In many microcomputers the information required to perform the operations is contained in groups of eight bits called a byte.

Each byte is removed from memory and placed in a storage register until it is needed. This allows the memory to go on with other business without waiting for computer operations to be performed. When the eight bits of the byte are to be loaded into the storage (or memory buffer) register, the memory supplies the logic levels to the eight lines feeding the D input of each of the register stages. A clock pulse is generated by the memory (or memory control) to tell the register to accept the information, and the complete byte is "dropped" in

The Storage Register

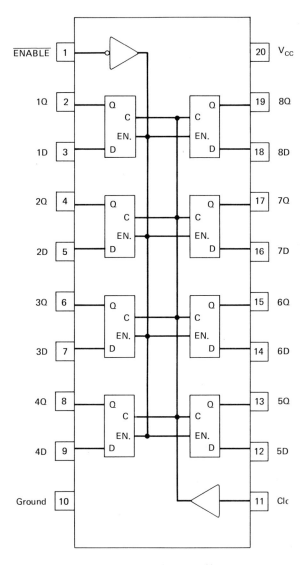

Figure 11-4 The 74LS374 buffer register

parallel into the memory buffer register. The outputs of the register may now be fed elsewhere in the computer to perform various functions, and the memory can proceed with its next step. The memory buffer register has provided the parallel in-parallel out storage register function, showing the practical application of IC prepackaged registers.

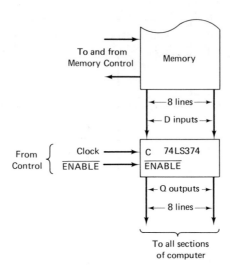

Figure 11-5 Buffer register application

11-2 THE SHIFT REGISTER

Basic Concepts

The shift register fits into the overall class of registers, since it may be constructed of FFs and is used to operate on binary data. *Shift registers move data,* usually from one stage of the register to an adjacent stage. The movement (shift) of data may be from left to right, right to left, or in both directions. Hence both *unidirectional* and *bidirectional registers* exist.

Shift registers differ from storage registers in that *adjacent stages are connected* to allow movement of data from stage to stage. Actually, the shift register also may serve as a storage register when proper controls exist. Data may be loaded into the shift register in parallel and read out of the register in parallel, just as in the storage register. However the same shift register can also accept data one bit at a time (serial), move it from stage to stage until all data is in the register, and then store the data until it is ready for readout. The data may be read out one bit at a time (serial) or all bits at once (parallel). Properly designed, then, the shift register can perform all four I/O operations: *serial-serial, serial-parallel, parallel-parallel,* and *parallel-serial.*

A common application of the shift register is seen in the block diagram in Fig. 11-6. An information display device, such as the keyboard control of a digital computer (similar to an electric typewriter in

The Shift Register

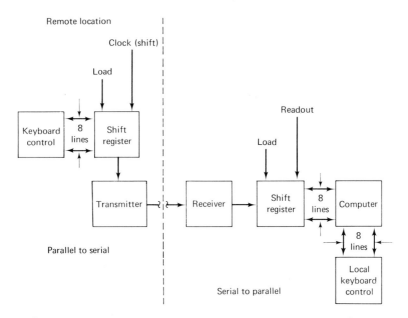

Figure 11-6 A shift register application to data transmission

form and operation), uses a multibit representation of each character to be displayed or entered. One particular representation, often used for remote computer control, employs an 8-bit code. Each alphabetic, numeric, or control character is represented by a group of 8 bits, all of which must be present at the same time (parallel). Transfer of data from one location to another requires eight separate lines, one for each bit. When sending and receiving locations are physically separated by great distances, however, the cost of multiline transmission becomes prohibitive.

Shift registers are used to allow transmission of the complete 8-bit character with only a single line. The 8 bits that result from actuating a key on the keyboard are stored in a shift register (8 stages), using parallel loading. As soon as loading is completed, the register is commanded to shift the data. One bit at a time is shifted from the register and converted for transmission. After the eighth bit is shifted out and converted, the register is ready to accept (in parallel) the next character.

At the receiving end, the 8-bit character is converted to logic levels and placed (one bit at a time) into another shift register. When all 8 bits are received, the information is read out to the receiving device and displayed. Thus, at the sacrifice of speed of transmission, the requirement for eight transmission lines has been reduced to one, and the cost of system implementation has been cut drastically.

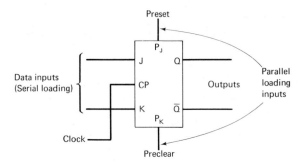

Figure 11-7 *J-K* flip-flop as a shift register stage

Implementation

Shift registers are implemented with the same kinds of FFs used in storage registers. Each stage in the shift register may be viewed as an individual storage register, capable of being loaded (in most cases) not only in the parallel mode, as shown in this section, but also in the serial mode. Due to the necessary interconnections required to move the data from stage to stage, parallel loading is accomplished in a slightly different manner than it is in storage registers. The asynchronous (direct SET and CLEAR) inputs are generally used for parallel loading, while the synchronous inputs provide serial-load capability. Figure 11-7 shows a typical *J-K* FF used in shift register applications.

Once again, the *J-K* and *R-S* FFs possess the same characteristics under the input conditions of most shift register operations and are functionally interchangeable. Although not as versatile, the *D* FF used in shift registers will be shown later.

Shift Registers at Work

SERIAL-SERIAL MODE

Perhaps the simplest of shift registers is the *serial in-serial out circuit* shown in Fig. 11-8. (The AND gates and READ OUT input should be disregarded at this time.) It consists of *D* FFs and requires a logic level that represents only the data to be moved into the register, and a shift input. The CLEAR input is assumed to be activated just prior to arrival of the first shift input so that all stages begin operation in the CLEAR condition.

The output of a *D* FF is the same as its input *after* the occurrence of the clock or shift input. Thus when the *D* input to a FF is HIGH, the *Q* output is HIGH following the next clock pulse. The same is true

The Shift Register

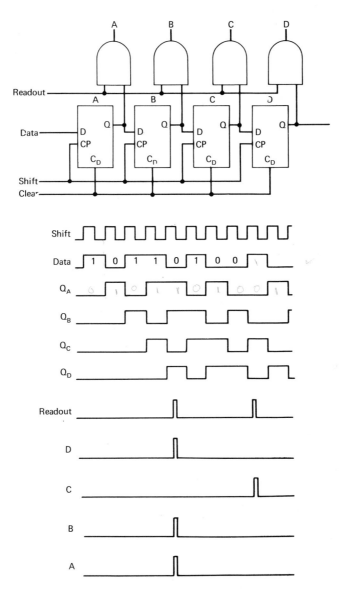

Figure 11-8 Shift register logic diagram and waveforms

of the LOW output. Note that the Q output of the A FF is the D input of the B FF. Whatever logic level is present at the Q output of the A FF appears at the Q output of the B FF upon occurrence of the shift pulse. Thus every shift pulse causes the information in the preceding FF to be transferred to the following FF.

The waveform drawing of Fig. 11-8 (disregarding the READ OUT and *D, C, B,* and *A* waveforms) does not really show the precise timing requirements that exist between DATA information and SHIFT information. Most FFs require that DATA information precede SHIFT information by a *specific time interval* (nanoseconds in high-speed FFs). This factor must be considered during logic design, but it is assumed when using waveform drawings. Although the drawing shows the DATA and SHIFT inputs changing at the same time, the required time differential is considered to be present. Thus when the SHIFT input goes HIGH to activate the FF, it uses the DATA level that was present just before the change in SHIFT input. The first HIGH-going transition of the SHIFT input occurs when DATA is LOW, and since all stages are in the CLEAR state, no change takes place. The second SHIFT input occurs with DATA HIGH, and FF *A* SETS, making its *Q* output HIGH. The DATA input has now been transferred to FF *A* from the input circuit. The third clock pulse occurs when DATA is LOW and FF *A* clears, making its *Q* output go LOW. However, at the same time that the SHIFT input was changing the state of FF *A,* it was also using the HIGH *Q* output of *A* to cause FF *B* to SET. The finite time that it takes to change the state of a FF is sufficient to enable FF *B* to change state before FF *A.*

Note that the DATA input has now been transferred from FF *A* to FF *B,* and the new piece of information that was present on the input line is now stored in FF *A.* Evaluation of the remainder of the operations in Fig. 11-8 will show that the incoming data is moved to the right one stage every time a shift pulse is provided. As soon as all of the data is loaded into the shift register (one bit at a time), it will be shifted out of the rightmost FF and disappear, unless the shift pulses are stopped. The usual mode of operation with serial in-serial out shift registers is to provide enough shift pulses to load the register and then stop shifting, until a time when the output data is needed. The shift pulses are then restarted, and the data feeds from the rightmost FF (one bit at a time), as long as shift pulses are present. Used in this manner, the register functions as a combination shift and storage register.

CIRCULATING REGISTERS

A special case of the serial-serial shift register is the *circulating register,* which is used in certain digital computer applications requiring *repetitive* use of the same information. In the serial-serial shift register, the information is shifted out the last stage and lost, while in the circulating register the last stage supplies its information *back* to the first stage to be recirculated. The information being recirculated is also available at the output of the register for use in the rest of the circuit.

The Shift Register

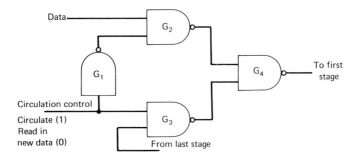

Figure 11-9 Circulating register input gating

Conventional serial-serial shift registers may be used as circulating registers by providing input gating similar to that shown in Fig. 11-9. Input to the first stage of the register may be either original data or the data from the last stage of the register, as selected by the circulation control input. If HIGH, the circulation control input enables G_3 and allows data from the last stage to be fed through G_4 to the first stage of the register. G_2 is disabled by the LOW output of G_1. Changing the circulation control input to a LOW level disables G_3 and is inverted by G_1 to enable G_2, thus allowing the original data to be read in.

SERIAL-PARALLEL MODE

The *serial in-parallel out register* may also be implemented by using gates connected to each of the FFs' Q outputs. Data is shifted in, until all desired information is stored. The READ OUT operation is enabled prior to the next clock pulse, and the information in each FF appears at the output of its respective output gate. No information is lost, and the register can continue its shifting operation as if nothing had happened. This operation is shown on the waveform diagrams in Fig. 11-8. It is representative of the type of register that would be used to convert the received serial data in the example of Fig. 11-6 to the required parallel data for machine operation.

PARALLEL-SERIAL MODE

Parallel in-parallel out shift registers are somewhat uncommon—they fit more properly into the storage register category. (Coverage of this configuration was provided in Sec. 11-1.)

J-K FFs are used to demonstrate the *parallel in-serial* out shift register in Fig. 11-10. This type of circuit could also be used in the remote computer control unit of Fig. 11-6 to convert the parallel keyboard data to serial form for transmission. The data to be entered is

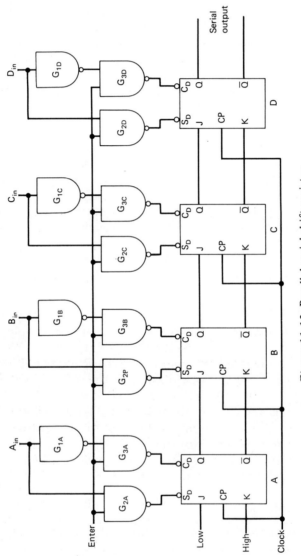

Figure 11-10 Parallel-serial shift register

provided at the *A*, *B*, *C*, and *D* inputs. When the "enter" pulse arrives, whichever of the two control gates becomes enabled for each FF allows a pulse to be developed, and each FF will be placed in the state represented by the inputs. The FFs do not have to be cleared prior to entering information, since both the CLEAR and SET inputs are capable of being activated by the input levels. Following entry of the data, clock pulses may be applied to cause the shifting of the data toward the output. If a 1 has been entered in a FF, its Q output is HIGH, enabling the following FF's J input. In a like manner, if a 0 has been entered, the following FF will have its K input enabled. Due to the interconnections, the J and K inputs cannot both be either 0 or 1, so the FF will respond to the next clock pulse by reproducing the state that was in the previous FF.

A Universal Register

Discussion of the four forms of shift registers should prepare us for the presentation of more complex prepackaged IC register circuits. The wide use of shift registers in digital equipment makes it feasible to provide packages that are usable in more than one operating mode. Figure 11-11a shows the 74LS295 IC, which is a 4-bit shift register with both serial and parallel data entry capability. It can shift data also to the right or to the left.

The operation of the 74LS295 shift register is indicated by the logic diagram of Fig. 11-11c. The element is composed of four clocked *R-S* FFs. The input of every stage can be switched between two logic sources by the mode control input. When the mode control input is HIGH, the inputs of the four stages are connected to the parallel inputs *A*, *B*, *C*, and *D*. When the mode control input is LOW, the inputs of the second, third, and fourth stages are connected to the outputs of the first, second, and third stages, respectively, thus forming a 4-bit shift register. The input to the first stage (with mode control LOW) is obtained from the serial input.

Conventional parallel-serial conversion is implemented with the 74LS295 shift register by causing the mode control to be HIGH for one clock period (to parallel load the register), then changing the mode control to a LOW level so that incoming clock pulses can shift the parallel-loaded data out via the *D* output. Serial-parallel conversion is realized by supplying enough clock pulses while the mode control is LOW to load all information in serial form. The clock pulses are then stopped so that output data can be made available by enabling the output control.

Shift registers are used not only to convert from serial to parallel form (and vice-versa) but also to *manipulate binary representations of*

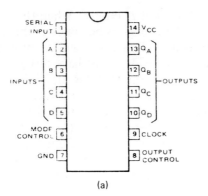

FUNCTION TABLE

MODE CONTROL	INPUTS		PARALLEL				OUTPUTS			
	CLOCK	SERIAL	A	B	C	D	Q_A	Q_B	Q_C	Q_D
H	H	X	X	X	X	X	Q_{A0}	Q_{B0}	Q_{C0}	Q_{D0}
H	↓	X	a	b	c	d	a	b	c	d
H	↓	X	Q_B†	Q_C†	Q_D†	d	Q_{Bn}	Q_{Cn}	Q_{Dn}	d
L	H	X	X	X	X	X	Q_{A0}	Q_{B0}	Q_{C0}	Q_{D0}
L	↓	H	X	X	X	X	H	Q_{An}	Q_{Bn}	Q_{Cn}
L	↓	L	X	X	X	X	L	Q_{An}	Q_{Bn}	Q_{Cn}

When the output control is low, the outputs are disabled to the high impedance state, however, sequential operation of the registers is not affected.

†Shifting left requires external connection of Q_B to A, Q_C to B, and Q_D to C. Serial data is entered at input D.
H = high level (steady state), L = low level (steady state), X = irrelevant (any input, including transitions)
↓ = transition from high to low level.
a, b, c, d = the level of steady state input at inputs A, B, C, or D, respectively.
Q_{A0}, Q_{B0}, Q_{C0}, Q_{D0} = the level of Q_A, Q_B, Q_C, or Q_D, respectively, before the indicated steady state input conditions were established.
Q_{An}, Q_{Bn}, Q_{Cn}, Q_{Dn} = the level of Q_A, Q_B, Q_C, or Q_D, respectively, before the most recent ↓ transition of the clock.

(b)

(c)

Figure 11-11 The 74LS295 universal register

Shift Counters

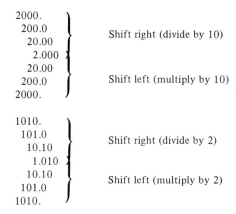

Figure 11-12 Shift multiplication and division

numbers. Multiplication and division by powers of 2 is performed on binary numbers in a shift register. Such operations may be demonstrated with decimal numbers (Fig. 11-12) so that the concept is more understandable. A movement of all digits in a decimal number 1 position (order) to the right is equivalent to dividing the number by a power of 10. Thus 2000 moved (shifted) one position to the right becomes 200.0. Similarly, left movement is equivalent to multiplication by a power of 10; 20.00 becomes 200.0 when a one-position left shift is accomplished. Since decimal numbers are represented in most digital computers by their binary equivalent, similar operations may be performed. Shift right results in division by a power of 2 instead of 10. Multiplication by a power of 2 results from shift-left operations. Thus a need for registers capable of *either* right or left shifting can be visualized.

Shift-right operations have been discussed in detail in all the shift registers examined. Left shifting of data is accomplished easily in the 74LS295. It is only necessary to make the mode control HIGH and connect the output of each FF to the parallel input of the previous FF (D output to C input, etc.). Serial data is entered at input D for left-shift operations.

11-3 SHIFT COUNTERS

Basic Concepts

Shift counters are a specialized type of *clocked counter*. The name is derived from the use of shift registers to perform the counting operation. This type of counter generally results in outputs that are easily decoded, and gating is usually not required between stages.

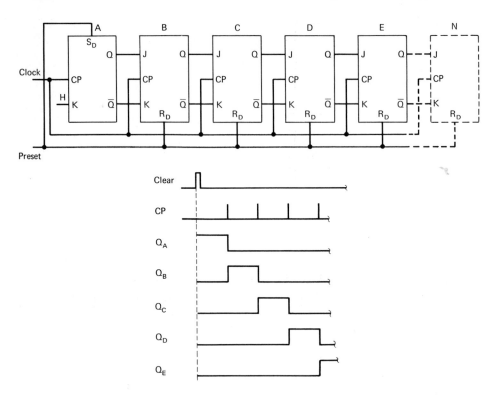

Figure 11-13 Basic shift counter

The concept of operation is very similar to the serial-serial shift register. A 1 is loaded into the first stage of the serial connected register and is shifted one stage to the right upon arrival of each clock pulse. No additional 1s are entered into the register, so the actual number of shift pulses (clock inputs) counted may be determined by the location of the stage in the register which contains the 1. Figure 11-13 shows the principles of operation of the basic shift counter.

This conceptual counter is, of course, impractical. It requires one stage for each count (a decimal counter would require ten stages), and no means is provided to restart the counter at the reference count without stopping the clock input and applying the CLEAR input.

The Ring Counter

By providing feedback from the last stage to the first stage in the shift counter (Fig. 11-14), *count recycling* is provided. A 10-stage decimal counter could then count from 0 through 9 and start the count at 0 again automatically. Ten stages are needed (A through M).

Figure 11-14 Ring counter

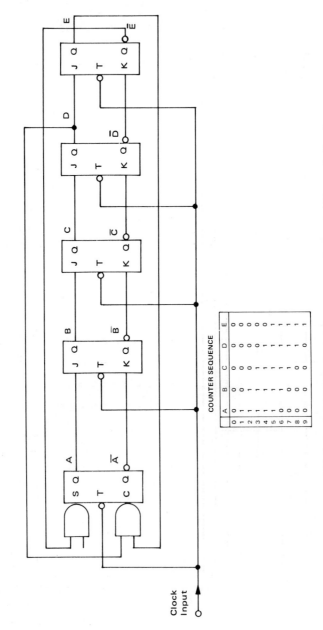

Figure 11-15 Switch-tail ring counter (courtesy Motorola Semiconductor Products)

A 1 is preset into the first stage of the register and shifted through one stage per count, until the final stage is reached. Note that the Q output of the final stage feeds the J input of the first stage and that the $\overline{Q}$ output feeds the K input. As long as the final stage is in the CLEARed state, the $\overline{Q}$ output is HIGH and the K input to the first stage is HIGH. Each time a clock pulse is received the first stage is returned to (or left in) the CLEARed state. When the propagating 1 reaches the final stage, the J input of the first stage is enabled, and the 1 transfers from the final stage of the register to the first stage. Ten counts are required for the propagating 1 to recycle, and the circuit of Fig. 11-14 functions as a decimal counter. Since the propagating 1 effectively travels in a circle, the counter of Fig. 11-14 is sometimes known as a *ring counter*. Waveforms for this circuit resemble those of Fig. 11-13. A count sequence table showing the state of each stage is used in place of waveforms.

The Switch-Tail Ring Counter

The requirement for ten FFs in the decimal ring counter can be reduced to five by the circuit shown in Fig. 11-15. This counter is sometimes called a *switch-tail ring counter* or a *Johnson Counter*. The output is inverted (taken from the complementary output) of the last stage, before it is fed back to the first stage. In addition, a feedback connection from the next-to-last stage is applied to the first stage to make the counter self-correcting and to eliminate the requirement for initial presetting in most applications. The count sequence table is shown in Fig. 11-15. Waveform development is left to the reader.

11-4 APPLICATION AND ANALYSIS

The use of registers in both digital computer and information transfer situations has already been discussed. Numerous other applications in many other types of equipment exist, but one specific use of registers easily demonstrates both multiple register application *and* analysis techniques.

The Block Diagram

A portion of a block diagram of a digital logic system is shown in Fig. 11-16. Such a form is common in large digital systems, since complete logic diagrams tend to be too large. The system diagram of Fig. 11-16 provides some clues to circuit functions. Note that the input is defined as "serial data in" and the output as "parallel data out." We might presume that the circuit is a serial-to-parallel data converter as discussed in Section 11-2. Without additional information, analysis of this register application is difficult however.

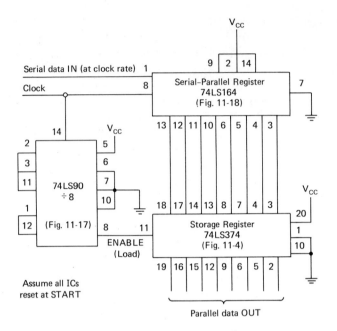

Figure 11-16 Digital system diagram

IC Specifications

The next step requires consulting specification sheets to determine the nature of each of the ICs. Figure 11-17 is the logic diagram and pin connections for the 74LS90 decade counter; Fig. 11-18 furnishes the same information for the 74LS164 parallel-out serial shift register. The 74LS374 buffer register has already been shown in Fig. 11-4.

Investigation of the 74LS90 counter shows that it consists of flip-flops and gates that provide a divide-by-two and a divide-by-five capability. External connections can be made to supply the divide-by-N capability. The 74LS164 register contains 8 flip-flops and associated gating which is connected to shift input data serially from left to right. Each clock pulse causes data at the input to be placed in the left-most flip-flop, while shifting data in the rest of the flip-flops one position to the right.

Determination of the actual logic system operation requires full knowledge of each of the ICs. Truth tables accompanying each logic diagram can be used during system analysis.

COUNTER ANALYSIS

For normal decade (BCD) counting, the Q output of flip-flop A is connected to the clock input of flip-flop B. Then the counter operates

Application and Analysis

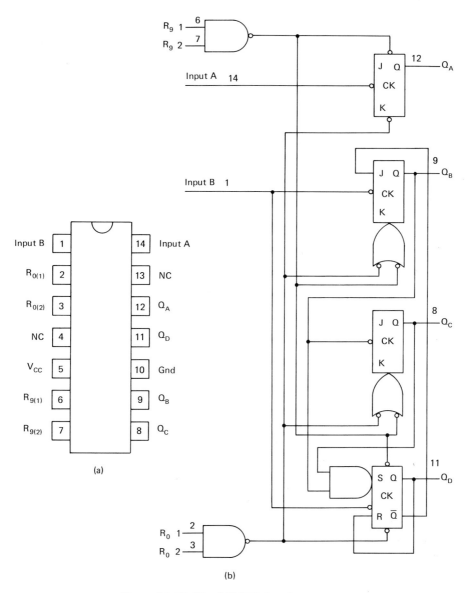

Figure 11-17 The 74LS90 decade counter

according to the BCD count sequence table of Figure 11-17. Verification of the BCD count sequence is left as an exercise for the reader.

When the external connections shown in Figure 11-16 are used, however, the count sequence is altered. Connection of the Q output of the D flip-flop to the Reset 0 gate causes all flip-flops to be reset when the Q output of the D flip-flop goes HIGH. This occurs at count 8, and

Count	Output			
	Q_A	Q_B	Q_C	Q_D
0	L	L	L	L
1	L	L	L	H
2	L	L	H	L
3	L	L	H	H
4	L	H	L	L
5	L	H	L	H
6	L	H	H	L
7	L	H	H	H
8	H	L	L	L
9	H	L	L	H

(c) BCD Count Sequence

Figure 11-17 The 74LS90 decade counter (cont'd)

the count sequence then becomes as shown in the modified count sequence table. The Q output of the D flip-flop is also connected to the ENABLE (Load) input of the 74LS374 register. Thus the 74LS374 receives a "Load" input every eighth clock pulse.

SHIFT REGISTER ANALYSIS

Operation of the 74LS164 shift register during data input is very much like the serial register discussed in Section 11-2. Input data is latched into the left-most flip-flop when the clock pulse occurs. At the same time, each of the following flip-flops stores the data previously in the preceeding flip-flop. Following occurrence of the clock pulse, data stored in each flip-flop is available at the individual Q outputs. No new techniques are found in the 74LS164 shift register. Previous discussions have explained shift register operations.

Operational details of the 74LS374 buffer register have been discussed in Section 11-1. As further analysis is shown, refer to Fig. 11-4 for details.

System Analysis

One of the most easily understood methods of system analysis uses the actual waveforms that occur. For example, modern logic analyzers allow simultaneous display of waveforms at many different locations in a system. Figure 11-19 shows the waveforms that would occur at the data input, clock input, ENABLE (load) input to the buffer register, and at the eight buffer register outputs. (It should be noted that most logic analyzers allow display of data in terms of 1s and 0s

Application and Analysis

(a)

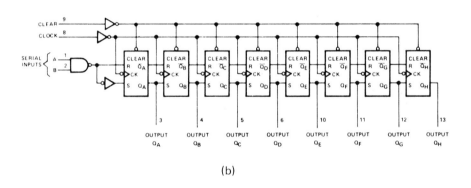

(b)

INPUTS				OUTPUTS		
CLEAR	CLOCK	A	B	Q_A	Q_B	Q_H
L	X	X	X	L	L	L
H	L	X	X	Q_{A0}	Q_{B0}	Q_{H0}
H	↑	H	H	H	Q_{An}	Q_{Gn}
H	↑	L	X	L	Q_{An}	Q_{Gn}
H	↑	X	L	L	Q_{An}	Q_{Gn}

H = high level (steady state), L = low level (steady state)
X = irrelevant (any input, including transitions)
↑ = transition from low to high level.
Q_{A0}, Q_{B0}, Q_{H0} = the level of Q_A, Q_B, or Q_H, respectively, before the indicated steady-state input conditions were established.
Q_{An}, Q_{Gn} = the level of Q_A or Q_G before the most-recent ↑ transition of the clock; indicates a one-bit shift.

(c)

Figure 11-18 The 74LS164 shift register

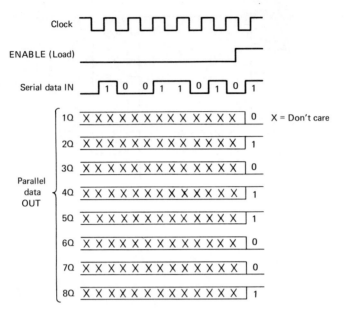

Figure 11-19 Digital system waveforms

instead of waveforms. Some technicians and engineers prefer the digital display over the waveform display.)

The waveforms of Fig. 11-19 assume that all ICs are reset before the occurrence of the first clock pulse and arrival of input data. It is also assumed that data arrives synchronized with the clock pulses. Such requirements are easily met with additional logic circuits. Note that the ENABLE (load) input to the buffer register and the buffer register outputs do not change until eight clock pulses have occurred. Thus eight bits of data have been fed into the shift register and shifted right. As the eighth data bit is entered, the counter supplies the ENABLE (load) input to the buffer register, and information present on each of the eight outputs of the shift register is transferred to the buffer register. The transferred information is now available at the buffer register outputs, and the counter and shift register can proceed to process the next eight data bits. Buffer register outputs can be examined for eight full clock pulse times before another group of eight bits are stored. A serial-to-parallel data converter thus has been analyzed.

If more detailed analysis is required, the waveforms at the outputs of the shift register can be examined as data is entered. The operation of the shift register can then be seen easily. Proper selection of points to be viewed with the logic analyzer makes analysis and troubleshooting much simpler.

11-5 THE FUTURE

Continuing advancements in semiconductor technology have resulted in great improvements in register design and applications. Now 1000 or more bits of static storage and shifting capability using latch-type circuitry are available in common IC packages. The use of data-storage techniques that do not depend on FF operation has resulted in decreased complexity and increased capacity. More than 65,000 bits of storage and shifting can be accommodated by the use of dynamic (circulating) techniques. This non-FF type of operation requires that the data be constantly circulated—or it will be lost. In contrast, the static type of register retains data without the constant presence of clock pulses to keep it recirculating.

Both of these high-density registers are of the serial-serial type, since parallel operation would require an excessive number of external connections. However the techniques of construction result in operating speeds greater than 25,000,000 operations per second, which makes up for the slower serial type of operation. Technological advances will certainly extend the upper limit in the very near future.

PROBLEM APPLICATIONS

11-1. Define the term register.
11-2. What are the characteristics of a storage register?
11-3. What are the characteristics of a shift register?
11-4. List the four classifications of register input-output methods and explain the concept of each method.
11-5. Discuss the applications of R-S, J-K, and D FFs to storage registers.
11-6. What is a buffer register?
11-7. List some advantages and disadvantages of prepackaged IC registers.
11-8. How do shift registers differ from storage registers?
11-9. How are shift registers similar to storage registers?
11-10. Discuss the application of R-S, J-K, and D FFs to shift registers.
11-11. What is a circulating register?
11-12. Draw the waveform diagrams that represent loading of the binary data 1010 into the register of Fig. 11-10. Show the same data being shifted to the right four times.
11-13. Explain the application of shift registers to the performance of arithmetic operations.
11-14. Draw the waveforms that show operation of the ring counter in Fig. 11-14.
11-15. Draw the waveforms that show operation of the switch-tail ring counter of Fig. 11-15.
11-16. Redraw the logic diagram of Fig. 11-8 using J-K FFs.

11-17. Identify the logic circuit shown below and document the steps of the analysis.

11-18. Identify the logic circuit shown below and document the steps of the analysis.

11-19. In microcomputer applications, information is handled in groups of eight binary digits (bits). The information appears all eight bits at once on eight separate lines called the data bus. Referring to the figure below, assume that the following events occur in sequence:

1. the data 10101010 is on the data bus.
2. the OUTPUT ENABLE (OE) input goes HIGH.
3. the clock input transitions from LOW to HIGH and back to LOW.
4. the data on the bus is removed.
5. the OUTPUT ENABLE (OE) input goes LOW.

Using the information for the 75LS374 shown in Figure 11-4, explain what would occur.

The Future

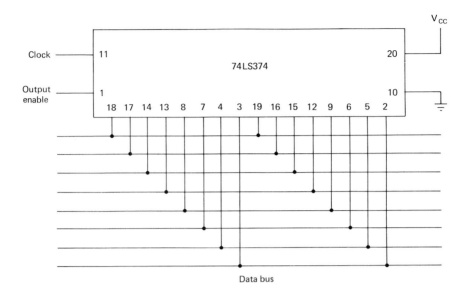

11-20. Discuss the future developments that may be expected in the design and application of registers.

12

MISCELLANEOUS LOGIC FUNCTIONS

You may have noted that in the two preceding chapters many of the counter and register applications were implemented with *packaged circuits.* It would seem reasonable that such packaged circuits would be used when complete functions may be obtained in sizes much smaller than the flip-flop (FF) implemented functions. The cost of implementation of the hardware by either method is similar. But the reduced manufacturing cost (less components to handle) plus the smaller physical size of the finished product generally leads to a preference for packaged integrated circuits (ICs) whenever possible.

Four prepackaged logic functions are explained in this chapter: *Exclusive-ORs, full adders, multiplexers,* and *memory elements.* Two other common prepackaged functions (decoders and code converters) will be discussed in subsequent chapters, due to their specific applications. It should be recognized, however, that the trend is toward putting *all* functions required for a specific application on a single chip. We continue to investigate less complex ICs and functions to expand our knowledge so that the multi-function LSI chips can be more easily understood.

12-1 THE EXCLUSIVE-OR OPERATION

Functional Description

The Exclusive-OR operation is a function that appears so often in digital circuitry that it has been assigned a special symbol. But, since it is not an independent operation, it can be expressed in terms of the

The Exclusive-OR Operation

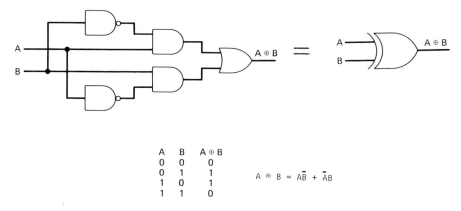

Figure 12-1 The Exclusive-OR operation

basic connectives: AND, OR, and NOT. The expression *A Exclusively-OR B* is written $A \oplus B$. This function is true ($A \oplus B = 1$) when *A* or *B* is true but *not* when both are true. Figure 12-1 shows the standard symbol for the Exclusive-OR operation, its truth table, algebraic expression, and a typical IC implementation of the operation. Many other logic implementations of the Exclusive-OR operation are used and will be discussed as they are encountered. A common IC (74LS386 for example) contains four Exclusive-OR functions per package.

The Equality (Exclusive-NOR) Operation

Many useful logic functions are derived from the Exclusive-OR operation. The expression for the Exclusive-OR circuit ($A\bar{B} + \bar{A}B$) shows that the nonequal cases may be determined by noting the logic level at the output. The output is 1, when *A* is 1 or *B* is 1, but not when both are 1 at the same time or when both are 0 at the same time. The mere addition of a logic inversion following the Exclusive-OR results in a circuit that can determine whether inputs are equal or unequal.

Negating the Exclusive-OR expression and simplifying the resulting expression $\overline{A\bar{B} + \bar{A}B}$ yields $AB + \bar{A}\bar{B}$. When *A* and *B* are both 1 or both 0, the logic value of the negated Exclusive-OR operation is 1. This is sometimes called the equality (*Exclusive*-NOR) operation. Details of the equality operation are shown in Fig. 12-2. The 74LS266 IC contains four Exclusive-NOR functions on a single chip.

Comparators

One of the most common applications of the equality circuit is found in digital computers. The values of variables are often compared, and the results are then used to determine future computer action. For

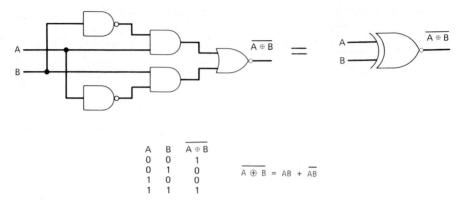

A	B	A ⊕ B
0	0	1
0	1	0
1	0	0
1	1	1

$A \oplus B = AB + \overline{A}\overline{B}$

Figure 12-2 The Exclusive-NOR (equality) operation

example: if A is less than B, the computer is instructed to perform a specific program step; if A is greater than B, it must perform a different step; if $A = B$, still another "branching instruction" is required. So far, we have discussed the logic circuitry that determines whether $A = B$ or $A \neq B$. But the determination of *relative magnitude* (i.e., $A<B$ or $A>B$) involves a logic circuitry called a *comparator*.

Actually, the Exclusive-NOR circuit is capable of furnishing *all three* of these comparisons if it is implemented like the logic circuit of Fig. 12-3.* As the truth table shows, $A>B$ is only 1 when $A = 1$ and $B = 0$; $A = B$ is 1 only when A and B are both either 0 or 1; and $A<B$ is 1 only when $A = 0$ and $B = 1$.

When comparators are used in computers, they must be capable of comparing multi-digit numbers, so the logic circuitry becomes quite complex. A comparator that determines the relative magnitude of two 3-bit binary numbers is shown in Fig. 12-4. The accompanying table lists circuit outputs for the magnitude relationships between A and B.

The function of each gate is listed as follows:

G_1 – determines equality of the most-significant digits in A and B.
 A HIGH output indicates $A_2 = B_2$.
G_2 – same as G_1, except A_1 and B_1.
G_3 – same as G_1, except A_0 and B_0.
G_4 – determines if $A_2 > B_2$ ($A_2 = 1$ and $B_2 = 0$). A HIGH output means $A_2 > B_2$.
G_5 – determines if $A_1 > B_1$ ($A_1 = 1$ and $B_1 = 0$) and $A_2 = B_2$. A HIGH output results if the gate is activated by these conditions.

*This is a relatively simple circuit, comparing only two single-digit binary numbers.

G_6 – determines if $A_0 > B_0$ ($A_0 = 1$ and $B_0 = 0$), $A_2 = B_2$, and $A_1 = B_1$. A HIGH output means that the gate has been activated by all of these conditions.

G_7 – ANDs the outputs of all three equality gates to furnish an overall HIGH indication that $A = B$. A LOW output signifies $A \neq B$.

G_8 – ORs the output of comparison gates G_4, G_5, and G_6 so that a HIGH output results if an $A_n B_n$ combination of $A = 1$ and $B = 0$ is present. A LOW output is interpreted as $A < B$.

The circuit operation of the 3-bit comparator is summarized in the following diagram.

G$_1$ output HIGH   G$_2$ output HIGH   G$_3$ output LOW – forces G$_7$ output LOW
G$_4$ output LOW   G$_5$ output LOW   G$_6$ output HIGH – allows G$_8$ output to
                                        be HIGH. Therefore $A > B$ due to $A_0 > B_0$

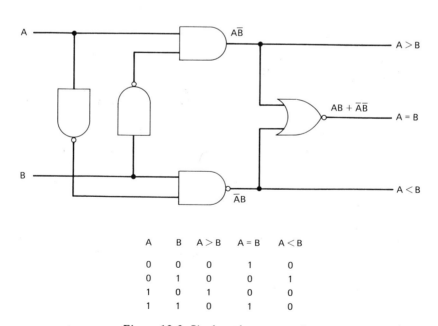

A	B	A>B	A=B	A<B
0	0	0	1	0
0	1	0	0	1
1	0	1	0	0
1	1	0	1	0

Figure 12-3 Single-order comparator

Figure 12-4 3-bit comparator

Parity

The constant movement of data both within digital machines and from machine to machine requires a very *high degree of accuracy*. For example, if the 15-bit computer word 000001000000000 represents the amount of $512, any error in transmission can dramatically affect the actual value of someone's check! (A single 0 changed to a 1 by "noise" during movement of the data—000001100000000—results in a value of $768.) The increasing reliability of solid-stage logic devices is helping to attain this high accuracy, but external influences in the transmission path commonly cause errors to develop. Duplication of the system—using a different transmission path and then comparing both sets of data—is another (expensive) way of increasing accuracy.

One of the most common and relatively inexpensive methods of ensuring data-movement accuracy is the addition of a standardizing or *parity bit* at the end of each word.

> Since all data words contain either an even or an odd number of 1s, *if the standard (parity) is an even number* of 1s, *it is said that* even parity is being used; if an odd number of 1s is required to meet the standard, the standardizing system is called odd parity.

Prior to movement of a data word, the *number* of 1s in the word is determined. For a 16-bit word odd parity system, if the number of 1s

The Exclusive-OR Operation

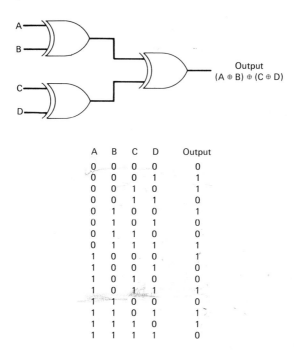

A	B	C	D	Output
0	0	0	0	0
0	0	0	1	1
0	0	1	0	1
0	0	1	1	0
0	1	0	0	1
0	1	0	1	0
0	1	1	0	0
0	1	1	1	1
1	0	0	0	1
1	0	0	1	0
1	0	1	0	0
1	0	1	1	1
1	1	0	0	0
1	1	0	1	1
1	1	1	0	1
1	1	1	1	0

Figure 12-5 4-bit parity detector/generator

is even, a 1 is placed in the final position. Thus a 16-bit word is formed, so the data is moved with an odd number of 1s present. When received, the number of 1s is again determined. If an odd number of 1s is present, it is assumed that the data word has been moved correctly. If an even number of 1s is detected, a warning is provided to show that the data is incorrect.

This type of parity system is effective only in detecting an *odd number of 0s that have changed to* 1s (i.e., 1, 3, 5, etc.). A simple, *4-bit parity detector/generator,* with its algebraic expression and truth table, is shown in Fig. 12-5.

Each Exclusive-OR gate is really an even parity generator for the two inputs. Parity generation for a *4-bit word* thus requires one Exclusive-OR gate for each pair of inputs, plus a third Exclusive-OR gate to compare the outputs of the first two gates.

The 4 bits of the binary word to be transmitted are applied to A, B, C, and D, respectively. Parity is checked, and if an odd number of 1s is detected (as it will be if the input is $A\bar{B} + \bar{A}B$), the output of the parity generator becomes 1. A 1 is then attached as a fifth bit to the transmitted word. If an even number of 1s is detected ($AB + \overline{AB}$), a 0 is attached. (The circuit that adds the parity bit to the transmitted message is not shown in Fig. 12-5.)

TABLE 12-1 Rules of binary addition

Digit A	Digit B		Sum	Carry
0	plus 0	=	0	0
0	plus 1	=	1	0
1	plus 0	=	1	0
1	plus 1	=	0	1

Parity systems capable of detecting and even correcting multiple errors in transmitted data are now in use, but they are beyond the scope of this text.

Binary Addition Using the Exclusive-OR Operation

The ability to perform *arithmetic addition* is an inherent requirement in digital computers. Binary addition is quite similar to decimal addition, but the addition operation is greatly simplified. In the decimal system all combinations of *ten* digits must be examined to construct an addition table; in the binary system, only *two* digits are used. The four rules of binary addition are shown in Table 12-1.

By arbitrarily assigning the *logic value* 0 to the *arithmetic value* 0 and the *logic value* 1 to the *arithmetic value* 1, the Carry column of Table 12-1 becomes identical to the logical AND truth table, and the Sum column matches the truth table for the Exclusive-OR operation (Fig. 12-1). The logic circuit that implements the Exclusive-OR operation may then be used to obtain the sum of two single-digit binary numbers, while any AND circuit will develop the required carry value. It should be noted, however, that only *single-digit* numbers may be accommodated by this maneuver. If multi-digit numbers are added and a carry from a previous order of addition is to be included, other means must be provided. Logic circuits that perform this type of operation (Full adders) are discussed shortly.

If a logic circuit designed to provide the sum of two numbers is so arranged that the carry operation may also be derived without additional circuitry, so much the better. Actually, the circuit of Fig. 12-1 is designed to provide both the sum and carry output, although the carry output is not available externally. (The Exclusive-OR output is, of course, the sum.) One of the inputs to the final NOR gate is AB, which is the carry expression. Another implementation of the Exclusive-OR operation is shown in Fig. 12-6. It provides both sum and carry outputs. Proof of the $A \oplus B$ output is left to the reader.

A circuit that performs all the logic operations required to add two binary digits is called a *half-adder*. It must have *two inputs*—one for

Arithmetic/Logic Circuits

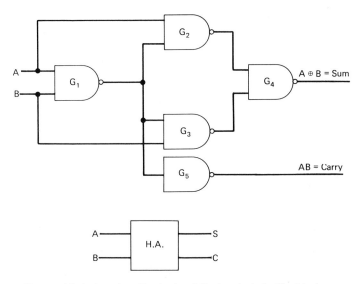

Figure 12-6 Another Exclusive-OR circuit (a half-adder)

each digit representation—and *two outputs*—one for the sum and one for the carry. It is called a "half-adder" because two half-adders are required to add two digits *and* the carry that may result from the addition of two less-significant digits.

Both Exclusive-OR circuits discussed so far fit this description and may be used as half-adders. The half-adder as a separate IC package is fast disappearing. In applications where the half-adder is required, it is usually implemented with packaged Exclusive-OR circuits and the necessary separate gates.

12-2 ARITHMETIC/LOGIC CIRCUITS

Although most arithmetic functions, such as addition and subtraction, adapt readily to digital computer applications, many of the circuits designed to implement these functions are also of value in other digital applications. Included in this category are *adders* and *multifunction assemblies*. Since the adder is the basis for many of the other circuits used in arithmetic functions, we shall discuss it first.

The Full Adder

During the explanation of the half-adder operation, we pointed out that additional circuitry was necessary to accommodate a third variable (the carry from an addition operation in a lower significant

Figure 12-7 Binary full adder

position). This requirement dictates a circuit with *three inputs, one sum output,* and *one carry output.* Again, by equating *arithmetic* 0 and 1 to *logic* 0 and 1, we can construct a table that describes the *full adder* function (see Fig. 12-7a).

The Boolean expression that results from the tabular description is extracted and subjected to simplification (Fig. 12-7b). A logic diagram (Fig. 12-7c) is then drawn to implement the equations for both the SUM and CARRY terms. All eight of the possible combinations of A, B, and C_i should be applied as inputs to the full adder and logic levels traced through the circuit to assure that the proper logic levels result at the SUM and CARRY outputs.

With proper modification of the numbers to be operated upon, adders may be used to perform other basic arithmetic operations. But since this text is not oriented toward the digital computer, the performance of subtraction, multiplication, and so forth is not discussed here.

In IC packages, the full adder appears in *single, dual,* and *quad* configurations. Some versions provide internal interconnections, while

Arithmetic/Logic Circuits

others make all (or at least the most important) inputs and outputs available externally. A very versatile, single-bit full adder (type 5480) is shown in Fig. 12-8. Gated complementary inputs are provided, along

(a) Package diagram

(b) Logic diagram

(c) 4-bit binary adder using 5480 packages

Figure 12-8 5480 single-bit full-adder

Figure 12-9 7483 4-bit full-adder

with both a true SUM output (S) and its complement ($\bar{S}$), and an inverted CARRY output. Input gating and complementing is accomplished by G_1 through G_6, inclusively. The remainder of the circuit functions as a conventional full adder. A 4-bit, binary full adder is also shown implemented with the type 5480 IC package. The complementary inputs allow easy implementation of the subtraction operation.

A 4-bit, binary full adder (type 7483) in a single IC package is shown in Fig. 12-9. It is very similar to modernized versions such as the 74283. While not nearly as versatile as the 5480 circuit, in applications where intermediate outputs are not required this full adder is quite satisfactory.

Parity Generation and Checking

As pointed out in Sec. 12-1, the Exclusive-OR operation is fundamental to the *parity-checking* concept. Whereas the Exclusive-OR gate is used to detect the existence of an odd number of 1s between two bits, the full adder can be used to detect the same relationship between *three* bits. The sum output of a full adder may be written as $S = A \oplus B \oplus C$, which describes its function as a parity checking device. Expansion of this equation gives us

Multiplexers

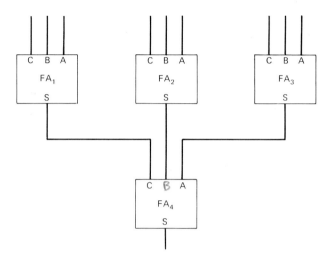

Figure 12-10 9-bit parity generator and checker using full-adder

$$A \oplus B \oplus C = (A \oplus B) \oplus C$$
$$= (A\bar{B} + \bar{A}B) \oplus C$$
$$= (A\bar{B} + \bar{A}B)(\bar{C}) + \overline{(A\bar{B} + \bar{A}B)}(C)$$
$$= A\bar{B}\bar{C} + \bar{A}B\bar{C} + (AB + \bar{A}\bar{B})(C)$$
$$\therefore \quad A \oplus B \oplus C = A\bar{B}\bar{C} + \bar{A}B\bar{C} + ABC + \bar{A}\bar{B}C$$

The diagram of Fig. 12-10 shows how four full adders may be used to generate parity for a 9-bit word. FA_1, FA_2, and FA_3 perform the $A \oplus B \oplus C$ operation on their inputs, and FA_4 then combines the results in its 3-input Exclusive-OR operation. Since each full adder has both a true and a complemented SUM output, both odd and even parity are easily obtained—if parity is *not* odd, it must be even!

12-3 MULTIPLEXERS

Another class of logic circuits is the *multiplexer*. The *multiplexer* is actually *a signal selector gating circuit that is the logic equivalent of a multiposition selector switch.* A number of input signals are fed into the multiplexer, and by using control inputs, one of the signals is selected and made available at the output. Multiplexers can be used in both digital and analog applications.

Switching

The basic concept of digital data selection is shown in Fig. 12-11. A single-pole, 2-position switch (Fig. 12-11a) is logically implemented with the increasingly popular *AND-OR-INVERT* (AOI) *gate* (Fig. 12-11b) and an inverter. AOI gates are available in all the common packaging forms, and in some cases, they may actually implement the standard logic operations with a reduction in cost and package count. The accompanying truth table defines the operation of the *2-input multiplexer*. The output of the multiplexer is either A or B, depending on the logic level at the control inputs, S_1 and S_2. When both controls are HIGH at the same time, the conventional OR operation is performed—this is usually not considered a multiplexer function.

(a) Single-pole, 2-position switch

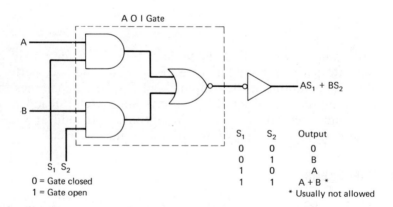

(b) Logic implementation

Figure 12-11 Multiplexer switching concept

As it becomes necessary to accommodate more inputs, the requirements for control signals can become restrictive. For example, if eight inputs must be switched by a multiplexer, eight control signals are necessary, and the number of connections for inputs and outputs exceeds the capability of a standard 16-pin IC package. This problem is solved by using *different combinations of control inputs* to perform selection. Eight different control signals may be derived from only three control inputs by using all possible combinations of 1 and 0 that exist with three variables.

The type 8232 8-input digital multiplexer (Fig. 12-12) uses this concept. Selection of I_2, for example, is made by an address of $A_2 = 0$, $A_1 = 1$, and $A_0 = 0$. G_2 is then enabled so that the information on input line I_2 is fed to the output NOR gate G_8. Both true and complemented outputs are provided and may be inhibited to provide a constant 0 at both outputs. The truth table accompanying Fig. 12-12 defines the operation of the 8232, which is much like the 74151 and 4512 ICs. Many other multiplexer I/O combinations exist, but all function under the general switching concepts just described.

An interesting application of multiplexers is shown in Fig. 12-13. A need often arises in both digital test equipment and computers to select *one of a number of different frequencies* that are related to a fundamental clock or timing frequency. Recalling from Ch. 10 that the simple binary counter divides its inputs by 2 is the clue to the operation of the *frequency selector* in Fig. 12-13. Two type 8281 4-bit binary counters (similar to the 7493 counters and its successors discussed in Sec. 10-1) are cascaded, thus providing a division ranging from a factor of 2 at the A_0 output of the first counter to a factor of 256 at the D_0 output of the second counter. The eight different frequencies are supplied to the inputs of the 8232 Multiplexer (or its modernized equivalents). Selection of the required frequencies is made by proper addressing of the A_0, A_1, and A_2 inputs, as shown in the table accompanying Fig. 12-13. Although only a simple switching application, this example shows the versatility of the multiplexer circuit and the imagination of the designer in applying the principles of logic.

Nondigital Applications

Multiplexer applications are not limited to digital circuits. *Analog information* may also be switched with a multiplexer, if the gates are properly designed. The enabling inputs are merely used to allow the analog information to pass. A specific example is *time-division multiplexing* (TDM). This technique sequentially samples each of the inputs

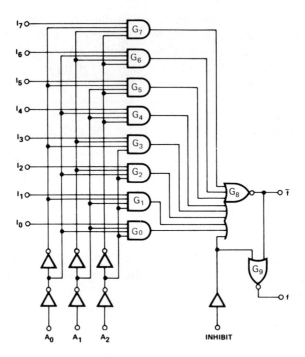

(a) Logic diagram

ADDRESS			DATA INPUTS								INH	OUTPUT		
A_2	A_1	A_0	I_7	I_6	I_5	I_4	I_3	I_2	I_1	I_0		f	8230 8231 $\bar{f}$	8232 $\bar{f}$
0	0	0	x	x	x	x	x	x	x	1	0	1	0	0
0	0	1	x	x	x	x	x	x	1	x	0	1	0	0
0	1	0	x	x	x	x	x	1	x	x	0	1	0	0
0	1	1	x	x	x	x	1	x	x	x	0	1	0	0
1	0	0	x	x	x	1	x	x	x	x	0	1	0	0
1	0	1	x	x	1	x	x	x	x	x	0	1	0	0
1	1	0	x	1	x	x	x	x	x	x	0	1	0	0
1	1	1	1	x	x	x	x	x	x	x	0	1	0	0
0	0	0	x	x	x	x	x	x	x	0	0	0	1	1
0	0	1	x	x	x	x	x	x	0	x	0	0	1	1
0	1	0	x	x	x	x	x	0	x	x	0	0	1	1
0	1	1	x	x	x	x	0	x	x	x	0	0	1	1
1	0	0	x	x	x	0	x	x	x	x	0	0	1	1
1	0	1	x	x	0	x	x	x	x	x	0	0	1	1
1	1	0	x	0	x	x	x	x	x	x	0	0	1	1
1	1	1	0	x	x	x	x	x	x	x	0	0	1	1
x	x	x	x	x	x	x	x	x	x	x	1	0	1	0

x = don't care

(b) Truth table

Figure 12-12 8232 8-input digital multiplexer (courtesy Signetics, a subsidiary of Corning Glass Works)

Multiplexers

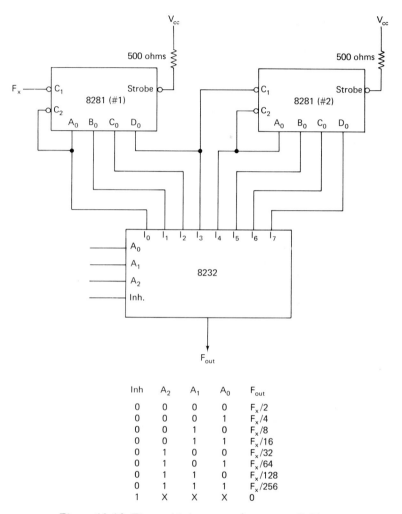

Inh	A_2	A_1	A_0	F_{out}
0	0	0	0	$F_x/2$
0	0	0	1	$F_x/4$
0	0	1	0	$F_x/8$
0	0	1	1	$F_x/16$
0	1	0	0	$F_x/32$
0	1	0	1	$F_x/64$
0	1	1	0	$F_x/128$
0	1	1	1	$F_x/256$
1	X	X	X	0

Figure 12-13 The multiplexer as a frequency divider

and provides a single output that contains a sample of the amplitude of each input. Voices and music may be transmitted from one location to another in this manner, so long as the sampling rate is at least twice the highest frequency in the signal being sampled.

Once the sample is obtained, the composite output is converted to a form suitable for digital transmission. Numerous transmission forms, such as pulse height, pulse width, pulse spacing, and pulse grouping, may be used. Upon reception of the coded information, it is applied to a *demultiplexer,* which extracts the information from each channel, separates it, and provides an output representative of the information

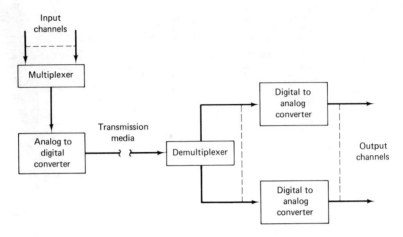

Figure 12-14 Non-digital multiplexer application

originally encoded at the transmitting end of the circuit. The techniques used for conversion to digital form are many and varied. The basic use of the multiplexer is such a system is seen in the block diagram in Fig. 12-14.

12-4 MEMORY ELEMENTS

Applications

Many digital systems require both *temporary* and *semipermanent data-storage capabilities.* Digital computers with their large-scale memory requirements also need smaller memory devices to increase their efficiency. The large memories are relatively slow, so smaller memories act as buffers between the rest of the computer and the large memory. When calculations are being performed, it is most inefficient to move the partial results to main memory. Small-scale memory devices are very useful in this application.

Data communication systems use small-scale memories to temporarily store groups of words, before passing them to the display device or to a computer's main memory. Test equipment used to verify digital system operation uses memory capability to store a sequence of events to be performed or a group of binary digits to exercise the system. Calculators need to store the variables to be operated on. Data display systems must accumulate sufficient data so that a complete line of information may be displayed at one time. Thus applications of *small-scale memory devices* are almost endless.

This section examines some of the common small-scale memory

devices used in digital equipment. Discussion will be limited to *semiconductor types of memories,* due to the almost universal application that these devices possess. In addition, they are available in the same manner as the other complex functions discussed in this chapter and fit readily into the prepackaged logic function category. Other modes of storage do exist—but usually at a prohibitive cost and size in terms of the relatively small amount of information that must be stored.

Classification of Small-Scale Memories

The most common of all semiconductor memory elements, the *flip-flop,* has already been examined in great detail. When more than one bit of information must be stored, FFs may be connected together to form registers, as we discussed in Ch. 11. In applications where no more than a few bits of information are to be stored, FF registers adequately fulfill this function.

When more than one group of bits is to be stored, however, FF registers tend to become physically large and expensive. Packages, called *random access memories* (RAMs), are available to fill the gap between single-group storage and computer main memories, which store thousands of groups of bits. *The* RAM *is a group of storage devices,* which may or may not be FFs, *that store binary data in a digital system when the system is not using that data.* Information is randomly written into or read out of each storage location, as required. The term "random access" is derived from this memory property. Availability of this type of logic circuitry allows temporary memory (with a greater storage capacity than that of a single register) to be distributed throughout the system.

Read-only memories (ROMs) are another form of small-scale information storage devices that are finding unprecedented applications in the digital field. *Anytime a combination of control variables or a fixed value is required from a fixed input, a* ROM *may be used.* Each storage location in the ROM is addressable, just as in the RAM. The information stored may be a complete trigonometric table, a predetermined message or sequence of bits to be used in a data communication system, or numerous other applications. ROMs can supply permanent storage of data that is to be used repetitively, thereby relieving other logic circuits of performing calculations to obtain the data each time that it is to be used.

Random Access Memories (RAMs)

One method of forming an IC memory cell is shown in Fig. 12-15. Many other types of memory cells are used, but this method is one of the easiest to understand. Each of the storage locations is a modified *R-S* FF, which requires both the X and Y inputs to be enabled before

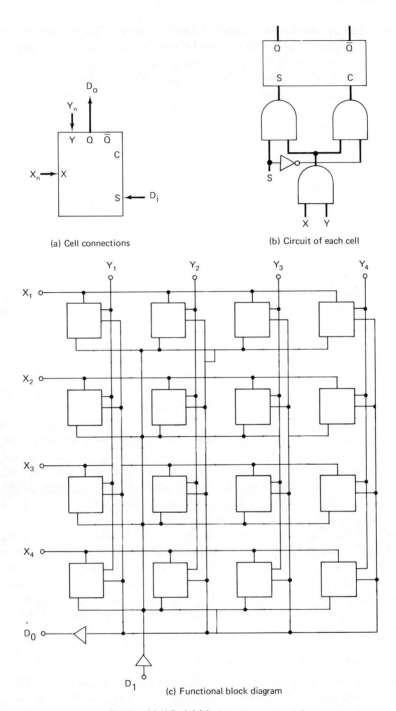

Figure 12-15 16-bit memory concept

the data input is active. The sixteen locations are arranged in an addressable 4 × 4 matrix. The desired bit location is selected by enabling the coincident X-Y address lines (hence the classification of "coincident selection") and by keeping the remainder of the address lines disabled. The data that was stored at the selected location is then read at the output terminals. If the data at the addressed location is a 1, the D_0 terminal is HIGH. If the data stored is 0, a LOW results at the D_0 terminal.

To write a 1 in a specific location, the desired storage location is addressed and the D_i input is set to a HIGH level. A 0 is written by setting the D_i input to a LOW level. This IC is capable then of accepting, storing, and reading out sixteen single-bit binary words. (A small storage capability IC has been chosen to explain RAM concepts more easily.)

Expansion to multiple-bit words merely requires an additional IC for each bit in the word. Equivalent address lines are paralleled, e.g., all X_1s connected together, so that when a location is addressed on one IC, the same location is addressed to the other ICs. Of course, the outputs and the inputs are not paralleled, since each bit in a word must be available separately, either for output or input. Thus a memory unit capable of storing sixteen 8-bit words could be constructed from 8 of the ICs being discussed.

Digital computers may require storage of many thousands of "words" in their memories. Techniques exist for this capability, but non-computer applications can function with much smaller memories. To simplify small-scale memory discussion, an early 256 bit memory IC housed in a standard DIP package (Fig. 12-16a and b) is used. The 2501 memory is fully decoded, thus permitting the use of the standard DIP package. Read/Write and Chip-select control modes are provided. A LOW logic level applied to the R/W control allows the information from one location in the memory to be made available within about 1 microsecond, following the arrival of the address code. Readout is nondestructive.

A HIGH level on the R/W input (in conjunction with an address code) causes the input data to be written into the selected storage location. Complete operation of the circuit is inhibited when the Chip-select input is HIGH.

Arbitrary size memories may be built by connecting appropriate numbers of 2501s together. Figure 12-16c shows a general-concept block diagram of a memory system containing (256 × N) words by M bits. For example, if the memory size were 4096 words by 12 bits, $N = 16$ and $M = 12$. Thus the number of 2501s required is $M \times N = 192$. The address inputs A_1 through A_8 are common to all rows. Inputs C_1 through C_n provide the column select and are wired to the Chip-select inputs of the 2501s. Extension of the techniques used in the 2501 have

362 Miscellaneous Logic Functions

(a) Package diagram (b) Functional block diagram

(c) Organization of 2501s into larger memory

Figure 12-16 2501 256 × 1 random access memory (RAM) (courtesy Signetics, a subsidiary of Corning Glass Works) resulted in single packages with a capacity of over 64,000 bits, and the end is not yet in sight.

Any of the units described so far perform one of two functions—a buffer storage device or a scratch-pad memory. *Buffer storage devices are used to compensate for differences in the rate of data flow.* Data is loaded into the memory device at one rate and read out at another rate. For example, in a digital computer, information is sometimes loaded into memory from punched cards. The rate of reading the punched card is hundreds of times slower than the rate at which the memory can store the data. By storing the punched card information in a buffer storage device, the main memory is free to perform other functions, until the buffer storage is ready to feed its information in for storage. It can then be interrupted for a short period of time, rapidly move the contents of buffer storage into its own memory, and go on with its other work, while buffer storage is being loaded with information from another punched card. Thus the buffer storage device allows the computer to more adequately use its time.

Scratch-pad memories are used for intermediate storage of data resulting from arithmetic or other signal-processing operations that have been accomplished in the main memory. Just as one uses a "scratch pad" to work arithmetic problems before transcribing them to a finished form, so does the scratch-pad memory provide a place to "work the problem" without cluttering up main memory.

The number of ICs required to implement the buffer storage device in the scratch-pad memory depends on the number of bits per computer word and the maximum number of words that must be stored. Their capacities range from as few as sixteen to thousands of bits, depending entirely on specific applications.

A noncomputer application of a (16 × 4) RAM is shown in the block diagram of Fig. 12-17. Almost all digital systems require some type of *word generator* to test system response. The simple programmable word generator is capable of providing sixteen 4-bit words at its output. Each word is loaded in the RAM by setting the data-entry switches to the desired bit pattern, selecting the location for storage by establishing the required address input combination (using the single-pulse switch), and depressing the R/W control switch. The RAM is changed from the READ to WRITE mode by the R/W switch action, and the logic data present at the WRITE inputs is placed in memory. As soon as all sixteen (or as many as are required) words are loaded, the word generator is ready for operation.

The Control Logic is the heart of the generator. Here the sequencing takes place, and clock signals are modified and distributed to per-

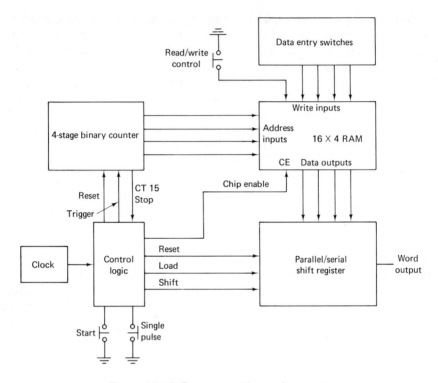

Figure 12-17 Programmable word generator

form their various functions. For example, if we desire to start generator operation by using the word stored at location 0, the binary counter must begin operation at the appropriate count. This can be assured by generating a RESET signal to the counter upon depression of the START switch. At the same time, the shift register is cleared. The RAM is enabled (C_E), and data present at the output of the RAM is made available to the shift register. A LOAD input is generated, followed by four shift commands, and the 4 bits of the first word are shifted out to the digital system being tested.

    A trigger pulse is now allowed through the control logic to the counter, and the counter advances one count. The second word is loaded into the shift register, shifted out, and the counter once more advanced. Each word is selected, loaded into the register, and shifted out. The recognition of count 15 in the counter is used to inhibit any further issuance of trigger pulses to the counter, and after shifting out the final word, the generator stops automatically. The digital system has now been exercised with 16 preselected test words of 4 bits each, and its response may be evaluated for proper operation.

Except for the clock and control logic, the entire word generator can be constructed from prepackaged logic functions already discussed. The 4-stage binary counter is similar to the 7493 binary counter of Sec. 10-1, and the shift register function can be implemented with an 74LS295 shift register (Ch. 11). The (16 × 4) RAM operates similar to the 8225 RAM discussed in this section. Clock and control logic implementation is accomplished with standard gates and FFs.

Read-Only Memories (ROMs)

A Read-Only memory is a system for storing information in a permanent form. In applications that require fixed data, the Read-Only memory offers advantages where speed, cost, and reliability are factors. In gross block-diagram form (Fig. 12-18a), a basic ROM has two types of inputs and one type of output. Operationally, the address of a word is placed on the address line, and the ENABLE input is activated. The contents of the addressed location then appear at the output.

The storage mode in ROMs is somewhat different than that in the RAMs. Whereas some type of semiconductor FF is often the storage medium in RAM, a simple semiconductor switch is sufficient for a ROM. An "open" switch may be either a logic 0 or a logic 1, depending on the technique used in the memory design. The ROM shown in the detailed block diagram of Fig. 13-23b uses an open switch as a logic 1 and a closed switch as a logic 0. The configuration of the switch at each bit position in the selected word is determined and then applied to individual buffer amplifiers for each bit of the output word.

ROMs may be obtained with any desired storage pattern or they may be programmed by the user. Some ROMs are not capable of being changed once data has been stored. This means that loss of power to the equipment or error in operations cannot destroy the contents. It also means that if new data is required, another ROM package must be used. Remember that once the ROM device has a program stored, it may not be changed. Mistakes with ROMs are costly.

If the cell must be erasable and rewritable, other techniques are used. Although different in the manner in which they are implemented, most of the erasable ROMs store data as an electrical charge. Complex principles of semiconductor physics are needed to explain the special charge storage phenomenon and are beyond the scope of this text. Data are placed in the ROM cell by an electrical current and retrieved by noting the effect of the stored charge on the electrical characteristics of the semiconductor cell. EAROMs (electrically alterable ROMs) can be erased by using an electrical current to remove the stored charges; EPROMs (erasable programmable ROMs) require the application of ultraviolet light for erasure.

Figure 12-18 Read-only Memory (ROM) (courtesy Motorola Semiconductor Products)

Modern ROMs contain up to 64,000 bits of storage, and the top limit is not in sight. Applications are seemingly endless. Tables of mathematical data may be stored for use in digital computers, calculators, and computing counters. Repetitive small-scale computer programs may be placed in ROMs to free main memory space for more important tasks. Digital display devices, such as those used with digital test equipment and computers, must generate individual characters (letters, numbers, symbols, etc.). ROMs are used to store the form of each of the characters, so it may be addressed and put to use without having to form it each time it is required. A code used to develop a digital data message may have to be changed to a different form for transmission. ROMs once again tend to be the most efficient method of translation.

Effectively, the ROM is a code converter, since it accepts the input information (the address code) and provides an output (the stored word) that is the input's equivalent in another form. ROM applications are shown in Ch. 14, where code converters are discussed in detail.

PROBLEM APPLICATIONS

12-1. What are the conditions under which the output of an Exclusive-OR gate is logic 1 or True?

12-2. Draw the symbol for an Exclusive-OR gate and show the Boolean expression used to describe its output.

12-3. What is an Exclusive-NOR gate? Show its symbol and its output expression.

12-4. What is another name for the Exclusive-NOR gate?

12-5. List the functions that a comparator performs.

12-6. Define odd parity; even parity.

12-7. Determine the output of the 4-bit parity detector/generator in Fig. 12-5 with the following inputs: 0000, 0101, 1010, 1110, and 1111.

12-8. What is a half-adder? a full adder?

12-9. Verify algebraically that the logic diagram of Fig. 12-6 performs the half-adder function.

12-10. Verify algebraically, that the logic diagram of Fig. 12-7 performs the full adder function.

12-11. Draw a logic diagram of a full adder using only NOR gates.

12-12. Develop a means of subtracting two numbers using the full adder of Fig. 12-8.

12-13. What is a multiplexer?

12-14. If f_x in Fig. 12-13 is 1 Mhz, what is f_{out} under the following conditions?

$$A_0 = 0, \quad A_1 = 0, \quad A_2 = 0$$
$$A_0 = 0, \quad A_1 = 1, \quad A_2 = 0$$
$$A_0 = 1, \quad A_1 = 1, \quad A_2 = 0$$

12-15. Define random access memory; read-only memory.

12-16. What is the usual storage media in small-scale semiconductor memory elements?

13

BINARY
CODES

By the time you reach this chapter, there should be no doubt that the *binary number system* is commonly used in digital devices. Arithmetic operations have been performed using binary numbers. Sequences of operations have been generated. Many of the simple decisions encountered in machine operation have been adapted to binary circuits.

One problem remains, however. Although a digital device may conveniently use binary representation, when it becomes necessary for humans to intervene, the information must be converted to more easily interpretable form. That is, *the input of information to the device* and *the output of information from the device must be in something other than conventional binary form.*

One way of satisfying this requirement is to assign a *unique combination of* 1s *and* 0s to each number, letter, or symbol that must be represented. Such representations are called *codes*. Many digital devices employ one form of code for input, other forms internally, and even a different form for output. We first discuss the codes that are used to represent numbers and then proceed to more complex codes that are used to represent letters, symbols, and decimal digits.

13-1 BINARY RELATED RADIX CODES

Three related number systems (*binary, octal,* and *hexadecimal*) may be used to represent numbers in digital equipment. Table 13-1 shows the relationship between these systems. The binary number system (with a

Binary Related Radix Codes

TABLE 13-1 Binary and related radix codes

Decimal	Radix 2 Binary	Radix 8 Octal	Radix 16 Hexadecimal
0	0	0	0
1	1	1	1
2	10	2	2
3	11	3	3
4	100	4	4
5	101	5	5
6	110	6	6
7	111	7	7
8	1000	10	8
9	1001	11	9
10	1010	12	A
11	1011	13	B
12	1100	14	C
13	1101	15	D
14	1110	16	E
15	1111	17	F
16	10000	20	10

radix of 2) was discussed in detail in Ch. 2, but positional notation, binary-to-decimal conversion, and decimal-to-binary conversion should be reviewed in preparation for our discussion of the other number systems.

The Octal System

Numbers are stored within a computer in binary form, but the storage area may be arranged in such a manner that the number can be easily represented in another number system. The *octal* (base 8) *number system,* for example, is closely related to the binary system. Representation of the basic digits of the octal system (0 through 7) is easily accomplished with three binary digits. In fact, if the binary number is separated into groups of three digits, the octal number is immediately obtained. Limiting each group to three binary digits provides equivalent octal digits from 0 to 7, which is the range of the octal system.

Example 13-1. What is the octal equivalent of 101_2?
Solution $101_2 = 1 \times 2^2 + 0 \times 2^1 + 1 \times 2^0 = 4 + 1 = 5_8$

Note the use of positional notation.

Example 13-2. What is the octal equivalent of 110101_2?
Solution   110 101
             6   5 = 65_8

Figure 13-1 Binary, octal, and hexadecimal number relationships

Conversion from octal to binary is the reverse process. The three binary digits that represent each octal digit are written, starting with the least-significant digit first.

Example 13-3. What is the binary equivalent of 65_8?
Solution   6   5
   110 101 = 110101_2

The relationship between the binary and octal number systems is diagrammed in Fig. 13-1. Octal-decimal and decimal-octal conversion is accomplished in different ways. Although not the most "scientific" way, conversion first to binary, then either to octal or decimal, depending on which direction the conversion is going, may be the easiest. A more organized method for conversion from any radix to radix 10 follows.

CONVERSION FROM ANY RADIX TO RADIX 10

1. Multiply the most-significant digit (MSD) of the chosen number by its radix and add the next most-significant digit.
2. Multiply this sum by the number's radix and add the next most-significant digit.
3. Repeat the process until all digits in the chosen number have been used. The final result is the converted number in radix 10.

Example 13-4. Convert 65_8 to a decimal number.
Solution   6   ┌─► 5 = 53_{10}
      x8   +
      ──
      48 ──┘

This procedure is the equivalent of the *double-dabble* method used for conversion from binary to decimal in Ch. 2.

The successive division method of Ch. 2 may also be used for whole number conversion from decimal to any other radix, provided the divisor used is the radix of the desired number.

Binary Related Radix Codes

Example 13-5. Convert 53_{10} to an octal number.

Solution
```
     LSD
 8|53  ↓
  8 |6  5
    0  6  = 65₈
       ↑
     MSD
```
$$= 65_8$$

Formal mathematical methods for number system conversions exist, but they will not be discussed in this text. However, methods of conversion using logic circuits are examined in Ch. 14.

The octal number system is used for digital data processing input and output devices. Ease of encoding and decoding numbers makes the octal system the ideal choice for "minimum hardware" input and output applications. However, the recent advances in semiconductor technology and IC availability are moving octal operations into the background. Converting binary to decimal and vice versa no longer requires extensive hardware and excessive space. As existing digital equipment is updated, octal operations will gradually disappear in favor of the more easily manipulated decimal input and output devices.

The Hexadecimal Code

Some digital computers handle numbers in groups of four binary digits. One code used to represent these sets of four digits is the *hexadecimal* (base 16) *number system.* Sixteen numbers are representable in the hexadecimal system, and additional symbols are required to portray digits greater than 9. Table 13-1 shows the set of symbols for the hexadecimal number system.

Conversion from binary to hexadecimal is conveniently performed by grouping the binary number into sets of four digits and determining the equivalent value using Table 13-1.

Example 13-6. What is the hexadecimal equivalent of 1010_2?
Solution $1010_2 = 1 \times 2^3 + 0 \times 2^2 + 1 \times 2^1 + 0 \times 2^0$
$= 8 + 0 + 2 + 0$
$= A_{16}$

Example 13-7. What is the hexadecimal equivalent of 10011010_2?
Solution 1001 1010
            9   A $= 9A_{16}$

The process is reversed when converting from hexadecimal to binary. The four binary digits representing each hexadecimal number are written, starting with the least-significant digit first.

Example 13-8. What is the binary equivalent of $9A_{16}$?

Solution 9   A
1001 1010 = 10011010_2

The relationship between the binary, octal, and hexadecimal number system is diagrammed in Fig. 13-1.

Hexadecimal-decimal and decimal-hexadecimal conversion follows the same procedures as those for octal-decimal and vice versa.

Example 13-9. Convert $9A_{16}$ to a decimal number.

Solution   9 ⟶ A* = 154_{10}
×16 +
144 ⤴

*A_{16} = 10_{10}

Example 13-10. Convert 154_{10} to a hexadecimal number.

Solution   LSD
16⌊154   ↓
16 ⌊9   10*
    0   9 = $9A_{16}$
         ↑
        MSD

*$10_{10} = A_{16}$

Note again that positional structure has been implicit *throughout* the discussion of all number system conversions.

The hexadecimal system will continue to find widespread use in the so-called character-organized digital computers. Engineers, technicians, and programmers who work closely with the computer find no difficulty in using this nondecimal number system. Conversion hardware is reduced in these cases without loss of utility. However, when information must be displayed for the non-computer-oriented user, the decimal system is employed.

13-2 BINARY CODED DECIMAL (BCD) CODES

General Characteristics of BCD Codes

One of the major problems facing the engineer, technician, programmer or any user of digital equipment (whether it is computer, control, or measuring equipment) is *the conversion of a code readily handled by man to a code readily handled by the equipment.* Most people can best work with the decimal code, while digital machines work most effectively with binary codes.

Binary Coded Decimal (BCD) Codes

The best solution to this problem is found in the *binary coded decimal* (BCD) *codes.* Ten unique symbols (the decimal number system) must somehow be represented by only two unique symbols (the binary number system). In other words, *ten unique combinations of two symbols* must be found. Previous discussion tells us that any possible combinations of the two binary symbols is an *integer power of* 2, i.e., 2^0, 2^1, 2^2, and so on. Thus some integer power of 2 must be used that will allow at least ten combinations of binary digits (bits) to exist. Ten is greater than 2^3 but less than 2^4, so four bits must be employed to obtain the ten combinations. This results in sixteen combinations, six of which are not used. Any ten of the combinations may be used, and the specific combinations used will result in a particular type of BCD code.

Each decimal digit, therefore, is represented by a combination of four binary digits. The decimal number 345 may be represented in BCD by 0011 0100 0101. In binary, 345_{10} is represented as 101011001. Note that only nine bits are required for the binary representation, while twelve bits are required for the BCD form. However, since it is easier to recognize the BCD form, the *extra three bits* is the price that must be paid.

BCD codes may be classified as either *weighted* or *unweighted.* The *weighted codes* follow a *positional notation structure.* Positional values do not necessarily have to be ascending powers of 2; they may be arbitrarily assigned. In *unweighted codes,* the digit positions do *not* indicate the relative value of the represented number.

8-4-2-1 Weighted BCD Code

The most common weighted BCD code is the natural 8-4-2-1 code. The most-significant digit possesses a weight of 8, the next most-significant digit a weight of 4, and so forth. Correlation between decimal digits and the 8-4-2-1 BCD representation is shown in Table 13-2.

TABLE 13-2 Weighted Binary-coded-decimal (BCD) codes

Decimal	Natural 8421	2421
0	0000	0000
1	0001	0001
2	0010	0010
3	0011	0011
4	0100	0100
5	0101	1011
6	0110	1100
7	0111	1101
8	1000	1110
9	1001	1111

All that is required to convert an 8-4-2-1 BCD number to a decimal number (or vice versa) is a slight familiarity with the binary number system. Remember, each decimal digit is represented by four binary digits, weighted in an 8-4-2-1 order.

Example 13-11. Convert 1001 0101 (BCD 8-4-2-1) to a decimal number.
Solution 1001 0101
              9   5 = 95_{10}

Example 13-12. Convert 95_{10} to a BCD (8-4-2-1) number.
Solution   9   5
             1001 0101 = 1001 0101

Some digital computers actually operate in a BCD mode, and arithmetic operations may be performed directly in BCD without conversion to binary equivalents. However, since this text is not computer oriented, we shall concentrate on applications commonly encountered in the digital measuring and display fields. A BCD counter was discussed in Ch. 10. Methods of encoding and decoding BCD numbers are discussed in Ch. 14.

2-4-2-1 Weighted BCD Code

Another weighted BCD code that is widely used in computers is the 2-4-2-1 code (see Table 13-2). In this case, the most-significant digit of the 4-bit representation of a decimal number has a weight of 2 rather than 8. The major advantage of the 2-4-2-1 codes is that it is *self-complementing.** Thus the 2-4-2-1 representation of 1 (0001) is the opposite of 8 (1110), and 0 (0000) is the opposite of 9 (1111). All BCD numbers represented in the 2-4-2-1 code have this self-complementing property. And since one method of subtraction in a digital computer uses the minuend *added* to the complement of the subtrahend, the self-complementing feature is quite valuable.

Gray Unweighted BCD Code

Unweighted BCD codes are used in both digital data processing and in data transmission and measuring applications. When it is necessary to convert a physical parameter (e.g., a shaft position on a motor), conventional codes become difficult to work with. Mechanical con-

---

*Any two numbers that add up to their radix minus 1 complement each other: that is, are opposite.

Binary Coded Decimal (BCD) Codes

TABLE 13-3 Unweighted BCD codes

Decimal	(xs3) Excess 3	Gray	2 out of 5
0	0011	0000	00011
1	0100	0001	00101
2	0101	0011	00110
3	0110	0010	01001
4	0111	0110	01010
5	1000	0111	01100
6	1001	0101	10001
7	1010	0100	10010
8	1011	1100	10100
9	1100	1101	11000

struction methods place overly restrictive requirements on the encoding device. For example, if the conventional 8-4-2-1 BCD code is used, the encoding device *could* show a change in as many as *four bit positions simultaneously* (0111 to 1000). The *Gray code* alleviates this problem. When transitioning from one number representation to the next, *only one bit position changes*. Table 13-3 shows the Gray code representation of ten decimal digits, while Fig. 13-2 shows a typical mechanical encoder using both 8-4-2-1 and Gray code methods. From these comparisons, it is apparent that the Gray code implementation of BCD is preferable, when minimum possibility of generating improper representation of numbers due to bit change is required. Also, since fewer bits must change for each advance, binary counters implemented with a Gray code tend to have less operating delay.

Since no direct relationship exists between binary digits and bit positions in such unweighted codes as the Gray code, conversion is not as easy as it is between weighted codes. So, when converting between Gray and decimal codes, the binary system is used as an intermediate representation of the number. Conversion procedures are detailed in the following lists.

CONVERTING FROM BINARY TO GRAY CODE

1. Write the binary number completely.
2. Starting under the binary number—one place to the right—place the binary number again. The least-significant digit now appears beyond the radix point and should be dropped.
3. Add the two numbers, disregarding any carry terms that then result.

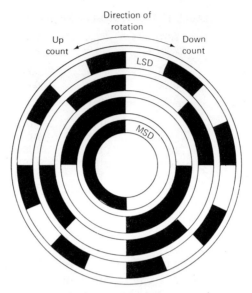

Dark areas represent conductive material and light areas insulative material. Electrical connection to the disc is via wire brushes. A complete circuit (logic 1) exists when a dark area is contacted; an open circuit (logic 0) exists when a light area is contacted.

(a) 8-4-2-1 BCD encoder

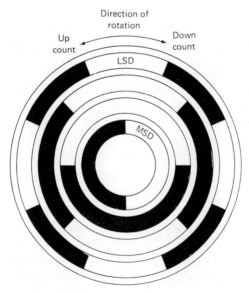

(b) Gray code encoder

Figure 13-2 Encoding binary codes: (a) 8-4-2-1 encoder, (b) Gray code encoder

Binary Coded Decimal (BCD) Codes

Example 13-13. Convert 93_{10} (1011101_2) to a Gray code number.
Solution   Disregard carry terms generated here
$$\begin{array}{l} \downarrow\downarrow \\ 1011101. = 93_{10} \\ \underline{101110.1} \text{ (drop)} \\ 1110011 \quad \text{Gray code representation} \end{array}$$

Note: This is equivalent to combining all adjacent digits of the binary number in an Exclusive-OR relationship, as shown in the following diagram.

CONVERTING FROM GRAY TO BINARY CODE

1. Write the Gray code number completely.
2. Bring the most-significant digit down and consider it the most-significant digit of the binary number being sought.
3. Add the most-significant digit of the binary number to the Gray code number in the next-significant digit column, neglecting any carry terms that result.
4. Continue the procedure until all Gray code digits have been used. The resulting number is the binary equivalent of the original Gray code number.

Example 13-14. Convert the Gray code number 1110011 to a decimal number, using the binary system.

Solution
$$1 \rightarrow 1 \rightarrow 1 \rightarrow 0 \rightarrow 0 \rightarrow 1 \rightarrow 1$$
$$\downarrow + = + = + = + = + = + =$$
$$1\,0\,1\,1\,1\,0\,1 = 93_{10}$$

Excess 3 (XS3) Unweighted BCD Code

Another unweighted BCD code often encountered in digital computer applications is the *XS3 code*. It is equivalent to the natural (8-4-2-1) BCD representation *plus* 3. Table 13-3 equates the XS3 code with decimal digits. Conversion from decimal to XS3 code may be accomplished either by converting to binary and then adding 3 or by adding 3 to the decimal number and then converting.

Example 13-15. Convert 9_{10} to an XS3 coded number.
Solution $9_{10} = 1001$
$$\frac{+\ 11}{1100\ (\text{XS3})}$$
or $9 + 3 = 12 = 1100\ (\text{XS3})$

Example 13-16. Convert 12_{10} to an XS3 number.

Solution	1	2	Decimal	1	2	Decimal
	0001	0010	Binary	+3	+3	Add 3
	+ 11	+ 11	Add 3	4	5	
	0100	0101	XS3	0100	0101	XS3

While each decimal digit is converted separately, XS3-to-decimal conversion is essentially the *reverse* process. Each digit is converted to its decimal equivalent, and 3 is then subtracted, as shown in Example 13-17. (The subtraction may also be performed at the binary level.)

Example 13-17. Convert 0100 0101 (XS3) to a decimal number.

Solution	0100	0101	XS3	0100	0101	XS3
	4	5		− 11	− 11	−3
	−3	−3	−3	0001	0010	Binary
	1	2	Decimal	1	2	Decimal
		12_{10}			12_{10}	

The XS3 code is *self-complementing,* just like the 2-4-2-1 BCD code. In addition, the proper carry is *always* generated, as though the XS3 number were in decimal form. Both of these characteristics make the XS3 code an excellent choice for arithmetic operations in digital computers, and since many digital measuring instruments interface directly with computers, the XS3 code is also found in this area of digital applications.

2-Out-of-5 Unweighted BCD Code

All of the BCD codes mentioned so far may be used *wherever number representations must be transmitted from one location to another* (either within the digital device or to a remote location). Each of the codes may also be implemented with one or more extra bits to establish a *parity check.* Another way to assure correct transmission is to use an *exact count code,* where the total number of 1s in each number representation is the same. The *2-out-of-5 code* (Table 13-3) is an example. A pseudoweighted BCD method (7-4-2-1) represents the ten decimal digits in the first four bit positions, while the final bit position is reserved for an exact count digit. The added digit in position 5 is selected so that each word has exactly two 1s and three 0s. When decimal numbers are transmitted in this form, all *single errors* are detected. Any character received with less or more than two 1s is in error.

Conversion to and from the 2-out-of-5 code is relatively straightforward due to the pseudoweighted format. However, the decimal digit 0 is represented in a nonweighted manner.

13-3 ALPHAMERIC CODES

The codes that we have discussed so far have represented only *numerical data*. But communicating information from one point to another necessarily includes other forms of data, such as *alphabetical characters* and *symbols*. Hence the total numbers of possible combinations concept of binary digits, called *alphameric representation,* is used to represent numbers, special symbols and alphabetical characters.

As you may have already noted, a direct correlation *may not* exist between a number and its representation in binary digits. The Gray code, for example, adequately represents the decimal digits, but the positions of the binary digits in the representation are unweighted. The most important consideration in its design was the single digit change from one representation to the next, *not* a consistent value for each bit position. Alphameric codes usually fall into the unweighted category, although in some codes a design pattern is noticeable.

5-Level (Baudot) Code

Four binary digits (bits) allow only sixteen different combinations, which are insufficient for alphameric representation. Up to 32 different characters may be represented by 5 bits, thus allowing all the letters of the alphabet to be represented. If 2 of the 32 bits are reserved for the purpose of selecting one of two separate "sets" of information, then a total of sixty-two available combinations is obtained.

A *5-level code,* developed by Jean Baudot (a French engineer) in the late nineteenth century, has been employed in data communication applications, using a typewriter-like device called a *teletypewriter.* The 26 letters of the alphabet are represented on the "lowercase" set of information, while numbers, punctuation marks, and special symbols are represented by the "uppercase" set.

The teletypewriter mechanism contains a group of switches that may be opened and closed by the depression of any key on the keyboard. Switch conditions may be translated into binary representation (e.g., 0 = open switch and 1 = closed switch). Actuation of each key then results in a specific combination of 1s and 0s that is representative of the selected alphameric character. The switches operate in sequence, following a beginning synchronizing indication and followed by a "stop" indication.

The Baudot code is shown in Fig. 13-3, correlated with a drawing of the perforated tape used with many alphameric code devices. The

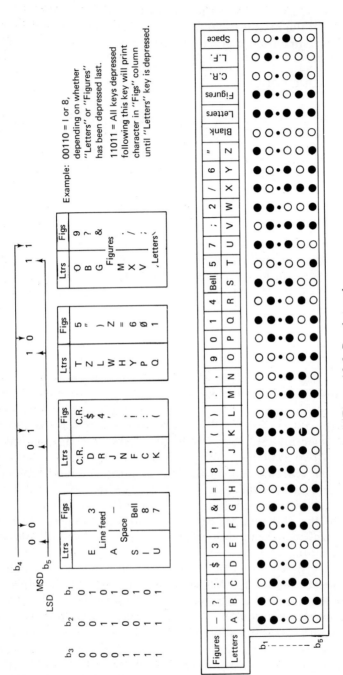

Figure 13-3 Baudot code

Alphameric Codes 381

black marks represent holes in the tape (1s), while the unmarked circles represent solid tape (0s). Information that is to be transmitted from one location to another is usually prepared by "punching" a tape; large amounts of data can then be transmitted without interruption. Devices that read the tape sense a hole or a lack of a hole at each of the five hole locations and generate the appropriate binary digit to represent the selected character.

A 5-level code is therefore *capable of representing all alphabetical and numerical characters,* as well as *several punctuation marks and special symbols.* However no means to determine the accuracy of the binary representation is possible with a 5-level code; that is, *no parity is used.* Of course, the code could be expanded to six bits to provide parity, but often it is necessary to communicate more than the minimal number of characters possible with five bits.

7-Level (ASCII) Code

Six-level codes have been used in the past, particularly in early computer applications. They have generally been replaced by 7-level codes. The use of 7 data positions results in a code having 128 possible combinations.

The most common *7-level* (expandable to 8-level) *code* is called the *United States of America Standard Code for Information Interchange* (ASCII) and appears in Fig. 13-4. It is also called the ANSI (American National Standards Institute) code or the Data Interchange code. Any of the 128 locations can be described by a unique combination of high-order (b_7, b_6, and b_5) bit locations and lower order (b_4, b_3, b_2, and b_1) locations. The ASCII code is quite versatile and is organized along logical lines. For example, note that the device using this code does not have to go beyond the two most-significant bits to determine whether the represented information is a *control* (both 0s) or a *character* (not both 0s).

Numerous variations of the ASCII code are in use, but most are compatible. Their differences lie specifically in the area of *names of the control functions* and *the actual symbol represented,* when neither letters, numbers, nor control functions are being used. (Note that all 128 locations are used.) Expansion to eight bits allows either parity checks or doubling of the number of characters that can be represented.

The ASCII code table of Fig. 13-4 is used in the following manner. First, determine the most significant digits b_7, b_6, and b_5 in the top row of the table. Then locate the least-significant bits b_4, b_3, b_2, and b_1 in the four columns on the left. The intersection of the column containing the most significant bits and the row containing the least significant bits is the character represented by the ASCII code. For example,

Binary Codes

b₇ b₆ b₅ \ b₄ b₃ b₂ b₁	000	001	010	011	100	101	110	111
0 0 0 0	NUL	DLE	SP	0	@	P	`	p
0 0 0 1	SOH	DC1	!	1	A	Q	a	q
0 0 1 0	STX	DC2	"	2	B	R	b	r
0 0 1 1	ETX	DC3	#	3	C	S	c	s
0 1 0 0	EOT	DC4	$	4	D	T	d	t
0 1 0 1	ENQ	NAK	%	5	E	U	e	u
0 1 1 0	ACK	SYN	&	6	F	V	f	v
0 1 1 1	BEL	ETB	'	7	G	W	g	w
1 0 0 0	BS	CAN	(	8	H	X	h	x
1 0 0 1	HT	EM	)	9	I	Y	i	y
1 0 1 0	LF	SUB	*	:	J	Z	j	z
1 0 1 1	VT	ESC	+	;	K	[	k	{
1 1 0 0	FF	FS	,	<	L	\	l	|
1 1 0 1	CR	GS	-	=	M	]	m	}
1 1 1 0	SO	RS	.	>	N	^	n	~
1 1 1 1	SI	US	/	?	O	_	o	DEL

(a) Basic 7-level ASCII code
$b_1 \ldots \ldots b_7$ Parity

(b) Tape sample showing 7-level code expanded to 8 level by Parity position

Figure 13-4 ASCII code

1000001 is equivalent to the letter A. The question mark (?) is represented by 0111111.

8-Level (EBCDIC) Code

One final example of alphameric codes is the *Extended Binary Coded Decimal Interchange Code* (EBCDIC). It is a full *8-level code* (see Fig. 13-5), since each letter, number, symbol, or control function is represented by two BCD numbers—the equivalent of eight binary digits. A number of similarities exist between EBCDIC and the other codes discussed in this chapter. Identification of control functions is made by checking the two most-significant digits, just like the ASCII

Figure 13-5 Extended binary-coded decimal interchange code; 8 and 9 level

Number	Punch	Letter	Punch	Letter	Punch	Letter	Punch	Char	Punch
0	0								
1	1	A	12,1	J	11,1			#	3,8
2	2	B	12,2	K	11,2	S	0,2	,	0,3,8
3	3	C	12,3	L	11,3	T	0,3	$	11,3,8
4	4	D	12,4	M	11,4	U	0,4	.	12,3,8
5	5	E	12,5	N	11,5	V	0,5	&	12
6	6	F	12,6	O	11,6	W	0,6	@	4,8
7	7	G	12,7	P	11,7	X	0,7	%	0,4,8
8	8	H	12,8	Q	11,8	Y	0,8	*	11,4,8
9	9	I	12,9	R	11,9	Z	0,9	¤	12,4,8

Figure 13-6 Hollerith code listing

code. One advantage EBCDIC has over other codes is that the identification of symbols is made easy by investigating only the two most-significant digits.

The EBCDIC is used when the *data transmission and reception devices must interface with a computer that operates predominantly in the BCD mode.* Parity is not provided in EBCDIC; the code most be expanded to nine levels in order to obtain this function.

Hollerith Code

Although the *punched card* is largely associated with digital computers, it should not be overlooked when discussing data representation codes. Letters, numbers, and some symbols are represented by *certain combinations of holes punched in a special card.* The most widely used code in punched cards is the *Hollerith code.* Each of the 80 columns carries one digit, letter, or symbol. Figure 13-6 shows a typical listing of the Hollerith code, which should be correlated with the punched card shown in Fig. 13-7.

Punch positions are arranged horizontally in twelve rows—the bottom ten for digits 0 through 9, plus two rows (numbered 11 and 12) along the top of the card for *zone punching.* Figure 13-7 illustrates how *each digit* (0 through 9) is coded with a *single punch.* Alphabetical *characters* require a single row punch *plus* a zone punch. Special characters use multiple combinations of row and zone punches.

When a punched card is read, each row in each column is investigated separately. Thus the Hollerith-coded punch card actually represents data with a *12-level code.* Due to the large number of digits, the Hollerith code is not usually approached in its binary form but rather through the listing of the rows and/or zones punched for each character.

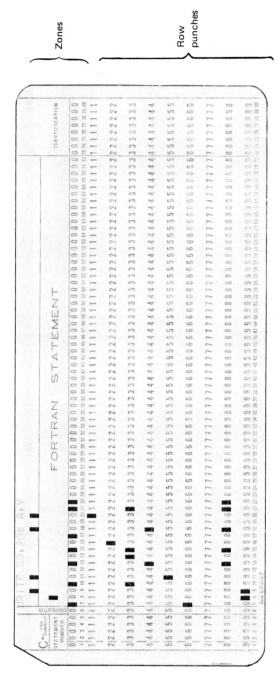

Figure 13-7 Hollerith-coded punched card

13-4 SUMMARY

Some of the common binary codes used to represent letters, numbers, and symbols have been discussed in this chapter. More important, however, has been the discovery that *information may be represented in binary form*. With this knowledge, *intelligence* can obviously be encoded, transmitted, decoded, and manipulated with logic circuits. Such capability has provided the impetus for the revolution in machine aids designed to improve our intellectual processes.

Many of the more complex codes used with digital equipment have not been covered in this text. Parity-checking methods may be extended to more than one bit per character—even to complete blocks of information (many groups of characters). All sorts of "exotic" error detection and correction schemes exist, but they are beyond the scope of this text.

In most cases that the digital logic technician will encounter, the codes will be listed in the technical information furnished with the digital device. If such information is not available, the principles discussed in this chapter should enable the technician to systematically approach the equipment and then determine the important characteristics of the codes, before proceeding with troubleshooting and repair activities.

PROBLEM APPLICATIONS

13-1. Discuss the relationships between the binary, octal, and hexadecimal number systems.

13-2. Convert the numbers 5_{10}, 76_{10}, and 348_{10} to (a) octal form, (b) hexadecimal form, (c) 8-4-2-1 BCD form, (d) 2-4-2-1 BCD form, (e) Gray code form, and (f) XS3 code form.

13-3. Why are codes necessary in digital devices?

13-4. Discuss the advantages and disadvantages of BCD codes versus straight binary codes.

13-5. What is a weighted BCD code?

13-6. What is an unweighted BCD code?

13-7. Discuss the relative advantages and disadvantages of weighted BCD codes versus unweighted BCD codes.

13-8. Develop a weighted BCD code (other than that used in this chapter) and list its representation of the ten decimal digits.

13-9. Develop an unweighted BCD code (other than that used in this chapter) and list its representation of the ten decimal digits.

13-10. Discuss the advantages and disadvantages of the 2-out-of-5 code.

13-11. What is an alphameric code?

Summary

13-12. What determines the number of bits necessary in an alphameric code?

13-13. Discuss the advantages and disadvantages of the 5-level Baudot code.

13-14. Show how the words "6-level alphameric code" would appear in the
   (a) Baudot code,
   (b) 7-level ASCII without parity,
   (c) 7-level ASCII with parity, and
   (d) EBCDIC without parity.

14

CODE CONVERTERS
AND
DISPLAY DEVICES

This chapter serves a dual purpose. The various codes used to represent alphameric characters in digital equipment were discussed in detail in Ch. 13. The need for conversion between codes was firmly established. A code must function efficiently within the digital equipment, but it must also be easily recognizable to the operator (i.e., it must be capable of being converted to a "human usable" code). To do this, it must be displayed in a manner that is easily recognized; hence *the second major purpose for code conversion is to actuate display devices. Both* the conversion between codes and the conversion of codes to actuate display devices will be discussed in this chapter.

The three major *character-display devices* are *cold-cathode direct-character, segmented,* and *dot-matrix units.* Each requires a specific conversion device for actuation, and each displays its information in a different manner. More complex devices capable of displaying complete pages of information are mentioned but are not discussed in detail.

As LSI and VLSI assemblies become more prevalent, conversion and interface capabilities are being incorporated directly on the parent chip. For example, many LSI/VLSI counter assemblies now connect directly to the display device without separate converters and display interface assemblies.

14-1 CODE CONVERTERS

Assuming that most digital devices operate internally in the *binary mode,* numerous requirements for code converters can be identified. Considering only numbers, Fig. 14-1 shows some of the possibilities.

Code Converters

Figure 14-1 Code conversion requirements

Input to the digital device may assume various physical forms, such as switches, punched paper tape, keyboard devices similar to typewriters, and magnetic tape. The actual process of converting the input information to binary is often accomplished by the input device itself. Of course, if the input device consists of switches (one for each binary order), no need for conversion exists, since the switch is either closed or open (representing a 1 or a 0). Output in binary usually takes the form of individual indicator lights (one for each binary order). Hence *no conversion is necessary, if both input and output are of a direct binary nature.*

Octal to Binary Conversion

Octal input is still used in certain special applications. When the input to a digital device is in octal form, it is necessary to convert to binary form so that the device may operate with the input information. *Mechanical switches,* constructed like those in Fig. 14-2, *provide direct conversion to binary form.*

Conversion of octal quantities in logic-level format requires the use of a *gating arrangement.* Seven inputs, representing octal digits 1 through 7, must be combined into three outputs, representing 2^0, 2^1, and 2^2 binary positions. Zero is recognized as the absence of all other

Figure 14-2 Octal switch (courtesy The Digitran Co.)

digits. A binary to octal conversion table (Fig. 14-3a) provides sufficient information to demonstrate such a circuit (Fig. 14-3b). For example, when the input representing 5_8 is activated and goes HIGH, it will cause the 2^0 and 2^2 outputs to go HIGH. No HIGH inputs are provided to the 2^1 gate, so its output remains LOW. Thus the number represented at the outputs is 101_2, which is equivalent to 5_8. (The remaining seven combinations should be verified by the reader.)

Hexadecimal to Binary Conversion

The octal to binary converter shown in Fig. 14-3b is expandable to hexadecimal capability by furnishing both *input lines* to represent each of the sixteen hexadecimal digits and sufficient *OR gates* to represent the binary equivalents.

Decimal to Binary Conversion

Decimal to binary conversion follows the same general concepts of octal to binary conversion. Figure 14-4 shows decimal to binary conversion of the decimal digits 0 through 9. In the decimal system, however, there is no *direct grouping relationship* with the binary system, as there is in octal and hexadecimal representation. The full count of the

Code Converters

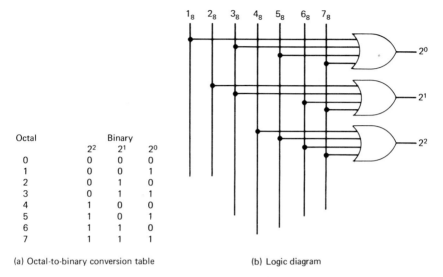

(a) Octal-to-binary conversion table (b) Logic diagram

Figure 14-3 Octal-to-binary conversion

decimal radix (10) is not a multiple of the binary radix (2), so no grouping can be arranged. This results in complex mechanical assemblies or electrical circuits, when large decimal numbers must be converted to binary.

Decimal to BCD Conversion

Many digital devices are designed to accept the *8-4-2-1 BCD codes* —some even perform operations with numbers directly in BCD format. The unwieldy logic circuit that may result from attempting to convert directly from decimal to binary may be simplified by using an intermediate conversion to BCD coding. Decimal to BCD conversion is accomplished *one order at a time* with four bits assigned to each decimal digit, as explained in Ch. 13. The encoding method shown in Fig. 14-4 is sufficient to perform logic-level conversion, while assemblies similar to the octal encoding switch of Fig. 14-2 are capable of performing the conversion mechanically.

BCD to Binary Conversion

Conversion from BCD to binary is not quite so easy, but it may be accomplished using a number of different techniques. Perhaps the simplest (and possibly the slowest) method of BCD to binary conversion is the *count comparison method* shown in the block diagram in Fig. 14-5.

Decimal	Binary			
	2^3	2^2	2^1	2^0
0	0	0	0	0
1	0	0	0	1
2	0	0	1	0
3	0	0	1	1
4	0	1	0	0
5	0	1	0	1
6	0	1	1	0
7	0	1	1	1
8	1	0	0	0
9	1	0	0	1

(a) Decimal-to-binary conversion table

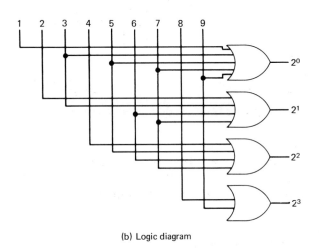

(b) Logic diagram

Figure 14-4 Decimal-to-binary conversion

The BCD number to be converted is loaded into a BCD down-counter; at the same time, a binary up-counter is reset. A Start input is provided to the clock control, and clock pulses are applied to both the BCD down-counter and the binary up-counter. Thus, *for each count that the BCD counter decrements, the binary up-counter increments.* An all-zero detector recognizes when the BCD counter reaches the zero condition, and clock pulses are terminated. The binary up-counter now contains the binary equivalent of the BCD number originally loaded into the BCD down-counter.

This method is quite slow, since an n-bit binary number will require 2^n clock pulses. Assuming a 10 megahertz clock rate (10,000,000 clock pulses per second), a count comparison method of conversion could take on the order of 3 milliseconds to convert a BCD number to its equivalent 15-bit binary number. Table 14-1 shows the actual count sequence for conversion of the BCD number 0010 0001 to its equivalent binary form 10101_2.

Code Converters

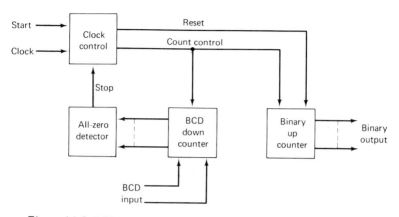

Figure 14-5 BCD-to-binary conversion—count comparison method

A faster method of BCD to binary conversion uses *full adders* to sum the binary expansions of all the digits in the binary number. The general concept of this type of conversion is shown in Examples 14-1 and 14-2, while implementation in block-diagram form appears in Fig. 14-6.

Example 14-1.   Convert BCD number 0010 0001 to its equivalent binary number.

Solution   Tens Digit   Units Digit
Weight   80 40 20 10   8 4 2 1
BCD   0 0 1 0   0 0 0 1
          $\underbrace{1 \times 20}$ + $\underbrace{1 \times 1}$ = 21_{10}
          10100   + 00001   = 10101_2

Example 14-2.   Convert BCD number 1001 1001 to its equivalent binary number.

Solution   1 0 0 1   1 0 0 1
       $\underbrace{1 \times 80} + \underbrace{1 \times 10}$ + $\underbrace{1 \times 8} + \underbrace{1 \times 1} = 99_{10}$
       1010000 + 1010 + 1000 + 0001 = 1100011_2

Binary to BCD Conversion

Following operations on the information within the digital device, the results must be made available in a form (such as decimal) that is easy to understand. Direct conversion of large binary numbers to decimal becomes just as unwieldy as did decimal to binary conversion. Again, the most common approach is to convert from binary to a BCD

TABLE 14-1 BCD-to-binary conversion—Count comparison method

	BCD				Binary
21	0010	0001	Start	0	00000
20	0010	0000		1	00001
19	0001	1001		2	00010
18	0001	1000		3	00011
17	0001	0111		4	00100
16	0001	0110		5	00101
15	0001	0101		6	00110
14	0001	0100		7	00111
13	0001	0011		8	01000
12	0001	0010		9	01001
11	0001	0001		10	01010
10	0001	0000		11	01011
9	0000	1001		12	01100
8	0000	1000		13	01101
7	0000	0111		14	01110
6	0000	0110		15	01111
5	0000	0101		16	10000
4	0000	0100		17	10001
3	0000	0011		18	10010
2	0000	0010		19	10011
1	0000	0001		20	10100
0	0000	0000	Stop	21	10101

code and then to decimal. Both forms of BCD to binary conversion have similar methods for the reverse conversion, but they will not be discussed here.

Perhaps the simplest binary to BCD conversion method uses a *read-only memory* (Fig. 14-7). The binary number serves as the address information to obtain the corresponding BCD code. Each possible binary number to be converted must have a BCD word placed in memory. Therefore this method requires a relatively large amount of storage. For example, a code comprised of ten bits requires 1024 memory words of 13-bit length. Other approaches exist for reducing the amount of read-only memory required. One of the most popular is to divide the binary number into parts, using small ROMs to store the equivalents of the now much smaller number, and then recombine the BCD equivalents in BCD adders. This technique can reduce the original storage requirement of 13,312 bits to as little as 576 bits.

One of the most popular binary to BCD conversion methods is the *ADD 3* approach. However this type of conversion from binary to BCD requires further investigation of BCD characteristics. Perhaps some simple additions will reveal the relationships between binary, decimal, and BCD systems.

Code Converters

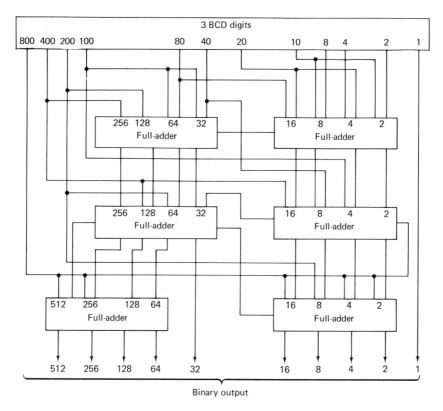

Figure 14-6 BCD-to-binary conversion—full-adder method

Example 14-3. Find the sum of BCD numbers 0011 and 0101.

Solution
```
 0011      3
+0101     +5
─────    ───
 1000      8
```

The addition of 3 + 5 in binary notation yields a valid sum in *either* binary or BCD form, since the BCD form consists of a group of four binary digits.

Example 14-4. Find the sum of BCD numbers 0110 and 0110.

Solution
```
 0110      6
+0110     +6
─────    ───
 1100     12
```

This sum is valid in binary form, but it is invalid in BCD form because, just as in the decimal system, 9 is the largest digit representable. A carry is required but not generated.

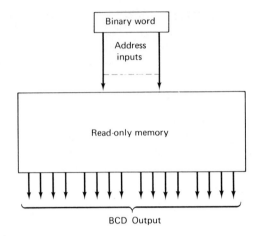

Figure 14-7 Binary-to-BCD conversion—ROM approach

Example 14-5. Find the sum of BCD numbers 1001 and 1000.

Solution
```
  1001      9
 +1000     +8
 -----    ---
 1 0001    17
```

Just as in Example 14-4, the sum is valid in binary form. In this case, a carry is generated for the BCD form, but the LSD (0001) is incorrect.

The problem in Examples 14-4 and 14-5 is that the decimal carry occurs at 10 and the BCD carry occurs at 16—a difference of 6. If the binary equivalent of 6 is added to the invalid sum and any overflow beyond four binary digits is added to the LSD of the next BCD group to the left, the sums will be valid BCD representations. In Example 14-4, 0110 + 0110 = 1 0010, which is the BCD representation of 12_{10}. Also, in Example 14-5, 1 0001 + 0110 = 1 0111, the BCD equivalent of 17_{10}. Thus it is not pure coincidence that 6 must be added to each sum 10_{10} or greater. These examples demonstrate that such action is necessary to maintain *the decimal structure when operation in BCD notation*.

As previously shown, any position (b_x) in a binary number has a decimal equivalent 2^x (i.e., $b_4 = 2^4 = 16_{10}$). The decimal equivalent of a binary number may therefore be obtained by multiplying x times the radix (2) of the binary system by itself.

$$2^4 = 2 \times 2 \times 2 \times 2 = 16$$

The decimal equivalent of a binary number (with the decimal equivalent expressed in BCD form) is the goal of binary to BCD conversion.

Code Converters

Multiplication by 2 is accomplished by shifting the binary number left (toward the MSD) one position.

$$\left.\begin{array}{l}0011 = 3 \\ 00110 = 6\end{array}\right\} 3 \times 2 = 6 \quad \left.\begin{array}{l}1000 = 8 \\ \rightarrow 10000 = 16 \\ \rule{0pt}{0pt}\text{—Not valid}\end{array}\right\} 8 \times 2 = 16 \quad \begin{array}{l}10000 \\ \underline{0110} \\ 1\,0110 = 16_{10}\end{array}$$

Valid
if ≥ 10, add 6

Note that the 10000 representation of 16_{10} is not a valid BCD number. As shown earlier, 0110 (6_{10}) must be added to the least-significant BCD group to form the proper representation. Actually, the BCD representation of 16_{10} is investigated, and if the least-significant BCD representation is ≥ 10, 6 is added. Another way to perform the same operation is to examine the BCD representation of the number before multiplication (left shift). If it is ≥ 5, add 3 and shift left. This is equivalent to checking for ≥ 10 and adding 6 *after* the left shift. The examination, ADD 3 (if necessary), and shift-left operations are continued until all binary digits are shifted. The binary number has now been multiplied by 2, the number of times required to convert it to its decimal equivalent, although the representation is in BCD form as required. Figure 14-8 shows conversion of two binary numbers to BCD form.

As Fig. 14-9 shows, this method may be implemented with *shift registers*. The binary number is inserted in parallel into a register with serial-output capability. The number is then shifted into a serial/parallel register in a serial manner. Examination of the three most-significant bits is accomplished with a combinational logic circuit. If these bits represent a number ≥ 5, the combinational circuit adds 3 and feeds the modified number back into the serial/parallel register via the parallel inputs. Following the next shift operation, the number is again examined, 3 is added if necessary, and another shift occurs. The control logic determines when to remove shift pulses, based on the total number of bits in the binary number to be converted.

The ADD 3 method may also be implemented directly with *gates* and/or *ROMs*. Such an approach results in a decreased conversion time but in an increased number of logic elements. Typical speeds for a 15-bit binary word and a 10 megahertz clock rate are 1500 nanoseconds for the sequential shift approach and 540 nanoseconds for the gate approach. Note that both values are well *below* the time required for the count comparison method of conversion.

BCD to Decimal Conversion

The remaining step in the sequence of decimal-in to decimal-out conversion is that from *BCD to decimal*. If the output device displays each of the decimal digits separately, then a means to convert from BCD

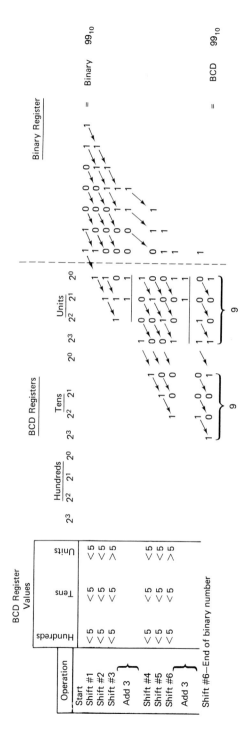

Figure 14-8 Binary-to-BCD conversion—ADD 3 method

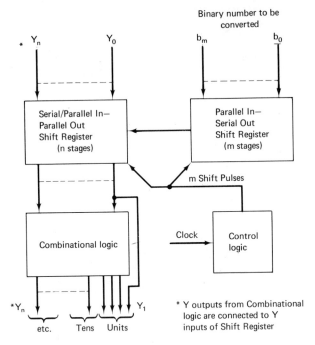

Figure 14-9 Implementation of "ADD 3" method

to a 1-of-10 outputs is needed. The actual display device* is unimportant from a logic standpoint, just so only *one* of the decimal digits is displayed at a time. As usual, conventional 8-4-2-1 BCD is assumed.

Figure 14-10 shows the general approach to 8-4-2-1 BCD to decimal decoding. Investigation of the logic diagram of the 7445A decoder and its truth table shows that the circuit meets the requirements of the preceding paragraph. In addition, all outputs are inactive for erroneous or not-allowed input combinations. The 7445A is similar to the 4028 and 74LS145 ICs in function.

An increasingly popular display is the *7-segment indicator,* discussed in Sec. 14-2. And, as in more and more digital equipment, the familiar 8-4-2-1 code is used to drive this device. The 8T05 7-segment decoder, which consists of the necessary logic to decode a 4-bit BCD code to 7-segment readout, is shown in Fig. 14-11. It is similar to the 74LS248 decoder. To provide an easily readable display, a *ripple-blanking input* is furnished to implement suppression of leading and/or

---

*Different display devices are discussed in subsequent sections of this chapter.

Code Converters

(a) Logic diagram

(b) Truth table

Figure 14-10 BCD-to-decimal decoder (Type 7445A) (courtesy Sprague Electric Co.)

trailing zeros. A lamp-test input is also provided, which forces all segment outputs HIGH. This allows the viewer to check the validity of the display presentation by testing the integrity of the display.

BCD codes other than the 8-4-2-1 version find applications in digital devices. For example, the *XS3 code* is often used when arithmetic operations must be performed. Its self-complementing characteristics greatly simplify some types of arithmetic operations. The XS3 code may be easily decoded to decimal, as shown in Fig. 14-12. No unusual circuitry is employed, and the operation is straightforward.

Binary to Octal Conversion

If the output is to be in octal form, a *binary to octal converter* is required. Such a decoder has three lines of binary input, representing the 2^0, 2^1, and 2^2 positions. Eight outputs are provided, only one of which is activated at a time. A typical prepackaged binary to octal decoder (type 8250) is shown in Fig. 14-13. The fourth input line (D) is used as an inhibit to allow use in larger decoding networks. Note that HIGH inputs are required, while the selected output is LOW when activated.

Code Converters and Display Devices

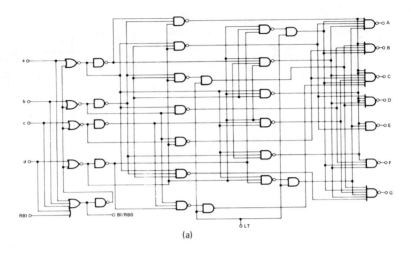

(a)

INPUTS							OUTPUTS							
INPUT CODE				LAMP TEST	RBI	BI/RBO	OUTPUT STATE							DISPLAY CHARACTER
d	c	b	a	LT		Note	A	B	C	D	E	F	G	
X	X	X	X	0	X	X	1	1	1	1	1	1	1	8
X	X	X	X	1	X	0 (Note 1 & 2)	0	0	0	0	0	0	0	BLK
0	0	0	0	1	0	0 (Note 2)	0	0	0	0	0	0	0	BLK
0	0	0	0	1	1	1	1	1	1	1	1	1	0	0
0	0	0	1	1	X	1	0	1	1	0	0	0	0	1
0	0	1	0	1	X	1	1	1	0	1	1	0	1	2
0	0	1	1	1	X	1	1	1	1	1	0	0	1	3
0	1	0	0	1	X	1	0	1	1	0	0	1	1	4
0	1	0	1	1	X	1	1	0	1	1	0	1	1	5
0	1	1	0	1	X	1	0	0	1	1	1	1	1	6
0	1	1	1	1	X	1	1	1	1	0	0	0	0	7
1	0	0	0	1	X	1	1	1	1	1	1	1	1	8
1	0	0	1	1	X	1	1	1	1	0	0	1	1	9
1	0	1	0	1	X	1	0	0	0	0	0	0	1	—
1	0	1	1	1	X	1	0	0	0	0	0	0	0	BLK
1	1	0	0	1	X	1	1	1	0	1	1	1	1	A
1	1	0	1	1	X	1	0	0	1	0	0	0	0	,*
1	1	1	0	1	X	1	0	0	0	1	1	1	0	L
1	1	1	1	1	X	1	0	0	0	0	0	0	0	BLK

*COMMA

X = Don't care, either "1" or "0".
BI/RBO is an internally wired OR output.

NOTE:
1. BI/RBO used as input.
2. BI/RBO should not be forced high when a, b, c, d, RBI terminals are low, or damage may occur to the unit.

(b)

Figure 14-11 Type 8T05 seven-segment decoder (courtesy Signetics, a subsidiary of Corning Glass Works)

Code Converters

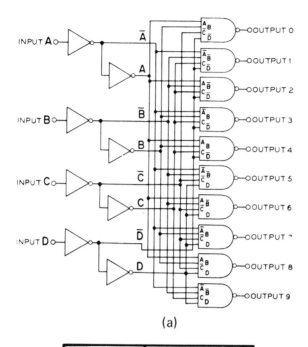

Figure 14-12 Excess 3-to-decimal decoder (Type 7443A) (courtesy Sprague Electric Co.)

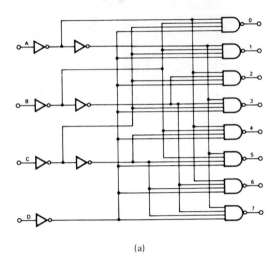

(a)

	INPUT STATE			OUTPUT STATES 8250							
A	B	C	D	0	1	2	3	4	5	6	7
0	0	0	0	0	1	1	1	1	1	1	1
1	0	0	0	1	0	1	1	1	1	1	1
0	1	0	0	1	1	0	1	1	1	1	1
1	1	0	0	1	1	1	0	1	1	1	1
0	0	1	0	1	1	1	1	0	1	1	1
1	0	1	0	1	1	1	1	1	0	1	1
0	1	1	0	1	1	1	1	1	1	0	1
1	1	1	0	1	1	1	1	1	1	1	0
0	0	0	1	1	1	1	1	1	1	1	1
1	0	0	1	1	1	1	1	1	1	1	1
0	1	0	1	1	1	1	1	1	1	1	1
1	1	0	1	1	1	1	1	1	1	1	1
0	0	1	1	1	1	1	1	1	1	1	1
1	0	1	1	1	1	1	1	1	1	1	1
0	1	1	1	1	1	1	1	1	1	1	1
1	1	1	1	1	1	1	1	1	1	1	1

(b)

Figure 14-13 Binary-to-octal decoder (Type 8250) (courtesy Signetics, a subsidiary of Corning Glass Works)

Binary to Hexadecimal Conversion

Hexadecimal conversion of a binary number is accomplished with a logic circuit *similar* to the binary to octal decoder. In the hexadecimal case, however, four inputs are provided, and only one of sixteen outputs is activated at a time. The 8250 one-of-eight decoder may be used, as shown in Fig. 14-14, to provide the one-of-sixteen operation. Such a scheme is expandable in groups of eight bits to provide up to a one-of-sixty-four capability.

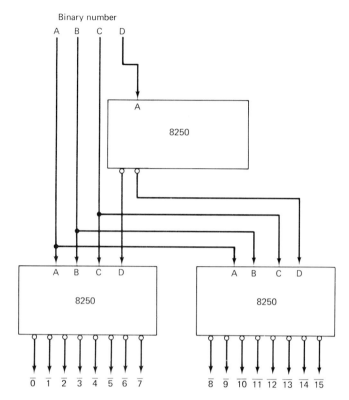

Figure 14-14 Binary-to-hexadecimal decoder using Type 8250 decoders

14-2 DISPLAY DEVICES

Cold-Cathode, Direct-Character Display

One of the first digital display devices was the *NIXIE* tube* (Fig. 14-15). It is classified as a cold-cathode, gas-filled glow tube. In simple terms, it may be compared to the neon bulbs seen in small orange-colored nightlights. Each of the "neon bulbs" is shaped like one of the ten decimal digits, and when properly activated, only one character glows at a time. The individual characters are so arranged inside the tube that there is a minimum of masking of characters which are in the rear. When one character is illuminated, it is difficult to see any of the nonilluminated numbers.

*Registered trademark of The Borroughs Corporation.

Figure 14-15 NIXIE tube (courtesy Burroughs Corp., Electronic Components Division)

The major advantages of the NIXIE tube included *the capability to quickly change from one character to another, long life,* and *relatively low power consumption.* High voltage was required to operate the NIXIE tube, but specially developed decoder/driver combinations performed the job quite satisfactorily. If there were any major disadvantages, they fell in the area of decoding complexity. Regardless of the code used within the display device employing NIXIE tubes, it had to be converted to a 1-of-10 output. Such a decoding requirement usually resulted in additional gate requirements.

NIXIE tubes were used in many digital applications where *numeric display* was required. Computer displays, automatic test equipment, digital clocks, and even everyday electronic test equipment used cold-cathode display tubes to indicate system conditions.

Segmented Displays

It is unnecessary to use ten separate elements to display the ten decimal digits. A close approximation of digits 0 through 9 may be had by breaking the digits into *segments* and activating only the segments necessary to display the digit. Seven segments quite adequately synthesize the decimal digits (plus fourteen alphabetic characters), as shown in Fig. 14-16a. When complete alphameric display is required, sixteen segments are used (Fig. 14-16b). The 16-segment display is capable of over 65,000 patterns, which allows great versatility in information display.

One of the most popular display device is the seven-segment display. Illumination of the individual segments is accomplished in a num-

Display Devices

Figure 14-16 Segmented display capability

ber of different ways. Light-emitting diodes (LEDs) as well as liquid crystals can be used to form segments. Other common techniques use vacuum fluorescent and incandescent segments. Each of these techniques is discussed below.

A light-emitting diode is a tiny solid-state device constructed much like a conventional solid-state diode. Depending on the LED materials, it can give off light ranging from yellow through red, and on into the infra-red part of the spectrum. Unfortunately, the LED itself is quite tiny (smaller than a speck of dust). Optical techniques must be used to spread the LED's light output so that it can be used for display purposes. Figure 14-17 shows a simple technique used in some segmented displays that employ LEDs.

Information in digital devices is not in the correct form for segmented displays. As noted in Figure 14-11, some type of code converter is needed to light the appropriate segments of the display. A typical application of LED displays may be seen in the common electronic digital clock. Figure 14-18a is a block diagram of one method of implementing such a function. The 60 Hz from the power line is divided by 60 (remember, counters perform the division function) to form 1 pulse per second. Another division by 60 provides 1 pulse per minute. One pulse per hour is provided by still another divide-by-60

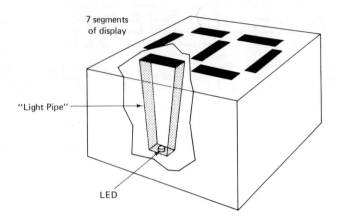

Figure 14-17 LED "light pipe" concept

counter. The actual count in each counter is decoded and applied to individual seven-segment displays via "drivers" to match the voltage/current requirement of the displays. It is not necessary to use a separate decoder/driver for each display, however. Multiplexers (see section 12-3) can sequentially sample each counter while simultaneously applying power to one display only. Sampling is performed so rapidly that

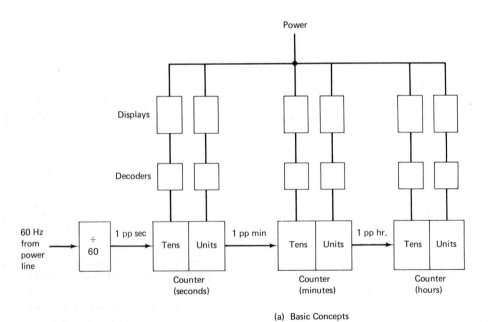

(a) Basic Concepts

Figure 14-18 Typical digital clock block diagram: (a) basic concepts

Display Devices

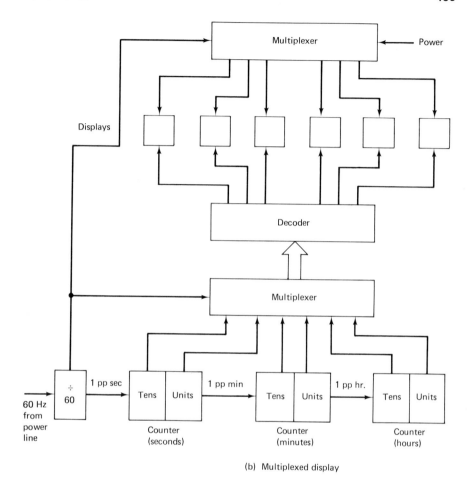

(b) Multiplexed display

Figure 14-18 (cont'd): (b) multiplexed display

the eye is fooled into thinking that the display is on continuously. The use of multiplexers allows only one decoder/driver to operate all displays. See Fig. 14-18b. It should be noted that modern electronic digital clocks are not constructed from separate assemblies. A single LSI assembly contains all components other than the displays. By combining the LSI assembly with displays on a small circuit board, a complete digital clock can be formed. Modern technique will soon furnish the complete clock built into the display device.

Liquid crystal displays (LCDs) do not emit light. (A liquid crystal is a chemical that maintains an orderly molecular structure even though it is in liquid form.) The basic principle behind a liquid crystal display is simple. A thin layer of transparent liquid crystal is placed between

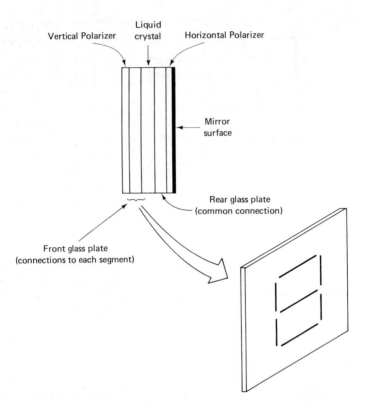

Figure 14-19 Liquid crystal display concepts

two sheets of glass. The sheets of glass are coated with an electrical conductor. While the rear electrode may be transparent, the front electrode is shaped in segments to form the desired character.

The display can be operated in a reflective or in a transmissive manner. In a transmissive display, the rear electrode is transparent; in the reflective display, the rear electrode is opaque, having a mirror-surfaced film.

Figure 14-19 shows a reflective LCD unit. As light strikes the unit, it first encounters a *vertical polarizing film.* This film acts much like polarizing sunglasses, aligning all light rays in the vertical direction. As the vertical light rays pass through the liquid crystal material, they are rotated another 90° because of the nature of the unenergized crystals. The now-horizontal light rays are passed by the horizontal polarizing filter and are cast back to the viewer by the reflector on the back of the unit. Thus the display appears clear because nothing has been done to impede the progress of the light rays.

Figure 14-19 shows that the front glass plate is arranged in the form of the typical 7-segment display. Electrical connections are brought out from each of the segments, which can be energized in much the same manner as the previously-discussed LED displays. When specific segments are energized, an electrical field exists between the energized segment and the common connection on the rear glass plate. Now the portion of the liquid crystal material between the energized segment(s) and the rear plate can no longer rotate the light rays, and those parts of the light rays passing through the liquid crystal remain vertically polarized. Hence they do not pass through the horizontal polarizing filter, and only the light rays that are horizontally polarized are reflected. Rays that are vertically polarized cause light to be blocked in the vicinity of the energized segments, and those areas of the display appear dark. Thus, by energizing the desired segments, specific digits can be displayed.

It should be noted that liquid crystal displays require the same type of code converters and drivers as did LED displays. Only a difference in operating characteristics must be provided, since LEDs and LCDs have different voltage and current requirements.

Segments can be formed by fluorescent or incandescent techniques also. In a fluorescent seven-segment display, each segment is a miniature fluorescent light. Incandescent displays often have a separate incandescent light for each segment. Optical techniques similar to those used with LED displays are employed to spread the light source into segment form. Some incandescent displays also have special segment-shaped lamps to form the displayed characters. The relative characteristics of various segmented displays and other display techniques are discussed in the summary of this chapter.

Dot Matrix Displays

Characters can also be formed for display by using a group of dots arranged in a matrix form. One method places single LEDs at each matrix location. For numeric display only, twenty-seven of the light-emitting diodes are arranged in a (5 × 7) rectangular matrix (not all of the thirty-five matrix locations are occupied by diodes); a twenty-eighth diode is offset at the lower left to serve as a decimal point. Figure 14-20a shows a typical solid-state numeric indicator, which includes decoding capability. The block diagram for the complete assembly appears in Fig. 14-20b. BCD input signals are decoded on the IC chip into ten signals, representing digits 0 through 9. These signals are then encoded into signals that drive an encoder that selects the proper combination of light-emitting diodes to display the required digit. The decimal point is activated by a separate input.

Figure 14-20 Solid-state numeric indicator diagram: (a) LED chip and decoder, (b) block diagram (courtesy Hewlett-Packard)

When complete alphameric display is required, a full (5 × 7) matrix of light-emitting diodes (LEDs) may be employed (Fig. 14-21a). The (5 × 7) matrix of LEDs, which make up each character, are X-Y addressable. This allows for a simple addressing, decoding, and driving scheme between the display module and the system logic.

To form alphameric characters, a method called *scanning* is used. Information is addressed to the display by selecting one row of LEDs at a time, energizing the appropriate diodes in that row, and then proceeding to the next row. After all rows have been excited (one at a time), the process is repeated. By scanning through all rows at least 100 times a second, a flicker-free character composed of discrete illuminated LEDs can be produced. Information may be moved from row to row of the display (*vertical scanning*) or from left to right (*horizontal scanning*).

Figure 14-21b indicates how (with vertical scanning) letters A and B would be formed by sequentially selecting the rows and energizing the correct diodes in each column. When row I is selected only columns 1B, 1C, 1D, and 2A, 2B, 2C, and 2D are energized. When selecting row II, columns 1A, 1E, and 2A, 2E are energized; the process is continued, as indicated by the solid squares in Fig. 14-21b, until all appropriate LEDs have been lighted. The cycle is repeated at a high rate, so the eye sees a continuous *flicker-free character.*

A typical vertical scanning circuit for addressing the display is shown in the block diagram in Fig. 14-21c. This particular scheme contains five display characters. Operation for other numbers of characters is similar. Circuit operation is as follows.

1. Coded 6-bit alphameric information is sequentially entered and stored in five 6-bit input storage buffers.
2. Next, with the input information stored in the input buffers, timing circuitry enables the ROM and input storage buffer number one, so its stored 6-bit code can be read by the ROM. (All other input storage buffers are disabled.)
3. The 6-bit input is decoded by the ROM, and the first row of character information is stored in the 5-bit output storage buffer number one. In other words, if the character A is being written, diodes 1B, 1C, and 1D of row 1 will be lighted (see Figure 14-21b); and this information is stored in the first output storage buffer.
4. Having completed the first loading operation, the timing circuitry now activates input storage buffer number two. (All other input storage buffers are disabled.) The coded character is read into the ROM, decoded, and the row information fed into output storage buffer number two. This operation is repeated until all five characters stored in the input storage buffers are sensed by the ROM character generator and the first line of character information is stored in the output storage buffers.
5. Next, the timing circuitry connects the top row driver, so current flows through and lights up all of the appropriate LEDs in the top rows of the five display characters.
6. The complete cycle is then repeated to decode and display row II of the LED matrix characters, then row III, on

(a)

(b) Formation of letters A and B with vertical scanning

Figure 14-21 Alphameric LED display operation (courtesy Hewlett-Packard)

Display Devices   415

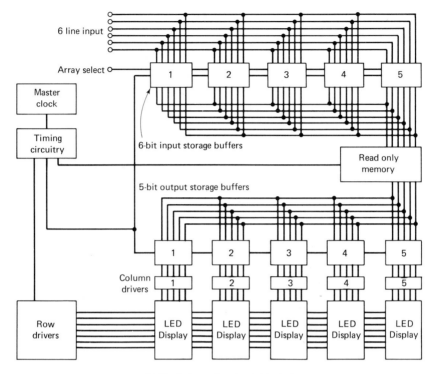

(c) Vertical scanning circuit, block diagram

Figure 14-21 (cont'd)

through row VII. Since the time to decode and load the output storage buffers is short in comparison with the time the display is lit, repeating the character scanning at a rate above 100 times/second gives a flicker-free alphameric character.

Cold-cathode, gas-filled display techniques have been modernized, and these improved techniques have resulted in some very versatile display devices. *Multicharacter assemblies,* using gas discharge techniques, have also been developed and are being used.

One of the most interesting gas-discharge displays is the so-called *plasma panel.* Characters are displayed in a dot matrix form, one of which is the familiar (5 X 7) form discussed earlier. The physical construction of one version of this type of display (SELF-SCAN* panel) is shown in Fig. 14-22a.

*Registered trademark of The Borroughs Corporation.

(a) Physical construction

(b) Cross-sectional view

(c) Block diagram

Figure 14-22 Plasma panel (courtesy Burroughs Corp., Electronic Components Division)

Although the following explanation is grossly simplified, enough of the general concepts are presented to determine its basic principles of operation. When the panel is energized, sufficient potential difference is established between the keep-alive cathode and the keep-alive anode to establish a glow discharge in the keep-alive grooves located in the rear glass cover. The physical construction of the keep-alive cathode and anode are such that charged particles can diffuse into the area of the reset cathode.

The scan anodes are maintained at a high dc voltage, and when the reset cathode is grounded, complete ionization of the partially ionized gas that has diffused from the keep-alive area occurs. A glow is then established in the seven rectangular areas of this single cathode, defined by the intersection of the rear of the cathode strip and the seven scan grooves in the rear glass cover. (This area is not visible from the front of the panel.) The first character-forming cathode is then grounded, immediately following removal of the ground from the reset cathode, and the glow is transferred. Each character-forming cathode has seven tiny holes (.030 inch in diameter) called *glow-priming apertures.* (These holes are small enough that the glow from the rear of the character-forming cathode does not show through, yet charged particles can diffuse through them into the display cavities.) The ground is removed from the first character-forming cathode and transferred to the second cathode. Glow discharge is transferred in this manner until the end of the display is reached. The reset cathode may then be grounded again, and the scan process started over. Thus *the glow discharge is started at the left and transferred one cathode at a time along the entire length of the panel.* Actually, the logic circuitry needed to perform the scan may be simplified by using a "3-phase" drive circuit and grounding every fourth cathode. Glow transfer occurs only between adjacent cathodes, so this technique still allows transfer of only one cathode at a time.

As the glow is scanned down the panel, a phenomenon called *glow-priming* occurs. Figure 14-22b shows a cross-sectional view of a panel, depicting one cathode with a glow discharge on the rear (scan side). The scan glow covers the rectangular portion of the scan side of the cathode located beneath each display cavity. The priming apertures have allowed the partially ionized gas to diffuse into the display aperture. If any of the display anodes are placed at a sufficiently high dc voltage, the partially ionized gas will become completely ionized, and a glow will occur in the display cavity. Characters are written on the viewing side of the panel by addressing the display anodes in synchronism with the glow-priming ionization present on the scan side of the cathode that intersects the point where the dot is to appear.

Figure 14-22c is a block diagram showing the general concept of "writing" information into the display panel. The clock input, which occurs at a frequency that allows the overall panel to be scanned approximately 60 times per second, feeds the 3-phase drive circuit to sequentially ground the cathodes that form the characters. A reset input is also provided to return the panel scan to the left-hand side. The character generator converts six bits of primary information and a clock pulse into a dot matrix character format made available column by column, seven bits in parallel. A counter section, using the count control and clock inputs, establishes the spacing between characters. The actual character information is provided by a read-only memory. Anode drivers are provided to interface between the logic level of the character generator and the high-voltage dc levels of the display panel. If information is to remain on the panel for display, external circuits must be provided to refresh the input to the character generator.

Panels using this or similar techniques may be used to display as few as 1 and as many as 256 characters. Display size of characters ranges from 0.25 to 0.5 inch in height. The future will surely see these maximums extended.

MISCELLANEOUS DEVICES

Two other methods of display, *projected-image* and *edge-lighted panels,* should be mentioned in passing. Both are still in use but are being phased out by the newer, more efficient displays. The projected-image display uses a separate light source and image for each character. The image is usually on film, and activation of the light source causes an enlarged image of the filmed character to appear at the face of the device. Edge-lighted panels have solid characters engraved on acrylic plastic panels, and illumination of the character is accomplished by activating a separate lamp for each panel. Both methods suffer from the decoding complexities and lack of versatility that plague the NIXIE tube, although the relatively low-operating voltage requirements tend to make them more applicable to standard IC decoder/driver packages.

Cathode-ray tubes (CRTs) should not be overlooked as potential display devices. In conjunction with logic circuits called *character generators,* they are finding widespread use as computer terminals and in other data display applications. No limit is placed on the types of characters that may be used, and CRTs promise to fulfill the need of displaying large amounts of data in a relatively small space. The only real disadvantage is cost. CRT displays are perhaps the costliest—although the most versatile—display devices in use today.

14-3 SUMMARY

The combination of code converters (decoders) and display devices has provided us with a means of communication between digital devices and humans. Selection of a specific type of display device depends on many factors. Generally, LED type displays, whether they be segmented or dot-matrix, are relatively bright except in direct sunlight. They do require a moderate amount of power for operation, however. When used in "plug-in" or rechargable battery-powered equipment, LEDs provide an excellent display.

LCDs function quite well in normal light, even in bright sunlight. They use very little power and are adaptable to portable equipment. However, they require an auxiliary source of light when used in the dark. Fluorescent and incandescent displays have special voltage/current requirements and seem to be slowly disappearing from the digital scene.

Gas-discharge displays are becoming more popular as additional information must be displayed. Although the associated electronics is complex, there are many applications where CRT displays are not warranted. For example, some computers, digital communication systems, and instrumentation systems require only a few lines of information. Gas-discharge displays fill this void.

PROBLEM APPLICATIONS

14-1. How many gates are required to convert a 3-digit octal number to binary form? Draw a logic diagram that will perform this operation.

14-2. How many gates are required to convert a 2-digit hexadecimal number to binary form? Draw a logic diagram that will perform this operation.

14-3. Why would it be impractical to convert large decimal numbers directly to binary form?

14-4. Discuss the count comparison method of BCD to binary conversion, noting its advantages and disadvantages.

14-5. Discuss how the count comparison method may be used to convert a binary number to BCD form.

14-6. What are the advantages of the full-adder method of BCD to binary conversion when compared with the count-comparison method?

14-7. Can the full-adder method be used to convert from binary to BCD? Discuss how it must be done.

14-8. Why is the ROM method often used to convert from binary to BCD?

14-9. What are the basic principles of the ADD 3 method of binary to BCD conversion?

14-10. Is the ADD 3 method of binary to BCD conversion applicable to BCD to binary conversion? If so, how is it performed?

14-11. Explain how the 7445A BCD to decimal decoder in Fig. 14-10 guards against improper outputs when erroneous binary input conditions exist.

14-12. What is the purpose of the D input of the 8250 decoder in Fig. 14-13?

14-13. Draw a block diagram showing a 1-out-of-32 decoder, using the 8250 decoder in Fig. 14-13.

14-14. Construct a logic diagram of a converter that will accept 8-4-2-1 input and provide XS3 output. Use the method described in Sec. 14-1.

14-15. Discuss how the method in Sec. 14-1 could be expanded to encompass numbers and special characters when converting from ASCII to EBCDIC.

14-16. Why are 16-segment displays sometimes used in preference to 7-segment displays? What are the advantages and disadvantages of each?

14-17. Discuss "dot-matrix" formation of alphameric characters.

14-18. What is a light-emitting diode?

14-19. Why is scanning used when displaying information with LEDs?

14-20. Discuss the future of plasma-panel and liquid-crystal display devices.

15

MICROPROCESSORS

15-1 INTRODUCTION

As the use of digital techniques became more and more popular, some difficulties began to emerge. The complexities of the functions being performed resulted in very large and complicated electronic devices. In fact, some functions were so complex that digital computers were required. Such functions as traffic light controllers, supermarket checkout, machine control, and the like required either a large-scale digital computer or hundreds of digital logic components. The use of such large-scale computers for these relatively insignificant applications was not appropriate. Even the minicomputers were "overdesigned" for such simple tasks. With the knowledge that semiconductor technology could put thousands of circuits on very small chips, designers approached the everyday applications of computers and logic circuits from a different viewpoint.

Why not replace the many "hard-wired logic" control and data processing circuits with a general purpose programmable digital computer? Then one computer could perform many tasks, requiring only a change in programming to convert from one task to another. By designing a computer specifically for applications that require small amounts of information storage and minimal mathematical capabilities, perhaps the semiconductor people could develop a low cost "computer on a chip that would be more suitable." After all, if they could make a calculator chip, why not a computer chip?

Figure 15-1 Single-board computer (courtesy INTEL Corp.)

The semiconductor people came through. Multi-function LSI chips replaced dozens of ICs. In fact, a computer that had taken a number of printed circuit cards to perform its functions now could be built on a single card (see Fig. 15-1). If one is willing to settle for small-scale control applications, the true computer on a chip exists today; Figure 15-2 is such a chip. It includes all of the circuits required to perform operations and route data. It stores its own program, interfaces with outside devices, and includes its own timing. And the end is not yet in sight. As semiconductor technology grows, more and more capability is being packaged in smaller and smaller units. As IC chips become more versatile, additional applications are generated, which require even more versatile chips.

Let's see if we can discover how microprocessors and microcomputers work. A *microprocessor* is, by generally accepted definition, *a LSI circuit assembly that contains about 70% of the computing capability of a very small digital computer.* Usually it has minimal capability to receive and put out information. (A device that has versatile input, output, and storage capability is usually called a *microcomputer.*) It should be noted, however, that the dividing line between microprocessors and microcomputers is becoming hazy. Technology is giving the engineers the ability to place a complete microcomputer on a single chip. When the term microprocessor is used, just remember that we could be talking about a device very close to a complete microcomputer. This chapter investigates the microprocessor and relates it to the digital logic functions already discussed (hardware) and the new concepts of *programmed lists of instructions (software).*

Figure 15-2 Computer on a chip (courtesy INTEL Corp.)

15-2 MICROPROCESSOR CONCEPTS

Understanding the microprocessor requires the recognition of two major factors. First, a microprocessor is a functional device that is capable of performing certain specific operations only. Second, the microprocessor is time-oriented. The interrelationship of the operations performed and the time required to achieve performance is of the utmost importance.

Figure 15-3 Microprocessor functional diagram

Functional Operation

Five basic functions are performed by an information-processing device, whether it be a large-scale digital computer or a microprocessor. These are:

1. the INPUT function, which interfaces between the outside world and the device;
2. the OUTPUT function, which interfaces between the device and the outside world;
3. the STORAGE (MEMORY) function, which stores information and instructions about what to do with the information;
4. the ARITHMETIC/LOGIC (ALU) function, which performs the mathematical and logic operations of the device;
5. the CONTROL function, which sequences and controls the device.

It should be noted that these are *functional* categories only, and it may not be possible to identify them with specific pieces of hardware or sections of the microprocessor chip.

The flow of information within a microprocessor is seen in Figure 15-3. Except in special cases, all information that the microprocessor has must be entered via the INPUT function. (In microprocessors and microcomputers, some of the information may be permanently contained within the device.) Depending on the type of information, it may be used either to tell the system what to do (*instructions*) or it may be used in arithmetic or logical operations (*data*). As an instruction, the information tells the processor where the next information is stored or what to do with that information. As data, it will be routed to the

Microprocessor Concepts

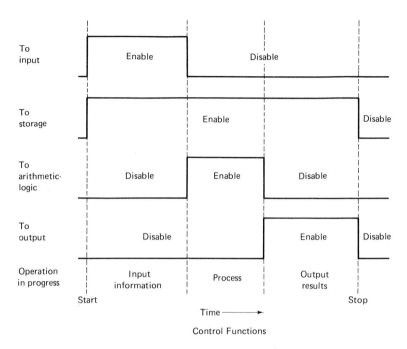

Figure 15-4 Gross timing diagram, microprocessor

ARITHMETIC/LOGIC function for processing and is then sent back to STORAGE to await further use. When called upon by the CONTROL function the OUTPUT function displays the results of the internal processing of information. The flow of information throughout the microprocessor is from INPUT to STORAGE to OUTPUT, with detours along the way to the ARITHMETIC/LOGIC function for processing operations and to the CONTROL function to determine what is to be done next.

Sequential Operation

Control signals, occurring at the proper time and in the proper sequence, are developed by the CONTROL function. The *timing diagram* of Figure 15-4 shows a very simplified version of the major control signals required to enable and disable the basic functions of Figure 15-3.

Information is accepted from the input device(s) during the period t_0 to t_1, when the input device(s) and STORAGE functions are enabled. It is during this time that the *list of instructions (program)* is input. The information to be processed may also be input at this time.

Instructions and data are placed in STORAGE and called upon as needed during the t_1-t_2 time period. With both STORAGE and ARITHMETIC/LOGIC functions activated, instructions and data are processed as required by the stored program. Results are returned to STORAGE. Upon completion of information processing, the ARITHMETIC/LOGIC function is disabled, and the OUTPUT function is enabled. The stored program then controls transfer of results to the output device(s) during t_2-t_3.

Information Representation

Each instruction must be presented to and stored within the memory in a form that the microprocessor understands. Since digital devices function only with binary inputs, both instructions and data are presented in binary form. Information must also be in a known format. In many microprocessor and microcomputers, *eight binary digits as a group* are used. This group is called a *computer word,* or *byte.* Computer words are input in the exact order in which they are to be used, and the microprocessor automatically selects the next byte in sequence unless told to do otherwise. The basic cycle of operation is one of retrieving the computer word from memory, evaluating it, obtaining the information at the proper location, performing the operation with the obtained information, and progressing to the next byte for another instruction.

The Bus

Although the microprocessor/microcomputer *could* be represented by the block diagram just shown, it may be easier to understand the microprocessor using a bus-organized functional diagram. *A bus is a group of circuit interconnections between all sections of the microprocessor.* It connects all sections to all other sections and effectively acts as a communication path between two or more of the microprocessor's sections.

A bus-organized microprocessor contains all of the functions shown in Figure 15-3. Instead of information flowing from one function to the next, however, each function is connected to all of the others. The CONTROL section decides which functional unit is to originate information to be used and which of the others is to receive that information. Figure 15-5 shows the general concept of a bus-organized device.

All of the functions are connected together by a bus. Some wires in the bus are committed to handling memory location (address) information only; others carry only data; still others transmit only control signals. Each function uses only the address, data, and control signals related to the specific operation being performed. For example, the

Microprocessor Concepts

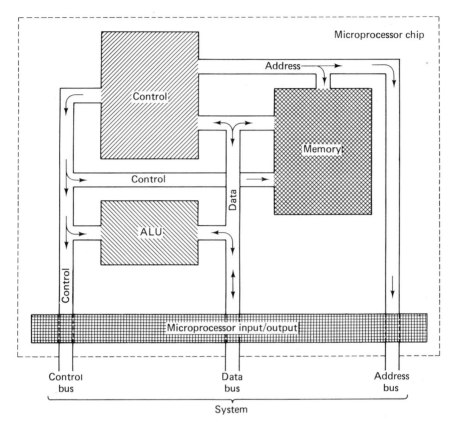

Figure 15-5 Bus-organized microprocessor

control bus supplies inputs to both the ALU and the on-chip MEMORY sections as well as to external units. *Address bus* information that originates in the CONTROL section is made available to both on-chip and external memory. All functional sections of the microprocessor are connected to the *data bus*. Therefore, whether data originates in the ALU, the MEMORY, or from external sources, it is available to all functional sections. The origination and destination of the data on this bus are determined by the CONTROL section of the microprocessor.

The simple application of gates, plus precise timing, keeps the information on the various lines in its proper bus. Figure 15-6 shows a simplified logic diagram of the busing concept. Each section of the microprocessor is connected through a gating network to each wire of the bus. In its simplest form, the gating network consists of hundreds of AND gates. One input of each gate connects to a specific bus line; another input connects to the timing source in the microprocessor; a

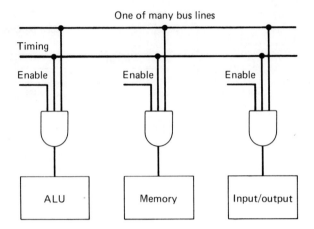

Figure 15-6 Bus logic diagram

third input connects to the device outputting the control operations. Information on a bus line is passed to a functional section of the microprocessor only when its AND gate is enabled, and at the time when the timing is correct.

It is very important to recognize the microprocessor as merely a collection of digital circuits, most of which are not "connected" in the conventional manner. The circuit connections are a function of the *computer software programs.* Most circuits in the microprocessors are connected by gates that are enabled only to perform certain desired operations. Thus the person working with the microprocessor must learn a new "wiring" skill—that of computer programming. "Wiring with software" is discussed later in this chapter.

15-3 THE MICROPROCESSOR AS HARDWARE

The microprocessor may be viewed either as a collection of hardware or as a set of operations that can be performed without regard as to how they are done. This section discusses the hardware; Section 15-4 examines the software of a typical microprocessor.

Figure 15-7 shows how a hardware-oriented person might view the microprocessor. Each of the major functions are still uniquely identified as in Fig. 15-5, but it should be noted that portions of the MEMORY section are shown attached to other sections. Also, the INPUT/OUTPUT sections are shown in three different locations. These changes are necessary to explain more adequately the detailed hardware operation of a microprocessor.

The Microprocessor as Hardware

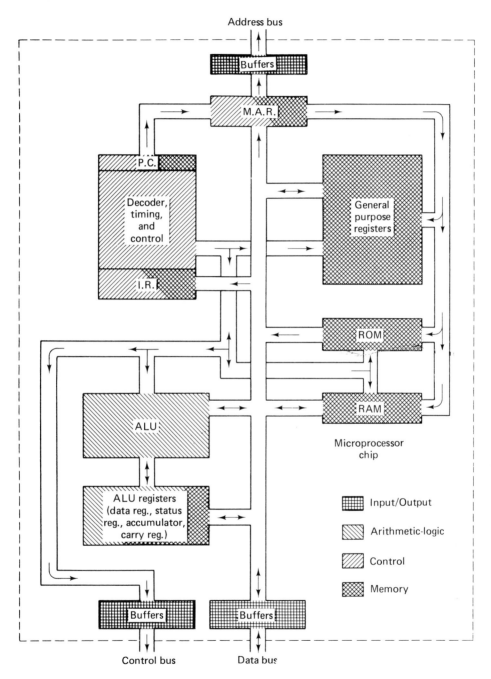

Figure 15-7 The microprocessor as hardware

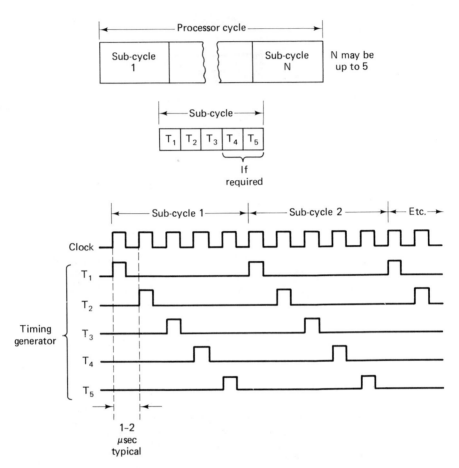

Figure 15-8 Processor cycle and timing

The CONTROL Function

A microprocessor operates in a repetitive sequence of operations while performing its functions. This sequence of operations is called the *processor cycle*. It is the *sequence of operations required to obtain information from storage and to perform the operations required by that information*. Each processor cycle has one or more *subcycles*. Furthermore, each subcycle consists of up to five *timing periods,* which are the smallest increment of operation available to the microprocessor. Figre 15-8 shows the timing and processor cycle relationships.

The first subcycle "fetches" information from memory, considering this information to be an instruction. During the first timing period of the first subcycle the CONTROL section addresses memory in order to obtain information. During the second timing period the next mem-

The Microprocessor as Hardware 431

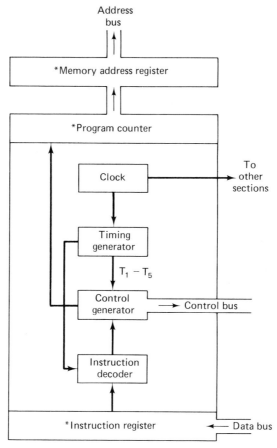

Figure 15-9 Microprocessor control section

ory location address is identified. During the third timing period information at the address noted in the first time period is moved to a temporary storage register. During the fourth and fifth timing periods information is moved between registers.

During the second and following subcycles, the same procedure is followed, except that the information obtained is considered to be data. Most processor cycles require up to three subcycles to perform their operations. However, some complex operations may require more than five subcycles.

Figure 15-9 is the functional diagram of a typical CONTROL section. The clock output, shown in Fig. 15-8, is routed to the other functional sections of the microprocessor to determine its basic operating speed. *Timing for each subcycle is developed by the timing generator.* Both the clock and timing generator contain the IC equivalents of counter, registers, and gates.

The *instruction register* receives a byte of information from the data bus and temporarily stores it while the *instruction decoder* determines what action is represented in the specific arrangement of 1s and 0s. Decoding takes place only during the subcycle when it is known that information on the data bus is an instruction. The timing generator is responsible for enabling the instruction decoder. A network of gates can be used for the instruction decoder.

Control signals for the rest of the microprocessor originate in the *control generator.* Here the outputs from the timing generator and the instruction decoder are combined to develop the required control signals. The control generator also oversees the operation of the program counter. Since the *program counter* must *contain the address of the next location in memory to be used,* it must be loaded with a given address, incremented, or decremented according to program instructions. The control generator is well suited for these operations, since it is aware of each operation to be performed. IC equivalents of gating networks usually make up microprocessor control generators. The program counter is usually implemented with counters or registers. One more register—the *memory address register*—is commonly found in microprocessors and it *contains the address of the location currently being processed.* Input to the memory address register is usually obtained from the program counter.

The STORAGE (MEMORY) Function

Those sections of the microprocessor that are considered as "*memory*" include the *register stack, random-access memory (RAM),* and *read-only memory (ROM).* Figure 15-10 shows the general organization of a microprocessor on-chip memory.

The *register stack* is merely a *collection of temporary storage registers;* these may be classified as special-purpose registers and general-purpose registers. *Special-purpose registers,* such as the *instruction register* and the *program counter,* are discussed as they are functionally used in other sections of the microprocessor. *General-purpose registers,* or "scratch-pad" registers, are used principally for the *manipulation and temporary storage of data.* These registers are used as a person would use pads of paper to write down notes or the intermediate results of an arithmetic operation. Storage registers and gating circuits are used to implement these functions.

The actual use of general-purpose registers is best discussed from a software viewpoint, in Section 15-4. The selection of a general-purpose register to be used and the operation to be performed is accomplished

The Microprocessor as Hardware   433

Figure 15-10 Microprocessor on-chip memory organization

by the CONTROL section. The byte in the instruction register is decoded by the instruction decoder, and if a register operation is indicated, certain bits in the instruction byte are selected and decoded by the REGISTER SELECT function of the register stack. The REGISTER SELECT function is enabled only when the register operations are

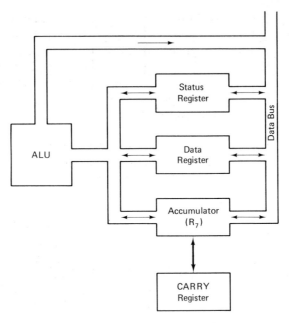

Figure 15-11 Microprocessor Arithmetic/Logic section

indicated. Thus any of the registers and any of the register operations may be selected.

RAM and ROM used in a microprocessor differ little from the general concepts of memory devices discussed in Chapter 12. When an address appears on the address bus, that location in memory is accessed. The CONTROL section determines what is to be done at that location. If a READ instruction has been decoded, the information at that location appears on the data bus. A WRITE instruction allows information on the data bus to be written into memory at the addressed location. To simplify our discussion, only 192 8-bit RAM locations and 64 8-bit ROM locations are used. Many more than 192 8-bit RAM locations and 64 8-bit ROM locations are commonly available in modern microprocessors.

The ARITHMETIC/LOGIC Function

The ARITHMETIC/LOGIC section (Figure 15-11) performs the data modifications required to accomplish both *logical and arithmetic operations*. A collection of gates called the arithmetic/logic unit (ALU) does the actual data manipulations. Dedicated registers from the register stack temporarily store *operands (quantities on which mathematical operations are to be performed)* during ALU operations.

Four registers are used during ALU functions. Because most arithmetic and logical operations require two operands, a register for each of the operands must be provided. One of the operands is temporarily stored in the *data register*. The second operand is stored in the *accumulator*. The accumulator also stores the results of the mathematical operation, which replaces the operand originally stored there. Inputs to the data register and the accumulator may come from either the data bus or the ALU.

Another register is needed to store any overflow from arithmetic operations. The *carry register,* a single-bit storage location, retains any overflow information so that it may be included in subsequent calculations.

In a microprocessor, arithmetic and data manipulation operations generate various "flags" (an indication of a specific condition) to define the results of the operation. For example, an arithmetic operation may generate a carry; the result may be −, 0, or +; and so on. One register, called the status register, may be dedicated to storing status information. The stored status can then be used to control conditional operations. For example, if the program detects a zero result in an arithmetic operation, it may branch to some portion of the program other than the next sequential step. The application of the status register is discussed with software in Section 15-4.

The ALU generally performs the following operations:

1. binary addition and subtraction,
2. Boolean logic operations, and
3. shift right and shift left.

Other mathematical operations, such as multiplication and division, are implemented by software programs. ALU functions are implemented by the IC equivalents of adders and gating circuits.

The INPUT/OUTPUT Function

Microprocessor input/output is generally used to isolate the microprocessor from outside influences. In many microprocessors, input/output is performed in the same manner as memory read/write. An address is placed on the address bus to represent the input/output device to be used. Data are then placed on the data bus, and the CONTROL section generates control signals to define whether an input or an output function is to be performed. Some microprocessors also allow serial (one bit at a time) input, but generally such operations are under program control.

The address and control buses in most microprocessors are one-way buses, transferring information to external devices from the microprocessor. Buffer amplifiers are used for this purpose. Two-way buffers are used on the data bus, because information must be both received and transmitted by the microprocessor.

15-4 THE MICROPROCESSOR AS SOFTWARE

All of our discussion to this point has been toward understanding the microprocessor from its hardware operations. Now the time has come to consider the microprocessor almost entirely from a software standpoint. In this section the software procedures of a microprocessor are explained and integrated with the hardware capabilities.

General Concepts

In many microprocessors, information is stored and manipulated in *groups of 8 binary digits* called *bytes*. As many as three bytes may be required to define an instruction and other information adequately. Fortunately, a large number of instructions can be represented by 8 bits (256 to be exact). Many instructions require reference to memory locations for information to be processed, loading of a specific numerical value, identification of a particular input/output device, and so on. Generally two-byte instructions are adequate for use with small memories or for loading small numerical values. When large memories or large numerical values are used, three byte instructions are common. Examples of all three types of "instruction words", which are computer words carrying only instructions, are shown in Figure 15-12.

To execute any program, the microprocessor must constantly refer to storage locations in the register stack, RAM and ROM. *Reference to storage locations is called addressing,* and numerous *addressing modes* are encountered in microprocessors. Examples of various addressing modes are seen in this section. Figure 15-13 depicts some of the simpler modes discussed below. *Direct addressing* tends to be the simplest mode. The instruction being executed contains both the command and the actual address to be used. Most microprocessors require at least one byte (for memories up to 256 bytes) or perhaps two bytes (for memories up to 64K bytes) of address in addition to the command byte. *Register addressing* refers to a mode that defines locations specifically in the register stack. When the register stack is small (on the order of eight registers or less), the register address may be included as an integral part of a single-byte instruction. *Immediate addressing* is the term used when the data to be used are included in the instruction byte. This mode is most often employed when constants are to be used in a program and, depending on the value of the constant used, may require up to three bytes. Many other addressing modes may be used,

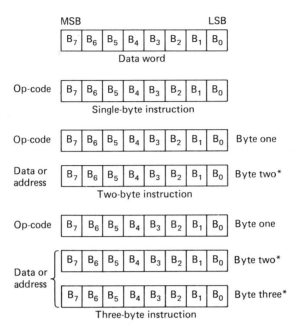

*Bytes two and three must immediately follow their op-code in memory

Figure 15-12 Instruction formats

but to understand the basic ideas of software capability it is not necessary to discuss the others. Information from the manufacturers of specific microprocessors explains all of the usable addressing modes for their devices.

As the programmer prepares the list of instructions for the microprocessor, it is necessary to know where all instructions and data are in memory. A *memory map* assists the programmer in the most efficient use of available memory. Table 15-1 is a memory map showing typical data stored at representative locations in a microprocessor. Purely arbitrary information is shown here, but actual data are used in a subsequent example. Note that both the location and the information available at that location are shown. The most primitive of memory maps is shown in binary form, since that is the only form that the microprocessor really recognizes. Octal, hexadecimal, and decimal equivalents are also shown in Table 15-1, and may be used depending on the programmer's background and manufacturer's technical data.

A programmer's primary concern with the microprocessor is that of applying its capabilities effectively. Since the microprocessor is

Figure 15-13 Addressing modes

merely a data-manipulation device, the programmer is most interested in those parts that perform a specific function and are available for use.

Figure 15-14 represents a programmer's viewpoint of a microprocessor. Here the registers are arranged to give the programmer maximum flexibility. Note that the general-purpose registers are separate from the special-purpose (dedicated) registers. This provides the programmer with an operational aid during programming. Each of the registers in Figure 15-14 have been discussed from a hardware standpoint in Section 15-3. From a software standpoint we now investigate the operations that can be performed with and between registers, plus those operations that allow data manipulation between other parts of the microprocessor. *NOTE: The complete list of instructions is called an instruction set.* We discuss only a small part of an instruction set here and leave the remainder to individual study.

The Instruction Set

The *instruction set is really the means of connecting one part of the microprocessor to another.* This is what makes a microprocessor so versatile. With just the change of one simple instruction, the complete function of a microprocessor can be changed.

The Microprocessor as Software

TABLE 15-1 Composite Memory Map

DATA				LOCATION			
(d) Decimal	(c) Hexadecimal	(b) Octal	(a) Binary	(a) Binary	(b) Octal	(c) Hexadecimal	(d) Decimal
85	55	125	01010101	00000000	000	0	0
109	6D	155	01101101	00000001	001	1	1
170	AA	252	10101010	00000010	002	2	2
204	CC	314	11001100	10111111	277	BF	191
51	33	063	00110011	11000000	300	C0	192
146	92	222	10010010	11111111	377	FF	255

Many instruction sets have their binary arrangement of 8-bit bytes arranged for easy recognition. The byte is converted to octal or to hexadecimal notation so that groups of instructions having similar characteristics are recognized easily. Such a method lends itself well to the instruction classification scheme that follows.

INSTRUCTION CLASSIFICATIONS

INFORMATION MOVEMENT INSTRUCTIONS

Group 0—General information operations. Instructions in this group allow *incrementing* and *decrementing* (increasing and decreasing value of contents) of registers, *loading* and *storing* registers, *shifting* contents of registers, and the like.

Group 1—Information transfer operations. Group 1 instructions *move information* from one location (called the source) to another location (called the destination) without modifying the information. Typical sources and destinations are registers, memory, and input devices on the microprocessor chip. Actually these instructions perform the function of connecting one location to another just as though they were wired together.

INFORMATION MODIFICATION INSTRUCTIONS

Group 2—Arithmetic/logic operations. The *contents* of certain registers may be *modified* by these instructions. Modification is accomplished by performing either arithmetic or logic operations on the contents of the registers.

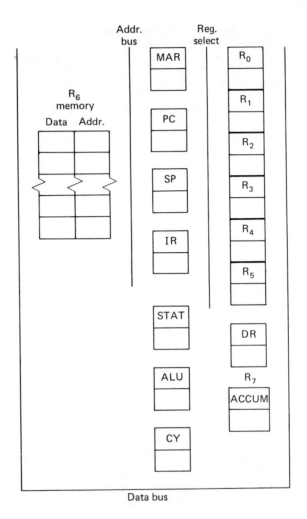

Figure 15-14 Programmers' view of the microprocessor

CONTROL INSTRUCTIONS

Group 3a—Program control operations. These instructions *transfer execution of a program* from its present location to some other location in memory. *Transfer may be conditional based on contents of the status register,* or *unconditional with no restrictions.* Furthermore, each transfer may be a returning transfer, in which the present address of execution is saved so that control may be returned, or nonreturning transfer which does not require saving the present address.

Group 3b—Processor control operations. Group 3b instructions perform operations directly on the microprocessor and are few in number. Typical instructions are PROCESSOR HALT, ENABLE/DISABLE interrupts, and NO OPERATION.

Figure 15-15 shows the general organization of bytes in an instruction set. Note that the two most significant digits of the byte define the instruction by group number as indicated above. The next two groups of 3 bits have varying functions, depending on their group assignment. For example, in group 9, the DDD bits define the destination register; the NNN bits indicate the specific operation to be performed on that register (an example of this is given shortly). Definitions for each combination of letter designations in the four groups of the instruction sets are given in Table 15-2.

Some elements from the microprocessor's instruction set are given in Table 15-3. A complete instruction set is quite comprehensive, and is not required for an initial understanding of its use. A complete manual for a microprocessor programmer would include the full instruction set summary and a complete description of every instruction. Such information as the number of subcycles used during execution of an instruction, the addressing modes, status flags affected, and so on, is commonly included.

Using the Instruction Set

Now that we have a set of instructions to interconnect all the electronic circuits in the microprocessor, let's see what it takes to actually provide those interconnections. First, a *list of instructions* for the operations to be performed (called a *program*) must be developed so that the microprocessor will know what to do. Writing a computer program is a demanding task. The computer programmer must understand the problem being programmed with great clarity; problem analysis must be complete, for no errors can be tolerated. The programmer's job can be separated into five separate tasks: (1) analysis, (2) organization, (3) coding, (4) testing, and (5) documentation.

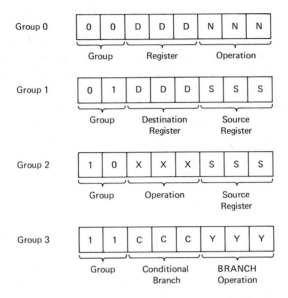

Figure 15-15 Instruction byte organization

TABLE 15-2. General Instruction Formats

Group 0	0	0	D	D	D	N	N	N
Group 1	0	1	D	D	D	S	S	S
	1	0	X	X	X	S	S	S
	1	1	C	C	C	Y	Y	Y

DDD & SSS—Destination and Source Registers

0	R0	
1	R1	
2	R2	
3	R3	
4	R4	
5	R5	
6	R6	(Memory—treated as register and requires address)
7	R7	(Accumulator)

NNN—General Information Operations

0	Not used
1	Immediate operations (Register pairs)
2	Load & Store (Register pairs)
3	Increment/Decrement (Register pairs)
4	Increment (Single register)
5	Decrement (Single register)
6	Immediate operations (Single register)
7	Register shifting/miscellaneous

TABLE 15-2 General Instruction Formats (cont'd)

XXX—Arithmetic/Logical Operations

0	Add
1	Add with carry
2	Subtract
3	Subtract with borrow
4	Logical AND
5	Logical EXCLUSIVE OR
6	Logical OR
7	Compare

CCC—Conditions of Branch

0	was not zero (NZ)
1	was zero (Z)
2	did not generate a carry (NC)
3	generated a carry (C)
4	generated odd parity (PO)
5	generated even parity (PE)
6	was positive (P)
7	was minus (M)

YYY—Branch Operations

0	Conditional return
1	Miscellaneous
2	Conditional JUMP
3	Miscellaneous
4	Conditional CALL
5	Miscellaneous
6	Special A/L Operations
7	Special unconditional JUMPs

During the *analysis phase* the programmer must *develop a concise statement of the problem, analyze the facts,* and *write the mathematical expressions for problem solving.* The key to a good program is an algorithm—a foolproof, step-by-step procedure that takes into account all possible situations, including exceptions. The *algorithm, then, is an unambiguous procedure that always leads to an answer (or at least an indication that an answer cannot be determined) in a finite number of steps.*

Let's assume that we have a requirement to assure that no extraneous information is stored in memory prior to beginning normal computer operations. One way to do this is to place all zeros in all memory locations. The general steps required to "initialize" memory are (1) place all zeros in the first location to be cleared, (2) repeat until all locations are cleared, and (3) recognize completion and stop to await further instructions.

TABLE 15-3. Partial Instruction Set

Instruction	Binary	Octal
Group 0		
MVI M	00110110	066
MVI R4	00100110	046
MVI R7	00111110	076
INR R4	00100100	044
DCR R4	00100101	045
Group 1		
MOV R1, R5	01001101	115
MOV M, R2	01110010	162
MOV R2, M	01010110	126
MOV R5, R1	01101001	151
MOV R4, R7	01100111	147
Group 2		
ADD R3	10000011	203
ANA R3	10100011	243
ORA R4	10110100	264
ADC R2	10001010	212
CMP R4	10111100	274
Group 3		
JZ	11001010	312
JMP	11000011	303
JC	11011010	332
RET	11001001	311
JPO	11100010	342

A possible algorithm that performs initialization follows:

1. Load the starting *address* in register 1.
2. Load the stopping *address* in register 2.
3. Move all zeros to the location in memory addressed by register 1.
4. Compare the present address with the address stored in register 2.
5. If both addresses are the same, stop and wait for more instructions.
6. If both addresses are not the same, increase the address in register 1 by one.

The Microprocessor as Software 445

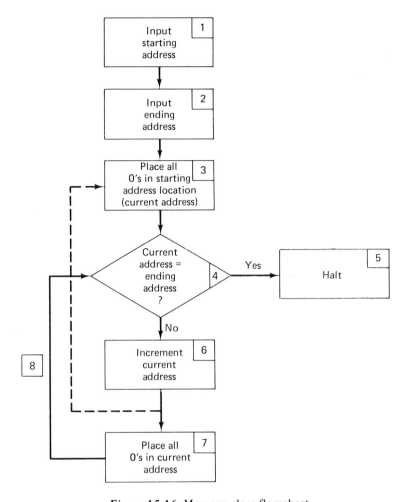

Figure 15-16 Memory clear flowchart

7. Move all zeros to the next address.
8. Go back to step 4.

Modifications to the algorithm often occur as organization, coding and testing are accomplished. The programmer should be alert to necessary changes as the work progresses.

During the *organization phase, each step of the algorithm is investigated,* and *a flowchart is developed.* Figure 15-16 is one way that a programmer might initially construct an initialization algorithm. Each block on the flowchart is numbered to correspond with steps in the original algorithm. When the information is put in flowchart form it becomes apparent that steps 3 and 7 are performing the same function. Therefore the flowchart can be modified as shown with the dotted line.

The *coding phase* requires knowledge of each of the instructions in the instruction set as the *flowchart is converted to the list of instructions called the program.* Although the function of the program is to place zeros in all memory locations bounded by the starting and ending addresses, the operation of the program revolves around the comparison of current and ending addresses. An instruction (or instructions) is required that compares the two addresses and takes specific action when they are equal. The programmer must also know the register assignments so that minimum movement of information is required by the final program. With this information and constraint, the following sequence of events can be listed:

1. Load starting address into R4.
2. Load ending address into R7.
3. Load all zeros into address specified in R4.
4. Compare contents of R4 with contents of R7.
5. Branch to HALT if they are equal.
6. If not equal, increment R4.
7. Branch to step 3.

Now all that remains is to convert these events into actual program steps. The completed program is shown in Table 15-4. Note that octal notation is used and correlated with both the binary and the mnemonic codes of Table 15-3.

As the *testing phase* is entered, *the program is compared with the flowchart* to assure that all operations are accounted for and are in the correct order. It is then advisable to *"walk through" the program* with sample data before inputting the program to the microprocessor. Quite often mistakes are detected, thus saving considerable time and cost. Once the programmer feels that the program will do its job, it is entered into the processor and its operation verified. If difficulties are encountered, then the programmer must evaluate the indications received and correct the discrepancies.

Every program should be thoroughly *documented* with as much background information as possible. The *rationale behind the algorithm* is extremely useful when changes must be made to a program, as are the various levels of *flowcharts* that have developed as the program has been organized. If *intermediate programs* have preceded the final program, these should be a part of the documentation, since they are a record of the thinking of the programmer. Finally, the *actual results of executing the program* should be an integral part of the documentation package.

The Microprocessor as Software 447

TABLE 15-4 Program to Initialize Memory

| Memory Address | | | Memory Contents | |
Binary	Octal	Mnemonic	Binary	Octal
00000000	0	MVI R_4	00100110	046
00000001	1	300	11000000	300
00000010	2	MVI R_7	00111110	076
00000011	3	377	11111111	377
00000100	4	MVI M	00110110	066
00000101	5	000	00000000	000
00000110	6	CMP R_4	10111100	274
00000111	7	JZ	11001010	312
00001000	10	014	00001100	014
00001001	11	INR R_4	00100100	044
00001010	12	JMP	11000011	303
00001011	13	004	00000100	004
00001100	14	HLT	01110110	166

Assume starting address 192_{10} (300_8) ending address 255_{10} (377_8)

15-5 SUMMARY

The reader should be fully aware by now that a microprocessor is a complex device in terms of both hardware and software. Once a person develops proficiency with these two operations however, no problem is too large to handle. More and more tasks formerly performed with logic circuits such as gates, flip-flops, counters, registers, and the like, are now handled easily by microprocessors. The digital technician/ engineer must use the knowledge of basic digital circuits and the techniques of software development to remain current in the field. Many fine texts exist in the microprocessor field for those wishing to pursue this special technology. Consult the bibliography for suggestions.

QUESTIONS

15-1. What is a microprocessor?
15-2. How does *hardware* differ from *software*?
15-3. List the five functional sections of a digital computer and explain the purpose of each.
15-4. What is a byte?
15-5. Using Fig. 15-3, show the flow of information through a digital computer.
15-6. What is the purpose of the "bus" in a microprocessor?
15-7. Discuss the three types of busses found in a microprocessor and explain their uses.

15-8. Define *processor cycle*. Explain what happens during each of the subcycles.

15-9. How do the *instruction register* and the *instruction decoder* work together?

15-10. What is the function of the *control generator*; the *program counter*; the *memory address register*?

15-11. Why are both *special-purpose registers* and *general-purpose registers* used in the *register stack*? Explain.

15-12. What is the difference between RAM and ROM?

15-13. List the four registers generally found in the ARITHMETIC/LOGIC section and discuss the function of each.

15-14. Discuss two different types of *addressing* found in modern microprocessors.

15-15. What is an *instruction set* and how is it used?

15-16. List the five phases of program construction and discuss each.

16

DIGITAL APPLICATIONS

As a closing gesture it is only fair that the reader be given a look at the new world of digital logic that has been discussed throughout this text. It is in this new world that the theorems of Boolean algebra, the logic maps, and the truth tables come together with gates and flip-flops in real ICs to form real, functional, helpful systems—not just combinations of gates or groups of flip-flops in packages.

Digital techniques are being used practically everywhere. It is difficult to name *any* business-related field where digital techniques have not made their mark. A few of the broad areas of digital applications are covered in this chapter; just enough to encourage the reader to find more.

If the question is *"why* are digital techniques being used?" the answer is easy—almost every application of digital techniques has resulted in faster, more accurate performance of tasks with more easily understood results. The everyday working world is starting to feel the impact of digital methods; the conversion to digital logic is just beginning.

16-1 TELEMETRY

Definitions

Telemetry is a *process that is used to measure data at a location remote from the data source.* The word telemeter is of Greek origin, derived from "tele" meaning far off, and "metron" meaning measure,

As mentioned in Ch. 7, a telemetering system acquires data at a relatively inaccessible location, encodes, transmits, decodes, and displays the data at the location where it is to be used. Figure 7-7 is expanded in Fig. 16-1, and the general telemetry system is discussed in greater detail next.

Transmitting System Description

For the purpose of clarity, the general telemetering system to be described here is divided into two parts: the transmitting system and the receiving system. This section discusses the equipment that is situated at the location where information is gathered, processed, and transmitted.

TRANSDUCER SUBSYSTEM

A typical space vehicle required hundreds of measurements to determine the operations of the propulsion, airframe, guidance, control, electrical, and perhaps biomedical (if living organisms are aboard) systems. The *physical quantity, property, or condition to be measured is converted to an electrical current or voltage by a transducer.* Transducers take many forms, some of which are:

1. resistive units whose resistance varies with temperature or physical strain;
2. thermocouples whose voltage/current output varies with temperature;
3. moving coil devices whose voltage output and/or phase varies with angular rotation;
4. acceleration sensing devices whose output is a function of the rate of change of motion;
5. pressure-sensing devices whose resistance or voltage output changes with pressure differences; and so on.

Each transducer may have a different basic output range, and it is necessary to *scale* the output of each so that individually or as a group they will be within the range of the remainder of the telemetering system. *Signal conditioners* are used for this purpose. These signal conditioners serve either to reduce the transducer output by attenuators or increase it by amplifiers so that all inputs to the analog multiplexer are in the same range.

MULTIPLEXER-PROGRAMMER SUBSYSTEM

In a typical space vehicle the *multiplexer-programmer subsystem combines, times, and encodes analog and digital data into a composite signal which can be transmitted to a receiving location.* A *timer,* con-

Figure 16-1 A general telemetering system

sisting of counters (Ch. 10) and gates (Ch. 3 and Ch. 4), is the control point of this subsystem. It develops the timing needed to sample each analog and/or digital input, converts analog inputs to digital values, loads the individual bits into the memory (Ch. 12) in the form of digital *words,* combines the words into *frames,* and supplies the total information as a series of synchronized frames to the *transmitter subsystem.* Digital inputs to the multiplexer-programmer are supplied to the *digital multiplexer* (Ch. 12), where, under control of the timer, they are sampled periodically and placed in memory. Switch positions and discrete outputs from guidance and control equipment that can be used to perform space vehicle positioning are examples of digital inputs.

Each of the conditioned transducer outputs is fed to the analog multiplexer, which also samples at the proper time under control of the timer. Analog multiplexers function in a manner similar to digital multiplexers, except that their outputs are nearly exact replicas of the magnitude of their inputs.

The *analog-to-digital converter* changes each analog signal from the analog multiplexer into a group of binary digits whose binary value is equivalent to the analog value. All analog and digital information is stored in memory until a transmittable group (one or more frames) is available. The combined information from the space vehicle is then transferred to the transmitter subsystem for transmission to the display location.

TRANSMITTER SUBSYSTEM

Information presented to the transmitter subsystem is in digital form, which is not easily transmitted via wire lines or radio methods; however binary 1s and 0s can be converted to a change in signal amplitude, phase, frequency, or pulse width, then supplied to the transmitter, Conversion is accomplished by the *code converter.* The *transmitter prepares the data for transmission* so it can be sent out either by wire lines or radio.* The output of the transmitter, then, is a signal that represents, in a given period of time, the combined measured parameters of the space vehicle.

Receiving System Description

At the data display location, transmitted information is received, recovered, separated into its individual measurement channels, and displayed or recorded as required.

---

*Transmitter operation is non-digital and therefore is not discussed in this book.

RECEIVER SUBSYSTEM

The *receiver* performs the necessary selection and amplification of the transmitted signals. In space vehicle telemetry, the received signals may be very weak, and complex receiving equipment may be required to recover and amplify the information satisfactorily. A *detector* converts the received signal into digital form, separating the transmitted timing and data for use in the demultiplexer/decoder.

DEMULTIPLIXER/DECODER SUBSYSTEM

In the demultiplexer/decoder subsystem the received and separated data is reconstituted for storage and display. The extracted timing is used to shift each unit of information into the shift register (Ch. 11), and to operate the various controlling counters (Ch. 10). As data is shifted into the shift register, the synchronizing pattern that accompanies each word and/or frame is recognized, and the counter is synchronized to indicate which frame or word is being received. The word counter gates each word into the storage register, where it is made available to the display devices (Ch. 14) and recorders. The next word may be shifted in while the existing word is being displayed. The display of information varies between the coding systems. Analog data is converted by the digital-to-analog converter and fed to display devices, whereas digital data is directly displayed. Recorders are commonly supplied so that received data may be stored for future use.

DISPLAY SUBSYSTEM

The end result of the data put into the general telemetering system is seen at the display device. For example, analog measurements of physical parameters inside or outside of a vehicle somewhere in space are displayed on gauges and meters so that vehicle operation may be closely monitored, second-by-second. Switching operations and commands to steering rockets are displayed instantly—as they happen. As this occurs, various recorders and plotting devices retain such telemetered information input for future study and incorporation into design improvements. There is no doubt that the ability to measure physical parameters remotely by use of telemetry has been especially influential in our rapid progress in the exploration of space, and many land-based, commonplace but necessary applications hold such promise as well.

Applications of Telemetry

A partial list of telemetry applications must include:

1. protection systems, such as fire and burglar alarms;
2. weather balloons;
3. natural gas pipeline monitoring;
4. highway and air traffic monitoring;
5. remote buoys gathering oceanographic information;
6. biomedical monitoring of patients;
7. industrial control systems, like chemical batching, sewage effluent discharge, electrical power distribution, and so on almost ad infinitum.

Only the user's imagination may limit the use of telemetry in the modern world.

16-2 DATA COMMUNICATION

Definitions

The term *data communication* implies the movement of information in some type of machine language. Data communication is usually concerned with machines that exchange information via some link, whether it be wires, radio-waves, or optical beams. The machines involved may be a simple as a teletypewriter or as complex as a computer, and the data communicated may be simple business records like the water company bills, or complex trigonometric calculations that concern the operation of a manned space vehicle.

Data communication is allied to telemetry in that information is transmitted in both cases; however, in data communication, the volume and speed of transmission are usually much lower. Many data communication devices operate at regular audio frequencies (frequencies that may be heard by the human ear) via regular voice-grade telephone lines. Thus high-speed operation becomes difficult due to the limited frequency response characteristics of such lines.

Data communicated from one machine to another is in the form of 1s and 0s and, as in telemetry, must be converted. The techniques mentioned in the transmitter subsystems section are also used in data communications, but the device used in data communication has a special name—*modem*. The term modem is a contraction of the words *modulator/demodulator, which defines the process that is used to prepare data for transmission.* The 1s and 0s of the message to be transmitted are used to modulate (change) some characteristic—phase, frequency, or amplitude—of a continuous carrier signal furnished by the transmitting device. At the receiving end, the signal is *demodulated* (the modulation is removed) and used to actuate the display device. Detailed discussions follow.

Data Communication   455

The Modem

A simple modem is shown in Fig. 16-2 in block diagram form. Frequency shift (FS) transmission is used, whereby the frequency of the output signal is different for a mark (1) than it is for a space (0). The FS transmitter of Fig. 16-2 contains an *oscillator* to generate the basic carrier frequency, and a *dc keying switch* to shift the carrier frequency when a mark is to be generated. Remote turn-off of the carrier may be accomplished if desired. The basic carrier frequency in the Fig. 16-2 example is about 30,000 Hz (cycles per second). This basic frequency is supplied to the modulator, where a non-varying frequency is supplied from the RF oscillator and demodulator to translate the 30,000 Hz FS signal to approximately 1700 Hz. Frequency translation is used to reduce distortion at high data rates and to generate a stable transmitted signal. An *output amplifier* increases the power of the FS signal to a level compatible with telephone-line operation, and a *transmitter filter* is used to reduce interference.

At the receiving end, the incoming signal is fed to a *preamplifier* in the RF oscillator and demodulator. Here it is amplified and translated to a higher frequency in the *modulator* for reasons similar to those previously mentioned. A *receive* filter removes *extraneous components of the signal*, and the signal is sent to a *limiting amplifier* in the *FS receiver*. Variations in signal amplitude are removed in the *limiting amplifier* and *mark-and-space frequencies* are detected in the *discriminator*. If the incoming signal falls below a preset level, *signal-squelch* is provided to *disable the discriminator. Loss of carrier signal will actuate the carrier detector alarm*. Next, an output circuit converts the detected marks and spaces into the dc levels needed to operate the display device.

One important point should be made at this time. Note that very little digital logic is used in the modem; other electronic techniques borrowed from the communications field are applied in order to accomplish digital communication. However, the digital technician need not be alarmed for many modems are constructed with replaceable assemblies, just as digital units. The instruction manual provides techniques, which, used in conjunction with provided test points, usually enable one to repair malfunctioning units quickly and completely. The use of LSI and VLSI assemblies is further simplifying the troubleshooting task. Complete modems are now available on a single IC.

Applications of Data Communication

Current data communication applications include:

1. the interconnection of law-enforcement data files;
2. computer-assisted instruction, where classroom terminals are used to communicate directly with computers;

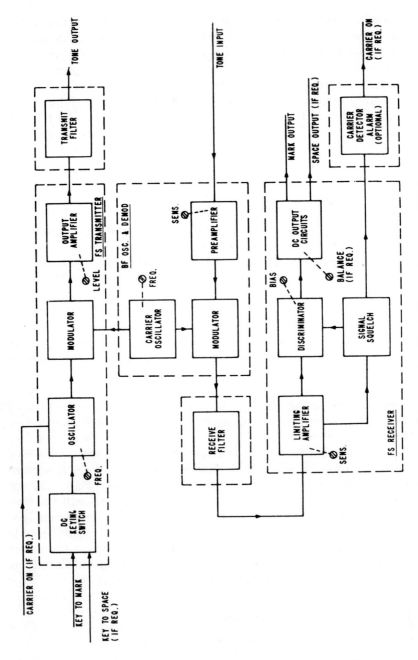

Figure 16-2 Data modem block diagram (courtesy RFL Industries, Inc.)

3. remote stock-brokerage offices;
4. computer time-sharing;
5. medical data-bank interconnection;
6. confirming travel reservations via remote-to-central computer;
7. verification of credit card accounts.

These are but a few of the present-day applications of data communications. The future leaves much to be discovered and applied in this field.

16-3 DIGITAL TEST EQUIPMENT

Digital testing equipment is different from other types of test equipment in that it directly displays in decimal form the parameter being measured or the signal being supplied. A digital voltmeter shows the voltage being measured in the types of displays discussed in Chapter 14, whereas a conventional moving coil voltmeter shows the voltage being measured by the position of a pointer on a scale.

Both conventional and digital equipment afford high accuracy, ruggedness, convenience, and utility. In addition, the digital measuring and testing equipment offers the special advantages of speed and direct numerical readout. These devices do have a high initial cost, are excessively influenced by ac fields, and are, in some cases, more difficult to maintain. Except for their high initial cost, the other two disadvantages can be bypassed by good design and application practices.

The General Digital Instrument

As indicated in Ch. 10, the basis for most direct-reading digital instruments is the BCD counter. The count in each decade is decoded and displayed in terms of the parameter(s) being measured. Figure 10-18 shows the general concepts of a device that counts events per unit time, such as pulses per second, cycles per second, and the like. Each decade in the BCD counter has an indicator associated with it, and the number of decades in the counter is determined by the number of significant figures to be displayed. Typical decade counters have been shown in Figs. 10-13 and 10-14. Decoders are discussed in Ch. 14.

Because a digital counting device is the basis for most digital measuring instruments, it becomes necessary to convert the input parameter(s) being measured to units that may be counted. Time-dependent waveforms, such as the frequency of pulses and ac voltages, are directly measured with the counter. (The counter has internal shaping circuits to convert input waveforms to pulses that may be counted.) The measurement of other parameters, such as *magnitudes* of voltages, currents, and light intensities, requires conversion to time-dependent, counter-compatible parameters. In other words, *some type of analog-to-digital conversion must be performed.*

Analog-to-digital converters are a subject unto themselves, and only the simple ramp-type converter will be discussed here. The concept of digital-measurement of dc voltage magnitudes was shown in Fig. 10-19 and is expanded and modified in Fig. 16-3 to demonstrate the use of digital techniques in combination with analog circuits.

The Digital Voltmeter

Note that the major functions exist in both Fig. 10-19 and Fig. 16-3, although the greater detail of Fig. 16-3 makes the operation more evident. The voltage to be measured is applied to the input terminals, where it is either manually or automatically magnitude-adjusted to fall within a predetermined range that is compatible with the *level* or *zero comparator devices.* The amount of division or amplification needed to establish this compatible range determines the position of the decimal point on the digital display.

The two comparators are used to determine the number of clock pulses that are provided to the counter. Magnitude-adjusted input is compared with an internally generated ramp signal in the level comparator; *a reference voltage* (0 volts, or ground) is compared with the ramp in the *zero comparator.* The level comparator pulses when the magnitude-adjusted input and the ramp signal are equal in magnitude; the zero comparator pulses when the ramp is equal to 0 volts. Typical comparator waveforms are shown in Fig. 16-3b.

During the time the ramp is changing from +10 volts through 0 volts to –10 volts, the input (magnitude-adjusted) is being examined. As shown in Fig. 16-3, when coincidence between the ramp and the level comparator input occurs, a pulse is generated which starts the clock and simultaneously enables the counter. The ramp continues downward, and when the zero comparator is enabled by coincidence between ground (0 volts) and the ramp, the generated pulse stops the clock and disables the counter. Thus the counter operates for a period that depends on the difference between the input voltage and zero volts.

The polarity of the input voltage is determined by a circuit, called the *polarity detector,* that determines the relationship between the input coincidence pulse and the zero coincidence pulse. In Fig. 16-3, the positive input causes the input coincidence pulse to occur first, so the polarity detector provides an input to the digital display that illuminates the + sign. Occurrence of the zero comparator pulse first indicates that the input voltage is –, so the – indicator is illuminated.

A *sample control circuit* determines the number of samples to be observed each second by controlling the frequency of ramp occurrence.

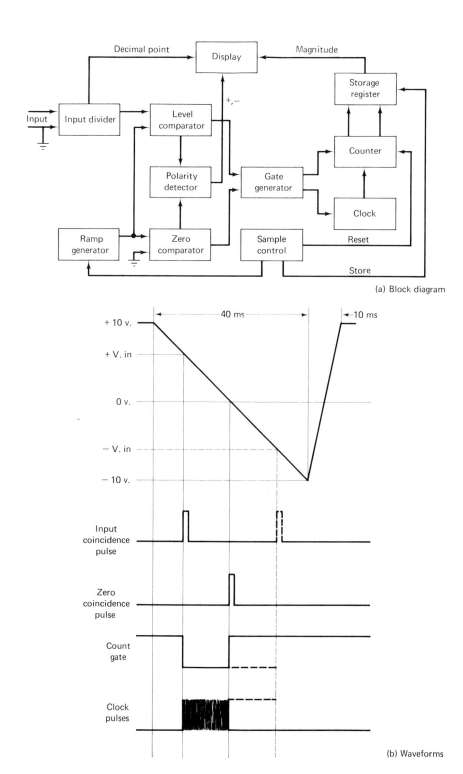

Figure 16-3 A digital voltmeter

The counter is reset following each sample period, but the contents have already been transferred to the *storage register* by the *store pulse,* so the counter is free to establish a new count if necessary.

Digital voltmeter resolution (precision) is determined by the relationship between the slope of the ramp and the frequency of the clock. For example, if the ramp changes at the rate of 500 V/sec, and clock frequency is 500 kHz, the total count displayed corresponds to input in millivolts.

Many other techniques for measuring electrical parameters are in use. Resistance, current, phase angle, capacitance, inductance, and the like, all fall easily within the scope of digital measurement. However, one common denominator exists: the parameter *must* be converted from analog to digital form. There are many excellent texts on this subject that may be consulted if more detailed information is desired.

16-4 MICROWAVE OVEN CONTROL

Functional Operation

Adaptation of microcomputer principles to high-sales-volume household devices is seen in Figure 16-4. The microwave oven has become a common appliance, and the microprocessor and associated supporting hardware and software make it a versatile addition to the kitchen.

Control of the microwave oven is accomplished using a 4-bit microcomputer contained on a single chip. The only additional hardware required is that necessary to tailor the input/output circuits to the microcomputer chip circuits. Review of Figure 16-4 will show that all traditional computer functions are present. *Input* and *output* devices interface the microcomputer with the outside world. *Memory* and *arithmetic/logic* functions are easily identified as part of the microcomputer chip, while the remainder of the chip accomplishes the *control* function. It should be noted that both RAM and ROM exist as part of the microcomputer chip, thereby making applications limited only to the software that is stored in ROM. Actually, this microcomputer chip is a standard item programmed by the manufacturer to function as a microwave oven controller. The same chip, with different software in ROM, is used in many other applications.

The complete controller acts as a special-purpose digital computer, performing its control functions as directed by the ROM-stored program. Software implements cooking by supplying seven separate timers (four cooking timers, a delayed-start timer, an alarm timer, and a time-of-day clock). All timers are "adjustable" from the control panel. The timing related to the actual operation in process is available on the time

Figure 16-4 Microwave oven control (courtesy Texas Instruments)

display, while the mode of operation is shown on the status display. Temperature cooking is also available by use of a temperature-sensing probe. Self-test programming is included for the convenience of service personnel.

The *control panel* is used to enter all numerical information (time-of-day, cycle duration, cooking speeds, etc.), select specific cycles, and reset alarms and errors. When a key on the control panel is pressed, the resultant switch closure is encoded by the *encoder* in the *input interface*. A 4-bit word is generated by the interface and routed to the microcomputer for processing. Timing and temperature probe signals are also converted to 4-bit words and switched by the *multiplexer* under program control for use by the microcomputer.

Subfunctions within the microcomputer operate in much the same manner as those functions in the microprocessors and microcomputers discussed in Chapter 15. All inputs to the microcomputer result from actions in the *input interface,* and all outputs are fed to the *output interface* as either control signals from internal *buffers* to actuate *power* and *status drivers* or time-related information from an internal *encoder* to be decoded and displayed on the *time display*. Input information is processed in the *ALU* and *accumulator register* under control of the preprogrammed *ROM*. Results of processing are routed to *latches* via either the accumulator or the *Y register* and thence to the *output buffers* and *encoder*. The *X register* and *RAM* are used for routing and temporary storage of program information and data.

Detailed Operation

An example of a "software" timer is seen in Figure 16-5. Each of the timers could have been implemented in hardware, but it will be seen that the software timer is also quite practical. The scheme employed is much like the delay routines used in earlier discussions of serial input/output.

A common subroutine that can be used to set any of the timers is assumed to be available. Such a routine is quite simple and may be seen in outline form in Figure 16-5a. Although it is possible to perform the timing functions directly in minutes and seconds, the set routine used in this discussion converts minutes and seconds into a total number of seconds and stores it in a temporary storage location that is mnemonically named "TS." The timer routine will constantly refer to the number stored in TS.

Actual use of the timer routine is accomplished by depressing the START button on the control panel of the microwave oven. The first step in the routine (Figure 16-5b) determines whether TS is 0. If TS is 0, no timing is necessary during this cooking sequence, and the routine

Microwave Oven Control

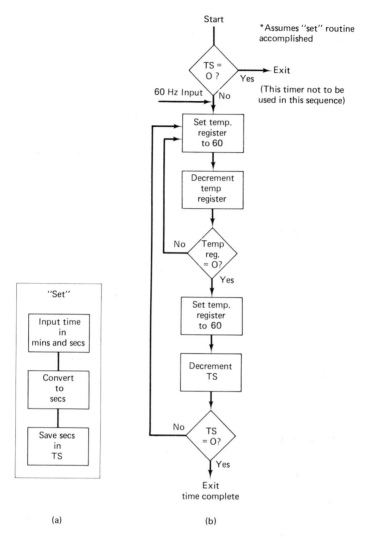

Figure 16-5 Timer flowchart

exits to the next timing sequence. If TS is not 0, however, a temporary register is loaded with the decimal number 60 and the 60-Hz power-line frequency is applied to the input of the timer routine. Each time the 60-Hz input generates a timing pulse (60 times per second), the temporary register is decremented by 1 and its contents compared to 0. If the temporary register has not yet reached 0, the routine returns to the decrement step to await the next 60-Hz input. If 0 has been reached (occurs only once each second), the temporary register is reset to 60 and TS is decremented by 1.

TS is examined to determine if it has reached 0, and if not, control is returned to the temporary register-decrementing step to continue counting. If 0 has been reached, the timer has completed its function, and the routine is exited with an indication that cooking time has expired.

The microwave oven executes its five cooking programs in a fixed sequence, as shown in Figure 16-6. Each of the programs has an associated timer, and their times may be set in any order. However, when the oven door is closed and the START control is depressed, execution will proceed in the order shown in Figure 16-6.

Up to 60 minutes is available from the *defrost timer,* followed by an equal amount of *automatic temperature settling time.* During the defrosting period, the oven's magnetron (source of cooking heat) is cycled on and off at 15-second intervals to provide maximum defrosting with minimum cooking of the food. The 50% duty cycle is an integral part of the defrost timer's ROM-stored program.

Cook cycle 1 can exist for up to 60 minutes maximum, with selectable duty cycle from 100% to 10% in 10% intervals. Upon completion of cook cycle 1, *Cook cycle 2* is entered. Selectable duty cycle is also provided in cook cycle 2. However, maximum cooking time is limited to 30 minutes. A *broil timer* (up to 30 minutes) is also provided by the microwave oven controller. A separate heating element is used by the broil timer to brown the food. Upon completion of *all* cycles, an alarm is sounded to inform the operator.

A delayed start capability is also provided by the microwave oven controller. The microcomputer computes the time to start the cooking sequence by subtracting the total of all cooking times from the desired end time. Thus, a complete cooking program may be entered at any time during the day, the food placed in the oven, and the operator may leave the unit unattended. At the proper time, the oven goes into action and prepares the food so that it will be ready at the calculated time.

Temperature-controlled cooking is also a feature of the microwave oven controller. A temperature probe is placed in the food to be cooked, and the temperature of the probe converted by the microcomputer into a digital value that may be used to determine completion of the cooking cycle. The programmed temperature is only approximate, however, and the operator must depend on experience to obtain maximum results.

Any of the timer's contents are accessible during the cooking cycle. Depression of the appropriate control button will display the timer's contents for 3 seconds and the cooking cycle will continue with-

Microwave Oven Control

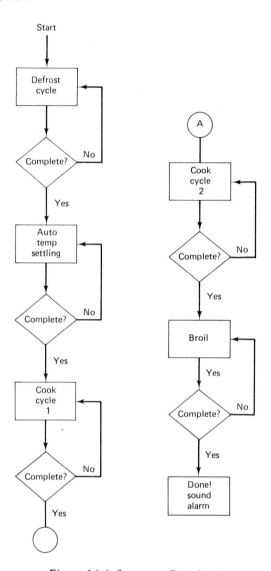

Figure 16-6 Sequence flowchart

out interruption. Furthermore, if the oven door is opened during cooking, the cooking cycle halts and the time remaining is displayed. When the door is closed, cooking resumes where it was interrupted upon depression of the START button. The program is lost only due to unanticipated power failure or use of the STOP button.

The microprocessor controlled microwave oven, then, has been perhaps one of the first consumer-oriented appliances to appear in the home. Microprocessor versatility has provided the homemaker with precise control of the cooking process. This is but one of the labor- and time-saving applications now available to the consumer. As the old saying goes—"The best is yet to come!"

16-5 SUMMARY

Each system, subsystem, or assembly described in this chapter should bring to mind the basic circuit elements that could be used to implement the function under discussion. For example, storage functions should recall the registers of Ch. 11 and the memories of Ch. 12. The counters of Ch. 10 should be visualized when timing and control are discussed. Selective control and routing of logic signals are obviously the domain of the gates of Chs. 3 and 4.

When these types of mental responses occur, you are on your way toward becoming a knowledgeable member of the digital logic profession. You will find an almost endless number of opportunities, depending only on your desires and background. Circuitry design, logic design, field engineering and maintenance, systems analysis and troubleshooting, and computer programming are but a few. WELCOME TO THE NEW WORLD OF DIGITAL LOGIC.

APPENDIX A

POSTULATES, THEOREMS, AND LAWS OF BOOLEAN ALGEBRA

POSTULATES

$0 \cdot 0 = 0$ (P–1)
$0 \cdot 1 = 0$ (P–2)
$1 \cdot 0 = 0$ (P–3)
$1 \cdot 1 = 1$ (P–4)
$0 + 0 = 0$ (P–5)

$0 + 1 = 1$ (P–6)
$1 + 0 = 1$ (P–7)
$1 + 1 = 1$ (P–8)
$0 = \bar{1}$ (P–9)
$1 = \bar{0}$ (P–10)

THEOREMS

$A \cdot 0 = 0$ (T–1)
$0 \cdot A = 0$ (T–2)
$A \cdot 1 = A$ (T–3)
$1 \cdot A = A$ (T–4)
$A \cdot A = A$ (T–5)
$A \cdot \bar{A} = 0$ (T–6)
$A + 0 = A$ (T–7)

$0 + A = A$ (T–8)
$A + 1 = 1$ (T–9)
$1 + A = 1$ (T–10)
$A + A = A$ (T–11)
$A + \bar{A} = 1$ (T–12)
$\bar{\bar{A}} = A$ (T–13)

LAWS OF BOOLEAN ALGEBRA

The Laws of Identity	$A = A$,	$\bar{\bar{A}} = \bar{\bar{A}}$
The Commutative Laws	$AB = BA$,	$A + B = B + A$
The Associative Laws	$A(BC) = ABC$,	$A + (B + C) = A + B + C$
The Idempotent Laws	$AA = A$	$A + A = A$
The Distributive Laws	$A(B + C) = AB + AC$,	$A + BC = (A + B)(A + C)$
The Laws of Absorption	$A + AB = A$,	$A(A + B) = A$
The Laws of Expansion	$AB + A\bar{B} = A$,	$(A + B)(A + \bar{B}) = A$
DeMorgan's Laws	$\overline{AB} = \bar{A} + \bar{B}$,	$\overline{A + B} = \bar{A}\bar{B}$

COMMON IDENTITIES OF BOOLEAN ALGEBRA

$A(\bar{A} + B) = AB$

$A + \bar{A}B = A + B$

$(AB)(A + B) = AB$

$(\overline{AB})(A + B) = A\bar{B} + \bar{A}B$

$\overline{AB} + \overline{\bar{A}B} = AB + \bar{A}\bar{B}$

$(A + B)(B + C)(A + C) = AB + BC + AC$

$(A + B)(\bar{A} + C) = AC + \bar{A}B$

$AC + AB + B\bar{C} = AC + B\bar{C}$

$(A + B)(B + C)(\bar{A} + C) = (A + B)(\bar{A} + C)$

APPENDIX B

MIL-STD-806B EXTRACTS

AND. The symbol shown below represents the AND function.

The AND output is high (H) if and only if *all* the inputs are high.

Input		Output
A	B	F
L	L	L
L	H	L
H	L	L
H	H	H

OR. The symbol shown below represents the INCLUSIVE OR function.

The OR output is high (H) if and only if *any one or more* of the inputs are high (H).

Input		Output
A	B	F
L	L	L
L	H	H
H	L	H
H	H	H

STATE INDICATOR (Active). The presence of the small circle symbol at the input(s) or output(s) of a function indicates:

(a) *Input Condition.* The electrical condition at the input terminal(s) that controls the active state of the respective function.

(b) *Output Condition.* The electrical condition that exists at the output terminal(s) of an activated function.

A small circle(s) at the input(s) to any element (logical or non-logical) indicates that the relatively low (L) input signal activates the function. Conversely, the absence of a small circle indicates that the relatively high (H) input signal activates the function.

A small circle at the symbol output indicates that the output terminal of the activated function is relatively low (L). This small circle should never be drawn by itself on a diagram.

EXCLUSIVE OR. The symbol shown below represents the EXCLUSIVE OR function.

The EXCLUSIVE OR output is high if and only if any one input is high (H) and all other inputs are low (L).

Input		Output	
A	B	F = A (H) and B (L) or B (H) and A (L)	
L	L	L	
L	H	H	
H	L	H	
H	H	L	

MIL-STD-806B Extracts

AND	OR	\multicolumn{3}{c}{Table of Combinations}
		A	B	X
A─⊃─X (A,B no bubbles)	A─⊃○─X (NOR)	H H L L	H L H L	H L L L
A─○⊃─X (A bubble)	A─⊃○─X	H H L L	H L H L	L L H L
A─⊃○─X (B bubble)	A─⊃○─X	H H L L	H L H L	L H L L
A─○⊃○─X (both bubbles)	A─⊃○─X	H H L L	H L H L	L L L H
A─○⊃○─X─output bubble	A─⊃─X (OR)	H H L L	H L H L	H H H L
A─⊃○─X	A─○⊃─X	H H L L	H L H L	H L H H
A─○⊃○─X	A─⊃─X	H H L L	H L H L	H H L H
A─⊃○─X	A─○⊃○─X	H H L L	H L H L	L H H H

FLIP-FLOP. The FLIP-FLOP (FF) is a device that stores a single bit of information. It has three possible inputs,* SET (S), CLEAR (RESET) (C), and TOGGLE (TRIGGER) (T), and two possible outputs, 1 and 0.** When not used, the TRIGGER input may be omitted.

Polarity. The two outputs are normally of opposite polarity. A "1" is stored in the FF when the "1" output level is active and the "0" output level is inactive. A "0" is stored in the FF when the above condition is reversed.

States. The FF assumes the "1" state when an active signal appears at the S input regardless of the original state. It assumes the "0" state when an active signal appears at the C input regardless of the original state. It reverses its state when an active signal appears at the T input. There are several possible variations to normal FF operations, depending on the response of the device when active inputs are applied simultaneously to more than one input.

There is a growing tendency among amnufacturers to use parts of the IEEE 91-1973 Standard for graphic symbols. This standard modifies logic symbols in three ways. First, it allows the use of rectangular symbols and qualifiers in place of the MIL-STD-806B distinctive shapes. For example, an AND gate could appear in any of the following forms:

Second, a *polarity indicator* (⊳) has been added. The polarity indicator is used as shown below.

High level sensitive

ACTIVE HIGH inputs and outputs are indicated by the *absence* of the polarity indicator (⊳) or negation symbol (o).

---

*Other input designations such as D and J-K are commonly used. These inputs are defined in Ch. 8.

**The 1 output is commonly called Q, and the 0 output is commonly called $\bar{Q}$.

Low level sensitive

ACTIVE LOW inputs and outputs are indicated by the *presence* of the polarity indicator (◺) or negation symbol (o).

Third, a symbol has been added to differentiate between inputs that are sensitive to level changes and those that are sensitive to a transition (edge sensitive). Use of the dynamic (edge sensitive) symbol (▷) is shown below.

Low-to-High Edge Sensitive

EDGE SENSITIVE (Dynamic) inputs are indicated by the presence of the dynamic indicator symbol (▷).

High-to-Low Edge Sensitive

Logic diagrams using both standards can be expected.

APPENDIX C

EQUIVALENT LOGIC SYMBOLS

Standard AND gate symbol and equivalents

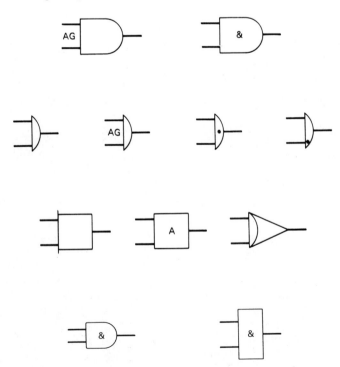

Standard OR gate symbol and equivalents

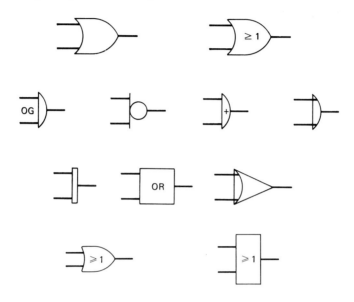

Standard NAND gate symbol and equivalents

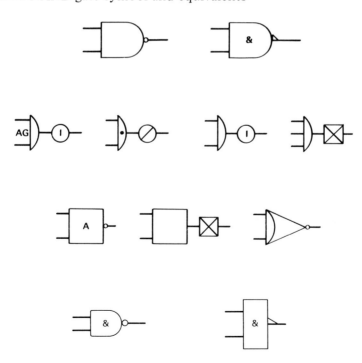

Standard NOR gate symbol and equivalents

Standard INVERTER symbol and equivalents

Standard AMPLIFIER symbol—common throughout symbology

*May not appear unless attached to another symbol.

Equivalent Logic Symbols

Standard FLIP-FLOP symbol and equivalents

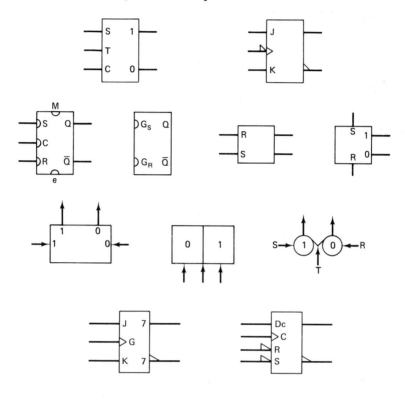

APPENDIX D

LOGIC FAMILIES

INTRODUCTION

In the past, schematic diagrams of assemblies used in digital logic systems were quite important. Now, however, as the level of integration becomes more and more complex, there is very little need for schematic diagrams. Of more importance are the characteristics of the general types, or "families" of ICs being used. Such parameters as what voltages and currents are expected, and what input/output characteristics are available are more appropriate. These general parameters, plus partial schematic diagrams to show the general concept of each of the "logic families" are provided here. If additional information about specific ICs or logic families is needed, the manufacturers' specification sheets should be consulted.

THE TTL FAMILY OF LOGIC CIRCUITS

TTL means *Transistor-Transistor-Logic,* and refers to the means by which inputs are combined in the logic circuit. Earlier logic devices used resistor-transistor and diode-transistor logic. TTL, however, has generally replaced the older RTL and DTL structures.

There are a number of sub-families within the TTL family of logic circuits. The regular TTL family is generally recognized by the prefix numbers 54 or 74. For example, a common TTL gate IC is the 7320

Logic Families

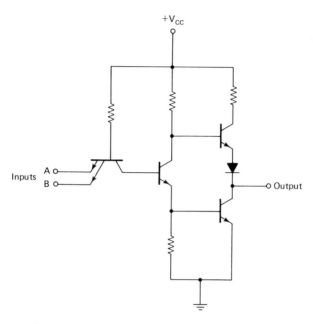

Figure D-1 Schematic diagram of simple TTL gate

dual 4-input NAND gate. (54XXX and 74XXX ICs perform the same functions. The 54 and 74 prefixes describe temperature ranges only.) This series of TTL ICs are widely available and are relatively inexpensive. The upper frequency of operation is around 35 MHz. Delay from input to output is about 10 nanoseconds and power consumption is near 10 milliwatts per gate. A logic 0 input/output is represented typically by about 0.1 volts, although values as high as 0.8 volts will be recognized as logic 0. A typical logic 1 representation is 3.6 volts, although values as high as 5 volts and as low as 2.4 volts are acceptable. Thus the regular TTL family is a very versatile, easy-to-use group of ICs. Figure D-1 shows a schematic diagram of a simple TTL gate.

Other TTL subfamilies provide improvements. The *low-power TTL* family, identified as 54/74LXXX, provide the same functions as the regular family. They consume only about 1/10 as much power, though. Such power reduction is, unfortunately, accompanied by a greater delay (about 33 nanoseconds) and lower frequency capability (near 3 MHz maximum). The *high-power family* (54/74HXXX) trades power for speed and frequency response. Consuming near 22 milliwatts per gate, the propagation delay is only 6 nanoseconds. Frequencies greater than 50 MHz can be used.

Figure D-2 Low power Schottky TTL ICS

Speed improvement is also accomplished by using *"Schottky" TTL*, identified as 54/74SXXX. These TTL ICs use special internal Schottky diodes to speed the propagation delay to only about 3 nanoseconds. Frequency response rises to about 125 MHz, but power requirements approach 19 milliwatts. Applying low-power techniques results in the 54/74LSXXX series. *Low-power Schottky TTL* ICs use only about 2 milliwatts per gate with delays near 10 nanoseconds. The upper frequency limit of operation is near 45 MHz. Figure D-2 shows low-power Schottky TTL IC concepts.

A new technique that can be applied to any of the TTL families is that of enabling or disabling the output of the circuit completely. In other words, a logic 0 state, a logic 1 state, and a disable state can exist. The disabled state places the output of the logic circuit in a high-impedance condition, which effectively disconnects it from the system. *Three-state logic* is becoming quite popular in computer applications where a number of devices must be connected to a common point. The three-state capability allows the enabling of only the device(s) required. Thus power requirements are reduced and operating speed is increased.

Logic Families

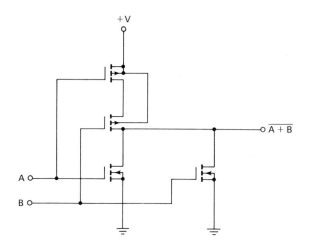

Figure D-3 Simple gate in CMOS form

THE MOS FAMILY OF LOGIC CIRCUITS

MOS (Metal-Oxide Semiconductor) logic ICs have a different manufacturing method than TTL ICs. It results in simpler, cheaper, smaller, and lower-powered logic assemblies. Unfortunately, MOS ICs are only about 1/10 (1 MHz) as fast as TTL. In lower speed applications, though, MOS logic is often the best choice.

Three types of MOS ICs are found in modern digital systems. The type is determined by how they are manufactured. *P-type* and *N-type* (*PMOS* and *NMOS*) both do the same job, but with different polarities of power supplies. NMOS is faster than PMOS because of the material used in construction. PMOS and NMOS ICs are found most often in high-density LSI/VLSI ICs such as the complete system on a single chip. Microprocessors and large-scale memories are typical examples.

Combining NMOS and PMOS results in *CMOS (Complementary MOS)*, a popular logic family. This is faster and uses less power then NMOS or PMOS, with an upper frequency limit near 5 MHz. CMOS is more complex to manufacture and uses more chip space than either NMOS or PMOS. However, it is easier to use and more versatile than the other logic types. It can operate with any supply voltage from as low as 3 volts to as high as 15 volts. A logic 0 is generally 0 volts and a logic 1 is the same voltage as the supply voltage. CMOS is used generously in portable equipment that requires long battery life. Figure D-3 shows a simple gate in CMOS form.

Figure D-4 ECL gating concepts

OTHER LOGIC FAMILIES

When extremely high speed is required of logic circuits, *ECL (Emitter-Coupled Logic)* is used. This is a variation of the same techniques used in TTL design and manufacture. Propagation delays are on the order of 1 to 2 nanoseconds, while the upper frequency limit is near 1000 MHz. Logic 0 and logic 1 levels are non-standard, but may be converted to standard TTL and CMOS levels quite easily. Often ECL ICs are used when high frequency requirements exist in a system, and TTL or CMOS are used in less critical parts of the system. Figure D-4 shows the concepts of ECL gates.

Another simple digital IC manufacturing technique is called I^2L (*I squared L*). Again it is a variation of standard semiconductor techniques and results in a simple, low power device. It does require one more manufacturing step when compared with MOS and is more costly therefore. Propagation delays are adjustable and are related to power requirements. If long delays can be tolerated (20-30 nanoseconds), then lower power can be used. This type of IC is quite versatile, but does require additional external components for adjustment.

Other IC techniques continue to appear. The reader should be on the alert for new technology, and use manufacturers' technical data generously.

APPENDIX E

NUMBER TABLES

Powers of Two

2^n	n	2^{-n}
1	0	1.0
2	1	0.5
4	2	0.25
8	3	0.125
16	4	0.062 5
32	5	0.031 25
64	6	0.015 625
128	7	0.007 812 5
256	8	0.003 906 25
512	9	0.001 953 125
1 024	10	0.000 976 562 5
2 048	11	0.000 488 281 25
4 096	12	0.000 244 140 625
8 192	13	0.000 122 070 312 5
16 384	14	0.000 061 035 156 25
32 768	15	0.000 030 517 578 125
65 536	16	0.000 015 258 789 062 5
131 072	17	0.000 007 629 394 531 25
262 144	18	0.000 003 814 697 265 625
524 288	19	0.000 001 907 348 632 812 5
1 048 576	20	0.000 000 953 674 316 406 25

Appendix E

Decimal-Binary and Binary-Decimal Conversion Tables

Decimal	Binary	Decimal	Binary	Decimal	Binary	Decimal	Binary
0	00000000	50	00110010	100	01100100	150	10010110
1	00000001	51	00110011	101	01100101	151	10010111
2	00000010	52	00110100	102	01100110	152	10011000
3	00000011	53	00110101	103	01100111	153	10011001
4	00000100	54	00110110	104	01101000	154	10011010
5	00000101	55	00110111	105	01101001	155	10011011
6	00000110	56	00111000	106	01101010	156	10011100
7	00000111	57	00111001	107	01101011	157	10011101
8	00001000	58	00111010	108	01101100	158	10011110
9	00001001	59	00111011	109	01101101	159	10011111
10	00001010	60	00111100	110	01101110	160	10100000
11	00001011	61	00111101	111	01101111	161	10100001
12	00001100	62	00111110	112	01110000	162	10100010
13	00001101	63	00111111	113	01110001	163	10100011
14	00001110	64	01000000	114	01110010	164	10100100
15	00001111	65	01000001	115	01110011	165	10100101
16	00010000	66	01000010	116	01110100	166	10100110
17	00010001	67	01000011	117	01110101	167	10100111
18	00010010	68	01000100	118	01110110	168	10101000
19	00010011	69	01000101	119	01110111	169	10101001
20	00010100	70	01000110	120	01111000	170	10101010
21	00010101	71	01000111	121	01111001	171	10101011
22	00010110	72	01001000	122	01111010	172	10101100
23	00010111	73	01001001	123	01111011	173	10101101
24	00011000	74	01001010	124	01111100	174	10101110
25	00011001	75	01001011	125	01111101	175	10101111
26	00011010	76	01001100	126	01111110	176	10110000
27	00011011	77	01001101	127	01111111	177	10110001
28	00011100	78	01001110	128	10000000	178	10110010
29	00011101	79	01001111	129	10000001	179	10110011
30	00011110	80	01010000	130	10000010	180	10110100
31	00011111	81	01010001	131	10000011	181	10110101
32	00100000	82	01010010	132	10000100	182	10110110
33	00100001	83	01010011	133	10000101	183	10110111
34	00100010	84	01010100	134	10000110	184	10111000
35	00100011	85	01010101	135	10000111	185	10111001
36	00100100	86	01010110	136	10001000	186	10111010
37	00100101	87	01010111	137	10001001	187	10111011
38	00100110	88	01011000	138	10001010	188	10111100
39	00100111	89	01011001	139	10001011	189	10111101
40	00101000	90	01011010	140	10001100	190	10111110
41	00101001	91	01011011	141	10001101	191	10111111
42	00101010	92	01011100	142	10001110	192	11000000
43	00101011	93	01011101	143	10001111	193	11000001
44	00101100	94	01011110	144	10010000	194	11000010
45	00101101	95	01011111	145	10010001	195	11000011
46	00101110	96	01100000	146	10010010	196	11000100
47	00101111	97	01100001	147	10010011	197	10000101
48	00110000	98	01100010	148	10010100	198	11000110
49	00110001	99	01100011	149	10010101	199	11000111

BIBLIOGRAPHY

Baron, Robert C., and Albert T. Piccirilli. *Digital Logic and Computer Operations.* New York: McGraw-Hill, 1967.

Bartee, Thomas C. *Digital Computer Fundaments,* 4th edition. New York: McGraw-Hill, 1972.

Boyce, Jefferson C. *Digital Computer Fundamentals.* Englewood Cliffs, N.J.: Prentice-Hall, 1977.

Boyce, Jefferson C. *Microprocessor and Microcomputer Basics.* Englewood Cliffs, N.J.: Prentice-Hall, 1979.

Burroughs Corporation, Technical Training Department—Defense, Space, and Special Systems Group. *Digital Computer Principles,* 2nd edition. New York: McGraw-Hill, 1969.

Churchman, Lee W. *Survey of Electronics.* San Francisco: Rinehart Press, 1971.

Davenport, William P. *Modern Data Communication.* New York: Hayden, 1971.

Dietmeyer, Donald L. *Logic Design of Digital Systems.* Boston: Allyn and Bacon, 1971.

Gasperini, Richard E. *Digital Troubleshooting.* Los Altos, CA: Movonics Company, 1975.

Hall, Douglas V. *Microprocessors and Digital Systems.* New York, NY: McGraw-Hill Book Company, 1980.

Hoernes, Gerhard E., and Melvin F. Heilweil. *Introduction to Boolean Algebra and Logic Design.* New York: McGraw-Hill, 1964.

Karnaugh, M. "The Map Method for Synthesis of Combinational Logic Circuits," *AIEE Proceed.,* 593: November 1953.

Ketchum, Donald J. *Applications of Digital Logic.* Farmingdale, N.J.: Buck Engineering Co., 1966.

Lancaster, Don. *TTL Cookbook*. Indianapolis, IN: Howard W. Sams & Co., 1974.

Lancaster, Don. *CMOS Cookbook*. Indianapolis, IN: Howard W. Sams & Co., 1977.

Levine, Morris E. *Digital Theory and Practice Using Integrated Circuits*. Englewood Cliffs, N.J.: Prentice-Hall, 1978.

Maley, Gerald A. *Manual of Logic Circuits*. Englewood Cliffs, N.J.: Prentice-Hall, 1970.

Maley, Gerald A., and John Earle. *The Logic Design of Transistor Digital Computers*. Englewood Cliffs, N.J.: Prentice-Hall, 1963.

Malmstadt, H. V., and C. G. Enke. *Digital Electronics for Scientists*. New York: Benjamin, 1969.

Mandl, Matthew. *Fundamentals of Electronic Computers: Digital and Analog*. Englewood Cliffs, N.J.: Prentice-Hall, 1967.

Marcus, Mitchell P. *Switching Circuits for Engineers*, 2nd edition. Englewood Cliffs, N.J.: Prentice-Hall, 1967.

Phister, Montgomery, Jr. *Logical Design of Digital Computers*. New York: Wiley, 1958.

Sifferlen, Thomas P., and Vartan Vartanian. *Digital Electronics with Engineering Applications*. Englewood Cliffs, N.J.: Prentice-Hall, 1970.

Thomas, Harry E. *Handbook of Pulse-Digital Devices for Communications and Data Processing*. Englewood Cliffs, N.J.: Prentice-Hall, 1970.

Tocci, Ronald J. *Digital Systems: Principles and Applications*. Englewood Cliffs, N.J.: Prentice-Hall, 1977.

Veitch, E. W. "A Chart Method for Simplifying Truth Functions," *Ass. Computing Machinery, Proceed.,* 127: May 1952.

———. *Application Notes*. Plainfield, N.J.: Burroughs Corp., 1971.

———. *Application Notes*. Phoenix, Az.: Motorola Semiconductor Products, Inc., 1972.

———. *Application Notes*. Sunnyvale, Ca.: Signetics (a subsidiary of Corning Glass Works), 1972.

———. *Application Notes*. Worcester, Mass.: Sprague Electric Company, 1971.

———. *Digital Logic Handbook*. Maynard, Mass.: Digital Equipment Corp., 1971.

Index

A

Abacus 2
Accumulator 435
Active/activating input, FF 233
Adder (*See* Full adder, Half adder)
Addition
 binary 119-120, 349
 logical 127
Addressing (microprocessor) 436
 direct 436
 immediate 436
 register 436
ALU (Microprocessor) 424, 434-435
Algebraic to logic conversion 118
Algebraic simplification procedures 124
 (*See also* Logic simplification, algebraic)
Algorithm 202, 443
Alphameric codes
 ASCII 381
 ANSI 381
 Baudot 379
 EBCDIC 382
 Hollerith 384
 5-level 379
 7-level 381
 8-level 382
Amplifier 63, 476
Analog device 3
Analog circuits analysis concepts 7-10
Analog/digital converter 452, 458

AND
 function 47, 60-62, 83
 analogy 60
 applications 61
 gate
  symbol 14, 469, 474
  truth table 17
AND-OR expression 157
AOI gate 353
Arithmetic/logic circuits
 Full adder 349-353
Associative law 90
Asynchronous 248, 279-287

B

Babbage, Charles 2
Base, number system definition 19
Baudot, Jean 379
BCD codes 154, 372-379 (*See also* Binary
 Coded Decimal (BCD) Codes)
Binary addition 119-120, 349
Binary Coded Decimal (BCD) Codes
 Excess 3 (XS3) 377
 general 154, 372
 Gray 155, 374
 2-out-of-5 378
 2-4-2-1 374
 8-4-2-1 373
Binary counter (*See also* Counters)
 analysis 40-41

487

488 Index

definition 279
in mixed function circuit 263
logic diagram 38-40
logic signal source correlation 76-77
state table 41
Binary counting 22-24, 30-32
Binary number system 15, 22, 25
Binary-to-decimal conversion 27-29, 483
Bistable circuit 15, 33
Bit (binary digit) 310
Boolean algebra 16, 47, 86 467
Boolean algebra identities 108-119, 468
Boolean algebra laws 86-108, 468
Boolean algebra postulates 52, 467
Boolean algebra theorems 52-55, 467
Buffer register 318-320
Bus (microprocessor) 426-428
    address 427
    control 427
    data 427
Byte 426, 436

C

Cathode-Ray oscilloscope (CRO) 17, 36,
    216-218, 270
Clear (reset) 33-37
Clock
  input 242
  pulse 242
Code converters 388-404
  BCD-Binary 392-393
  BCD-Decimal 397-401
  binary-BCD 394-395
  binary-hexadecimal 404
  binary-octal 401
  decimal-BCD 391
  decimal-binary 391
  hexadecimal-binary 390
  octal-binary 389
Codes
  alphameric 379-385 (*See also* Alphameric
    codes)
  BCD 372-379 (*See also* Binary Coded
    Decimal (BCD) Codes)
  binary related 368-372
Combinational logic
  analysis (*See* Logic analysis, combinational)
  applications 79-83
  definition 45
  operations 47
  troubleshooting (*See* Logic troubleshooting, combinational)
Combinational/sequential logic analysis (*See*
  Logic analysis, combinational/sequential)
Combined functions
  AND/OR/NOT 71-72
  AND/OR-OR/AND 68

multiple AND 66-67
multiple OR 65-66
Commutative law 87-90, 468
Comparator 344-345, 458
Computer, digital 310
Computer word 426
Conjunction 47
Constants 48-49
Control (microprocessor) 424, 430-432
Conversion, algebraic to logic 118-119
Conversion, number system 24-29, 369-378
  BCD (8-4-2-1)-decimal 400
  binary-decimal 27-29
    direct 28
    double-dabble 29
    successive multiplication 29
  binary-Gray 375
  binary-hexadecimal 371
  binary-octal 369
  decimal-BCD (8-4-2-1) 374
  decimal-binary (fractional) 27
  decimal-binary (integer) 24-26
    direct 25-26
    successive division 26
  decimal-hexadecimal 372
  decimal-octal 369
  decimal-XS3 378
  Gray-binary 377
  hexadecimal-binary 372
  hexadecimal-decimal 372
  octal-binary 369
  octal-decimal 370
  XS3-decimal 378
Converters, code 388-404 (*See also* Code
  converters)
Count decoding 79-83
Counters,
  applications 307-310
  asynchronous binary 38-42, 279-287
  bidirectional 284-287
  concepts 38-42, 279-281
  down 284
  up 38-42, 281-284
  shift 299, 329-332
    concepts 329-330
    ring counter 330
    switch tail (Johnson) 332
  special purpose 30-32, 293-299
    BCD 296-299
    biquinary 298
    concepts 293
    divide-by-N 293-296
    mechanical binary 30-32
    mod-5 299-302
    mod-17 303-307
  synchronous binary 287-293
    concepts 287-288
    down 289-290

Index **489**

    up 289
    up-down 291-293
Counting
    binary 22-24, 30-32
    decimal 19-22, 30

D

D flip-flop (*See* flip-flop, D)
Data (microprocessor) 424
Data communication 454-457
Decimal counting 19-22, 30
Decimal number system 19-22
Decimal to binary conversion 24-27, 484
DeMorgan's law 97-104, 468
Demultiplexer/decoder 453
Derived functions 73-76 (*See also* NAND and NOR)
Digital applications 4-7, 449-466
Digital circuits analysis concepts 11-15
Digital clock 309, 408
    analysis 407-409
    block diagram 408
Digital device 4
Digital test equipment
    digital voltmeter 308, 458-460
    general 307-309, 457
Digital voltmeter 308, 458-460
Disjunction 47-48
Display devices 405-418
    CRT 418
    edge lighted 418
    liquid crystal 409-411
    NIXIE 405-406
    plasma 415-418
    projected image 418
    segmented 406-411
    solid-state 406-414
Dissent circuit 219-224
Distributive law 91-97, 468
Dividers, frequency 281
Dual-rank flip-flop (*See* Master-Slave flip-flop)
Duality 106-107

E

Exclusive-NOR 111, 343
Exclusive-OR 109-110, 147, 342-343, 346, 470
Expansion input (gate) 303
Expression, Boolean algebra 56

F

Flip-flop, D 247, 314, 323
Flip-flop, general
    inputs 231

    master state table 253
    operation 34-38
    physical configurations 33
    rules of logic 472-473
    symbol 14, 33
Flip-flop, J-K 248-250, 289, 293-295, 314-315, 325-327
Flip-flop, MIL-STD 34, 231, 472
Flip-flop, R-S
    analysis 238-240
    applications 240-241
    clocked 241-242
    general 238, 472
    implementation 238, 280
    master-slave (dual-rank) 243-244
Flip-flop, T 244, 247, 280, 284, 291, 315
Flowcharts
    built-in indicators 206-210
    computer program 445
    concepts 201-205
    external indicators 210-216
    loops 205
    oscilloscope 216-218
    preliminary 204
    symbols 203
Full adder
    algebraic simplification 141-148
    analysis 183-195, 349-351
    general 120, 395
    mapsimplification 173-177
    output carry output 142, 146-148, 176, 193
    sum output 142, 145, 176, 193

G

Gate, definition 56
Gray code 155
Grouping signs 68-71

H

Half adder
    analysis 119-121
    general 144-145, 348-349
Hardware (microprocessor) 428-436
Harvard chart 151
Hexadecimal number system 371-372
    conversion 371-372
HIGH and LOW 17, 35

I

ICs (*See* Integrated Circuits)
Idempotent law 91, 468
Indicators (lamp) 37
Input/Output (microprocessor) 424, 435
Instruction (microprocessor) 424

Instruction 438-446
  decoder 432
  register 432
  set 438-446
Integrated circuits
  applications by type
    2501 361-362
    4512 355
    5480 351-352
    7443A 403
    7445A 401
    7476 303
    7483 351
    7493 282, 293-294, 365
    74151 355
    74176 294, 303, 306
    74197 303, 306
    74LS90 335
    74LS164 337
    74LS266 343
    74LS295 327-328, 365
    74LS374 318-319, 336
    8232 355-356
    8250 404-405
    8281 355
    8284 291
    8288 294, 297
    8290 298-300, 303, 306
    8291 303, 306
    8T05 402
  description 33, 101, 103
  packages 342
Inverter 63, 470

J

J-K flip-flop (*See* Flip-flop, J-K)

K

Karnaugh maps 154-156

L

Lamp drivers 37, 77
Latch circuit 228-230
Law of Absorption 104, 468
Law of Expansion 104, 468
Law of Identity 87, 468
Light-Emitting-Diode (LED) 407, 411-415
Liquid crystal display (LCD) 409-411
Logic analysis
  general 15
  combinational
    algebraic 183-195
    documentation 190
    example 261
    1s and 0s 193-194
    preparation 181-185
    procedure 182-183, 185-195
    truth table 189
    waveforms 194-195
  combinational/sequential
    identification and separation 259-260
    procedure 255-256
    recombination 266-267
Logic analyzer 216, 270-273
Logic circuit conversion
  NAND to AND-OR-NOT 137-139
  NOR to AND-OR-NOT 133-134
Logic diagrams 13
Logic operations
  definitions 45-48
Logic simplification
  algebraic
    from Boolean expression 134-135, 139-140
    from logic diagram 133, 136-137
    from word statement 130-132
    1-level 128-130
    2-level 130-135
    3-level 136-137
    4-level 137-141
  maps
    construction 156
    entering information 160-161, 178-179
    extracting information 162, 179
    from Boolean expression 161-168
    from logic diagram 168
    from truth table 173
    full adder 173-177
    general 16
    unplotted squares method 171
    2-variable 161-164
    3-variable 164-171
    4-variable 171-173
  procedures 177-178
Logic signal source 58, 76-78
Logic troubleshooting
  combinational
    example 219-224
    flow charts 201-219
    philosophy and concepts 199-201
  mixed functions
    built-in indicators 268-269
    external indicators 269-270
    general 267
    typical problem 274-277
Logical addition 127
Logical multiplication 92, 127
Logical product 127
Logical product of logical sums 156-157
Logical sum 127
Logical sum of logical products 156-157, 169-171

Index

M

Majority circuit 219-224
Map, Karnaugh 154-155
Master-slave flip-flop (*See* Flip-flop, R-S)
Maxterm 158, 161
Memory
  microprocessor 424, 432-434
  random access
    buffer storage 363
    microprocessor 432
    operation 359-362
    scratch pad 363
  read-only
    applications 366
    microprocessor 432
    operation 365
Memory address register 432
Memory elements
  applications 358, 363, 367
  classification 359
  coincident selection 361
  general 358
Memory map 437
Microcomputer 422
Microprocessor 421-447
MIL-STD-806B (*See* Symbols)
Minterm 157
Modem 454
Modulus, counter 293, 296
Multiplexers 353, 355-357
Multiplexer/programmer 450
Multiplication, logical 92, 127

N

NAND 73-74, 101-104, 189, 475
Negation 48, 470
NIXIE display tube 405-406
NOR 74-76, 99-101, 476
NOT circuit
  inverter 62
  symbol 470
NOT function 48, 62-65, 85
  analogy 62
  application 63-64
Number systems
  binary 22-24
  decimal 19-22
Number system conversions (*See* Conversions, number system)

O

Octal number system conversions 369-370
OFF state 34
ON state 34
ONE state 34
One shot 263-265
OR-AND expression 158
Order of operations 68-71
OR function 48, 57-59, 84
  analogy 59
  applications 58
OR gate
  symbol 469, 475
  truth table 48
Output, digital, definition 57

P

Parity 346-347, 351
Pascal, Blaise 2
Plasma panel display 415-418
Procedures
  algebraic simplification 126-128
  Boolean expression to logic diagram 89
  combinational logic analysis 181-183
  combinational/sequential logic analysis 255-256
  number system conversions (*See* Conversions, number systems)
Processor cycle 430
Product of sums, logical 156-158
Product, logical 127
Program counter 432
Program, computer 425, 436, 441
Programmer, computer 201, 436

Q

Q output (flip-flop) 35
Q̄ output (flip-flop) 35
Quine method 151

R

R-S flip flop (*See* Flip-flop, R-S)
Radix 19, 22
Registers
  analysis, block diagram 332
  carry 435
  circulating shift 324-325
  general 312, 339
  logic level analysis 334-338
  shift
    applications 321
    concepts 320-321
    implementation 322
    parallel-serial 325-327
    serial-parallel 325
    serial-serial 322-324
  stack 432

storage
  concepts 313-314
  count storage 318
  implementation of 314-315
  memory buffer register 318-319
Relay 58
Reset (clear) 33-37

S

Segmented displays 406-411
  decoding 407
  7-segment 407
  16-segment 407
Sequential logic
  algebraic representation 235-237
  nomenclature and symbology 230-233
  tabular representation 233-235
  waveforms 237
Set 33-37
Software (microprocessor) 436-447
Solid-state display (LED) 406-414
Standard product 156-161
Standard sum 156-161
State flow diagram 283
State indicator 470
State table 41, 235, 253
Storage (microprocessor) 424, 432-434
Strobe (input) 294, 303
Sum of products, logical 156-158
Sum, logical 127
Switching circuits 58
Symbols
  AND 14, 84, 469
  Exclusive OR 110, 470
  flip-flop 14, 232, 472
  MIL-STD-806B 15, 33, 469-473
  NOT 84, 470
  OR 84, 469
Synchronous 248
Systems concept 195-199

T

T flip-flop (*See* Flip-flop, T)
Telemetry
  applications 454
  general 197, 449-450
  receiving system 452-453
  transmitting system 450-452
Teletypewriter 379
Transducer 4, 450
Trigger (toggle) input 33
Truth table
  combinational logic
    analysis 83, 158-160
    applications 16, 50-51, 158-160, 189
  sequential logic 234-235

V

Variable 48-49
Veitch diagram 152-154
Venn diagram 151-152
Vinculum 71

W

Waveform analysis 79-82
Waveforms, general 17, 36-39
Word, computer 310, 318-320

Z

Zero state 34